KT-229-583

CONTENTS

Social Science Library
Oxford University Library Services
Manor Road
Oxford OX1 3UQ

WITHDRAWN
WITHDRAWN

University of Oxford
QEH Library

QC
925

WITHDRAWN

CLIMATE, WATER AND AGRICULTURE IN THE TROPICS

SECOND EDITION

I. J. JACKSON

Longman
Scientific &
Technical

Copublished in the United States with
John Wiley & Sons, Inc., New York

Longman Scientific & Technical,
Longman Group UK Limited,
Longman House, Burnt Mill, Harlow,
Essex CM20 2JE, England
and Associated Companies throughout the world.

Copublished in the United States with
John Wiley & Sons, Inc., 605 Third Avenue, New York, NY 10158.

© Longman Group UK Limited 1977
This edition © Longman Group UK Limited 1989

All rights reserved; no part of this publication
may be reproduced, stored in a retrieval system,
or transmitted in any form or by any means, electronic,
mechanical, photocopying, recording, or otherwise
without either the prior written permission of the Publishers or a
licence permitting restricted copying in the United Kingdom
issued by the Copyright Licensing Agency Ltd, 33–34 Alfred Place,
London, WC1E 7DP.

First published 1977
Second edition 1989

British Library Cataloguing-in-Publication Data

Jackson, I. J. Ian Joseph, *1939*
    Climate, water and agriculture in the
    tropics – 2nd ed.
    1. Tropical regions. Rainfall.
    I. Title.
    551.57'81213

ISBN 0-582-02159-6

Library of Congress Cataloging-in-Publication Data

Jackson, I. J. (Ian Joseph)
    Climate, water, and agriculture in the tropics / I. J. Jackson. –
2nd ed.
        p. cm.
        Bibliography: p.
        Includes index.
        ISBN 0-470-21069-9 (Wiley, USA only).
        1. Tropics – Climate. 2. Plant-water relationships – Tropics.
3. Water in agriculture – Tropics. I. Title.
QC993.5. J3 1988   87-36849
551.5'7'0913 – dc 19   CIP

Set in 10/12 pt Linotron 202 Ehrhardt

Produced by Longman Singapore Publishers (Pte) Ltd.
Printed in Singapore

10 PROBLEMS AND PRIORITIES 302

UNIVERSITY OF OXFORD
INTERNATIONAL DEVELOPME
CENTRE LIBRARY
QEH
DATE CATALOGUED 18.7.89

CLASS-MARK HD 350 JA

LIST OF ILLUSTRATIONS

PREFACE TO THE SECOND EDITION

Much has been written in the fields of tropical meteorology, hydrology and agriculture since the first edition. Hence a new edition seemed timely. While nearly all sections of the book have been extensively rewritten, several new sections have also been added. Current interest in the significance of El Nino and the Southern Oscillation prompted a new section (§ 2.8) and recent work linking rainfall variability to various phenomena suggested that a section dealing with this aspect would be useful (§ 3.7). Although in the first edition the question of drought received attention in various parts of the book, the continued importance of this topic has led to the introduction of § 3.4 which attempts to indicate the complexity of this aspect. In chapter 5, discussion of relationships between rainfall and evaporation in characterising areas suggested that an examination of climatic classification systems and their application, particularly to agriculture, was warranted (§ 5.4). The large volume of recent work on the impacts of land use and land management practices on the hydrological cycle has led to the original section (§ 9.3) dealing with this topic being split into three sections, using sub-headings from the old section as headings for § 9.4 and § 9.5. Finally, a concluding chapter has been added. This extends the discussion into areas such as appropriate experimentation, education and training and develops further the importance of the human factors introduced in § 1.4.

Comments on the first edition indicated that the references covering a range of disciplines were found to be useful, particularly by specialists in one area requiring an introduction to the literature in another field. Therefore, in the new edition, a deliberate attempt has been made to present a much more comprehensive indication of references which should serve not only as an introduction but also as a useful resource for more detailed study. The literature cited, together with a rather more detailed coverage of some topics in the book itself, will hopefully provide for more in-depth analysis while still serving a key purpose of introducing the interdisciplinary nature of the topic.

Most of the literature search for the new edition was carried out while I was a visitor in the Department of Geography, University of

Hawaii at Manoa. I am very grateful for the hospitality shown during my visit and for the useful discussions with a number of people in that Department as well as in the Departments of Meteorology, Agronomy and Soil Science, the Water Resources Research Center and the East-West Center. I also benefited considerably from discussions with staff in the Indian Institute of Tropical Meteorology, Pune, the Indian Meteorological Department in Pune and New Delhi, the Instituto Nacional de Meteorologia in Rio de Janeiro and the Jamaica Meteorological Service, Kingston. Thanks are due to Professor David Lea for his helpful comments on Chapter 10.

My thanks go to Denise Cumming and Megan Wheeler of the Department of Geography and Planning, University of New England who did most of the typing, with assistance from my wife, Jean. They coped with the horrendous task of dealing with the many bits of paper stapled to a photocopy of the first edition with enthusiasm and good humour. New diagrams were drawn by Rudi Boskovic. Thanks also to Steve and Chris Jackson for help with the boring task of reading the manuscript.

I. J. JACKSON
September 1987

PREFACE TO THE FIRST EDITION

There is a growing need for books dealing with the tropics, not only because of development in these areas but also because of increased interest in tropical developing countries and their problems from people in other parts of the world. The characteristics of tropical rainfall and high rates of evaporation exert a great influence on man's activities, particularly agriculture which is a mainstay of the economy in many tropical countries but also on water for domestic and industrial use.

This book examines characteristics of tropical rainfall and evaporation, together with their implications, especially related to agriculture, land use and aspects such as soil erosion and irrigation. Chapters 2 to 5 examine the characteristics and their broad implications. In Chapters 7 and 8, implications for plant growth, agriculture, soil and water conservation and irrigation are analysed in more detail. As an introduction to these two chapters, Chapter 6 presents a general review of the significance of water to plants and a discussion of the system through which rainfall becomes available to them. In the final chapter (Ch. 9) ways in which man alters rainfall and evaporation characteristics, either deliberately or inadvertently, are reviewed. This chapter also examines how man modifies the effects of the characteristics together with some of the results of this modification.

In a book of limited length, a complete coverage of all aspects of water resources in the tropics is impossible. Since in no sphere are climatic characteristics more important, the focus on agriculture and land use was considered justified. However, material relevant to aspects such as hydro-electric power generation and water transport is presented. For example, in § 4.4, rainfall-runoff relationships are examined and in § 9.3, ways in which land use affects variability in streamflow are discussed. This is important since financial restraints often limit the possibility of stream-flow control through reservoirs and many tropical countries must rely heavily on land-use measures to effect such control. In § 9.4, some effects of reservoir construction, which may be for a variety of purposes, on various aspects of the environment are analysed. The point is also made that soil erosion has implications for silting of reservoirs and navigation channels. Aspects related to water and health are also examined in Chapter 9.

Hopefully, the book will appeal to those with a general interest in the tropics as well as to those specifically concerned with aspects of development. A wide range of expertise is necessary to attack tropical development problems, this being especially true for water resources. Often, specialists in one area may not be sufficiently aware of aspects in other disciplines. Technical and scientific aspects of water resources involve hydrologists, meteorologists, agriculturalists and engineers, amongst others. The material here may broaden their knowledge. Planners, economists, geographers, sociologists, political scientists, administrators and others concerned with the wider aspects of development may gain an understanding of some of the physical aspects of water resources, especially in relation to agriculture and land use. The book should also be relevant to university courses concerned with aspects of tropical water resources.

I should like to thank Professor John Oliver for reading an earlier draft of the book and for making many valuable comments. My thanks are also due to Professor David Lea and Dr Ken Gregory for comments on individual chapters. I should like to thank colleagues in Liverpool, Dar es Salaam and New England for discussions which, however indirectly, contributed to the book. It is difficult to single anyone out, but perhaps my deepest debt is to Professor Len Berry for first stimulating my interest in tropical water resources. Much of the typing was done by my wife who also devoted a great deal of time and patience to the tedious business of sorting out material and checking references. Margaret Watson also helped considerably with the typing. Paul Branscheid, Bill Neal and Rudy Boscovic were responsible for drawing the figures.

I. J. Jackson

ACKNOWLEDGEMENTS

We are grateful to the following for permission to reproduce copyright material:

Edward Arnold (Publishers) Ltd. for fig. 2.8 and table 2.1 (Lockwood 1974); B. T. Batsford Ltd. for fig. 4.5 from fig. 3.9 (Hudson 1971); Blackwell Scientific Publications Ltd. for fig. 6.3 from fig. 7 (Cowan 1965); Cambridge University Press for fig. 6.2 from fig. 2.6 (Norman, Pearson & Searle 1984); College of Tropical Agriculture and Human Resources, University of Hawaii for tables 4.1 & 4.2 from tables 28 & 29 (El-Swaify, Dangler & Armstrong 1982); East African Community for figs. 4.9 (Lumb 1971), 5.10, 5.11, 5.12 and table 5.1 (Blackie 1965); Elsevier Science Publishers B. V. for fig. 3.9 from fig. 2 (Jackson 1982); Elsevier Science Publishing Co. for figs. 3.12, 3.18 (Griffiths 1972), 4.7 (Reich 1963), 4.12 (Sharon 1972) & 4.13 (Stol 1972) and tables 3.1, 3.2, 3.5 & 4.3 (Griffiths 1972); Controller of Her Majesty's Stationery Office for fig. 4.8 from fig. 3 (Lockwood 1967); Journal of Tropical Geography for figs. 3.19, 4.15 (Nieuwolt 1968) and 4.16 (Ramage 1964); McGraw-Hill Book Company for figs. 2.3, 2.5, 2.10 & 2.11 (Riehl 1965); Methuen & Co. Ltd. for fig. 5.4 (Davies & Robinson 1969); New Zealand Meteorological Service for fig. 2.9 (Ananthakrishnan & Rajagopalachari 1964); the author, H. Reihl for figs. 2.7 & 3.16 (Riehl 1954); Royal Meteorological Society for fig. 3.17 from fig. 3 (Hulme 1984) and table 4.4 from table 2 (Chaggar 1984); Royal Meteorological Society for figs. 2.1 (Palmen 1951), 2.14 (Coutts 1969), 2.15 (Lumb 1970), 3.20 (Kraus 1955) & table 2.3 (Soliman 1953); C. W. Thornthwaite Associates for figs. 5.5, 5.6 (Thornthwaite Associates) & 5.7 (Subrahmanyam 1972); UNESCO for fig. 7.1 (Slatyer, 1962) © UNESCO 1962; University of Chicago Press for fig. 5.2 from fig. 19, p. 66 (Sellers 1965); University of Tokyo Press for table 4.5 (Wada 1971); U.S. Department of Commerce for fig. 2.12 and table 2.1 (Gray 1968) and table 2.2 from table 2 (Crutcher & Quayle 1974); the author, Chye Lee Vun for fig. 4.1 (Lim 1969); Water, Engineering and Development Centre, Loughborough University for fig. 7.2 (Jackson 1986b); Westview Press for fig. 3.13 from fig. 15.2 (Benacchio 1983); World

Meteorological Organisation for table 9.1 from tables 4 and 5 (Munn & Machta 1979).

Whilst every effort has been made to trace the owners of copyright, in some cases this has proved impossible and we take this opportunity to offer our apologies to any authors whose rights may have been unwittingly infringed.

CHAPTER 1

WATER IN THE TROPICS

1.1 INTRODUCTION

Water is vital to life and development in all parts of the world. In some regions, the characteristics of water supply and demand pose few problems, but with increasing requirements for agricultural, industrial and domestic purposes, such areas are becoming fewer. A particular set of physical, economic, social and political factors combine to make water especially significant in the tropics.

1.2 EVAPORATION AND RAINFALL

High rates of evaporation and the characteristics of tropical rainfall pose particular problems. As will be seen in Chapter 3, in many areas rainfall is markedly seasonal in character, greatly limiting water availability at certain times of the year. At other times, these same areas may have excessive rainfall leading to a further set of problems. These wet season problems, such as flooding and soil erosion, are also shared by many areas which do not have a pronounced dry season. Flooding is an essential element of the agricultural system in some areas. There is also considerable variability from season to season and year to year. These temporal variations have a marked influence on water availability and hence on any form of human existence, growth or development.

Problems resulting from temporal variations in amounts are very much amplified by other characteristics of rainfall. Rainfall intensities tend to be high in the tropics, and a considerable proportion of the rain is concentrated in a comparatively small number of very heavy storms. As will be seen in Chapters 4 and 7, this means that much of the rainfall may therefore not be effective in the agricultural sense since it never becomes available to plants. Instead of contributing to the build-up of soil moisture reserves which can be drawn on in dry spells, surface runoff is considerable, creating problems of flooding and soil erosion. To the extremes on a seasonal basis referred to above, therefore, is added a short-term fluctuation between excess and deficiency. A dry, cracked landscape can be inundated and then return to its former appearance within a matter of hours, so that it is difficult to imagine that any rain has fallen for months. Tropical rainstorms are often very localised, creating marked differences in rainfall over short distances. As will be seen in Chapter 4, these differences can persist for considerable periods.

The problems due to rainfall characteristics are compounded by high rates of evaporation and transpiration. This results from the high availability of energy for evaporation in low latitudes and may conveniently be thought of as the 'evaporative demand of the atmosphere' discussed in Chapter 5. The effect is for water-loss rates from open-water and soil surfaces to be considerable and for vegetation and crop-water requirements to be high. In simple terms the result is that a rainfall amount more than adequate for agriculture or water supplies in temperate areas may be completely insufficient in the tropics.

1.3 THE HYDROLOGICAL CYCLE

The climatic elements referred to above create particular problems in the tropics, but an understanding of their impact rests upon a consideration of their place within the hydrological cycle, the circulation of water within the earth – atmosphere system (Fig. 1.1). The cycle can be briefly reviewed as follows.

Evaporation and transpiration add water vapour to the atmosphere. Following condensation and cloud formation, precipitation may occur. As precipitation falls some evaporation of drops occurs, the rest reaching the ground, a water surface or a vegetation surface. Of the precipitation reaching vegetation, some is held (intercepted) by the canopy and eventually evaporates again. Some will run down vegetation stems or drip from the canopy or be shaken off by wind and reach the ground. In the tropics, with the exception of hail, this precipitation is predominantly rainfall although the significance of other forms such as dew and 'fog drip' are considered in the book.

Of the water reaching the ground, some will infiltrate into the surface. The infiltration rate is extremely variable, depending upon factors such as soil type, condition and moisture content. In addition, the influence on infiltration of surfaces such as roads, paths and buildings is considerable. If rainfall intensity is greater than the infiltration rate then some water accumulates in surface depressions. When the depressions are full, water begins to move over the surface, usually towards channels. Not all water is channelled – sheet flow occurs, especially in sparsely vegetated areas. Slope and vegetation cover affect the rate of water movement over the surface and hence the time available for infiltration. Some of the water stored in depressions will evaporate and some will eventually be able to infiltrate into the soil.

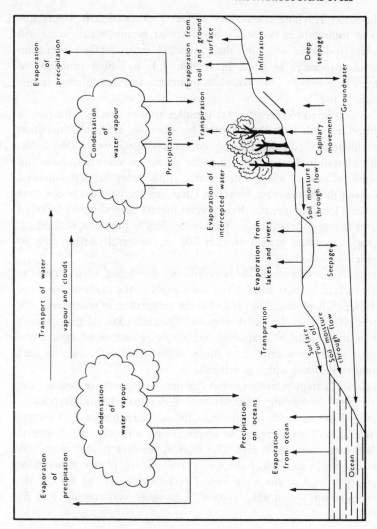

Fig. 1.1 The hydrological cycle.

Some of the water entering the soil may evaporate back into the atmosphere, and some will be taken up by plants and transpired. In the upper layers of the soil, sub-surface lateral movement of water (through-flow) may be considerable and may re-emerge on the surface and reach streams or lakes. Some water may percolate to deeper layers to become ground water where it may be held for very long periods. Ground water can move up over a limited depth by capillary action where it may evaporate or be taken up by plants and transpired. Some ground water, by lateral movement, may eventually seep into streams, lakes, seas or oceans.

Of the water reaching open-water surfaces, either directly as precipitation or indirectly as surface, sub-surface or ground water, some will evaporate and some seep into the ground. A more detailed discussion of runoff processes is given in section 4.4, including problems of distinguishing between the various components and the different terminologies in use.

This simplified picture of a very complex system indicates that people can interfere with virtually any part; for example, artificial stimulation of precipitation, reduction of evaporation, change in vegetation cover, modification of infiltration rates and rate of surface water movement and extraction of ground water. Interference at one stage has repercussions throughout the whole cycle, however. Changing the land use over a river catchment, for example, may have a great impact upon the path followed by water falling as precipitation, perhaps leading to problems of flooding, soil erosion, reduced dry season river flow or, conversely, having reciprocal beneficial effects.

Precipitation of various kinds represents the 'input' of water to the earth's surface in most areas. Even those parts of the earth whose water supply depends on rivers or ground water originating in other areas rely on precipitation in the latter regions. Consideration of precipitation alone is misleading in evaluating the water resources or agricultural potential of an area since it is the relation between it and the high evaporative demand which is critical.

However, simply relating rainfall amounts to evaporative demand can be extremely misleading, particularly in agricultural terms. Evaporative demand of the atmosphere influences the water requirement of a crop, and rainfall will, to a greater or lesser extent, influence the degree to which this requirement is met. The relation between rainfall characteristics and the water which becomes available to plants, functioning through that part of the hydrological cycle which may be termed the 'atmosphere–soil–plant water system' is, however, very complex (Ch. 6).

1.4 THE HUMAN FACTORS

Important as the physical factors are, it would be dangerous to divorce them from the range of human factors which unfortunately combine to make it very difficult for individual countries to overcome the problems posed by rainfall and evaporation. Financial and human resource limitations facing many tropical countries make it difficult to initiate schemes for the control of water resources to prevent flooding, soil erosion and to provide irrigation water. Even without such restraints, however, solutions to problems which may be applicable to the developed world and have received widespread acceptance may be inappropriate. Such solutions, to be effective, may demand levels of productive efficiency, use of fertilisers, seed selection and technical knowledge far beyond the reach of the peasant farmers who form the bulk of the population in many tropical countries. Large-scale multi-purpose schemes for the development of hydro-electric power, irrigation, flood control, recreation and fishing may not be suitable in an area of dispersed peasant farmers. A number of small-scale projects such as wells or small earth dams at the village level may be more appropriate. Unfortunately, such schemes are less likely to attract local or overseas aid since they lack prestige.

Technical developments may bring about social changes which are often overlooked and which can be for the worse. On the other hand, technical innovations often need to be accompanied by far-reaching social and institutional changes, such as land reform, in order to be effective. Without such changes new developments, even if effective in broad regional terms, may benefit a small number of the more wealthy rather than the mass of poor peasant farmers. There is therefore a need for research, not only into appropriate types of technical development of water resources but also into the resulting or necessary social changes which must accompany such development. In the past, technical developments tended to ignore the social and economic structure. Innovations need to fit existing structures or cause the minimum disruption.

Even when appropriate solutions to problems have been found, implementation may be difficult. Peasant farmers, with good reason, are very conservative and if change is necessary the way in which it is introduced is critical. To quote Dumont and Rosier (1969):

> The expansion of agricultural production in the countries of the Third World depends as much on the peasant communities' capacity for change and reaction to new ideas as it does on the use of new methods or the reclamation of new land.

The traditional systems have evolved over very long periods of time, taking environmental conditions, including periodic disasters, into account. The farmer therefore has very valid reasons for maintaining the *status quo* until he is convinced that change will be to his benefit. As will be seen in Chapter 8, considerable increases in yield may result from earlier planting than is traditional. The peasant farmer may plant later because he is well aware of the vagaries of the onset of the wet season and would rather be content with 'safe' but moderate or low yields than to run the risk of complete loss. Late planting may also be due to the inability of the farmer to work the hard ground at the end of the dry season prior to planting. Acceptance of new ideas such as early planting will depend upon relevant education, together with practical demonstration by means of model farms and plots. Unfortunately, all too often, even when farmers in an area adopt changes, factors such as a failure of the rains in the early years reinforce their distrust and they revert to traditional methods.

Education and ideas introduced from outside the local area by central government and the higher levels of administration may be viewed with suspicion, particularly in countries with a relatively recent colonial past. The successful dissemination of new ideas will depend on an effective political and administrative framework which allows the transmission of change from the central government to the local or 'grass-roots' level. Effective transmission of the views of the peasant farmer up through the hierarchy is just as important if any change is to be successful.

When innovations are introduced at the local grass-roots level, the people are perhaps more likely to be motivated and willing to engage in self-help projects such as the building of small earth dams. Such projects, requiring little capital input, are often more appropriate than high-capital, centralised development. In general, therefore, the success of changes to overcome problems resulting from the characteristics of tropical rainfall and evaporation will depend heavily on the degree to which social, economic and political factors are taken into account and the mechanism by which the changes are introduced. Different approaches to the provision of rural water supply in developing countries are compared by Bajard, Draper and Viens (1981). They pay particular attention to the importance of both physical and human factors in making appropriate decisions and the requirements for implementation and operation of developments.

Economic, social and political changes are often introduced from outside and beyond the control of the local people. Such changes can alter the impact of rainfall and evaporation characteristics (Ch. 3).

While water resource development and utilisation must take account of a wide range of factors, including social and political ones, planning must rest on a sound knowledge, understanding and assessment of the resource itself. In particular, the characteristics of tropical rainfall and evaporation, together with their links through the atmosphere–soil–plant water system need consideration because of their great significance. These characteristics and their implications for agriculture, land management and water resources are the subject of this book. This leads naturally to an analysis of how people adapt themselves to the characteristics and attempts to modify their impacts.

The limitations of data, resulting from problems associated with the measurement and analysis of rainfall and evaporation, are not always sufficiently appreciated. Some of these aspects will be considered, particularly in Chapters 3 to 5.

CHAPTER 2

ORIGINS OF PRECIPITATION

2.1 INTRODUCTION

While here interest centres on rainfall characteristics and their practical implications for water resources and agriculture, it would be a mistake to ignore the causes of precipitation. Such knowledge is vital to a correct appreciation of the nature of tropical rainfall – amounts, seasonal distribution, intensity, frequency of occurrence, variability, areal variations – which is necessary to understand its practical implications. Without such understanding, fundamental misconceptions may arise which can have disastrous results for development planning.

The limits of the tropics are to some extent arbitrary since the atmosphere is an entity and events in low latitudes cannot be divorced from those of higher latitudes. Links between the tropics and higher latitudes are examined by Harnack, Lanzante and Harnack (1982), Davis (1981), Arkin and Webster (1985). Reiter (1983) found significant teleconnections between precipitation surges in the equatorial Pacific 'dry zone' and mid-latitude airflow patterns. Limits defined by certain latitudes such as the tropics of Cancer and Capricorn, or 30°N and S latitude (the latter dividing the earth's surface into two equal parts) are especially meaningless since they do not take into account marked seasonal and longitudinal changes. Perhaps the simplest solution is to take the dividing line between easterly and westerly winds in the middle troposphere, the sub-tropical jet streams or the axes of the sub-tropical highs as rough guides. Such a division (Fig.2.1) allows for seasonal and longitudinal variations and there can be no doubt that there are particular characteristics to the weather equatorwards of these boundaries. The drawbacks in using latitudinal boundaries are illustrated by the fact that in late summer and autumn tropical hurricanes can affect south-eastern North America and areas of eastern Asia which would generally be considered to be outside the tropics. Conversely, the poleward margins of the 'tropics' in the cool season are affected by temperate-latitude weather systems. There are very great longitudinal differences in the areas coming under the influence of essentially tropical conditions. The south-east Asian monsoon system, to be discussed

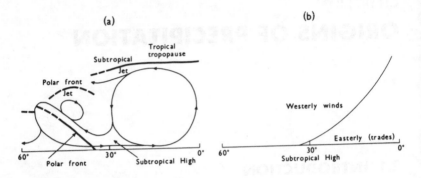

Fig. 2.1 (a) Cross-section of the tropical circulation (after Palmen 1951). (b) Cross-section of the westerly winds and trades.

later, provides a marked contrast with the situation in other regions of the same latitude.

2.2 EARLIER CONCEPTS OF SIMPLICITY AND UNIFORMITY

Before the Second World War, relatively small horizontal temperature and pressure gradients, together with the presence of large ocean areas, suggested considerable uniformity and simplicity of weather in the tropics to meteorologists from higher latitudes who were used to the great significance of fronts and considerable pressure gradients. Furthermore, these meteorologists and climatologists attempted to apply concepts and models from mid-latitudes. Inevitably, this led to attempts to adjust the limited weather information to fit assumed theories, a situation hardly conducive to the development of new ideas.

A simple concept of trade winds originating from a general belt of high pressure in the sub-tropics and blowing from the north-east in the northern hemisphere and the south-east in the southern hemisphere emerged. These airstreams met in an equatorial low-pressure trough where ascent of air took place and precipitation resulted. On this basis, a simple picture of tropical rainfall regimes emerged, with a seasonal migration of a rain belt following the movement of the sun. Thus, at the poleward margins of the tropics a single rainfall maximum following the summer solstice was to be expected, with a double peak at the equator in response to the passage of the overhead sun at the equinoxes. Viewed in terms of monthly averages of rainfall, a seasonal movement of a rain belt along these lines is evident in some areas, but there are great exceptions.

With greatly increased demands for meteorological data during the Second World War, the complexity of the situation began to emerge. While in higher latitudes the depression model and associated frontal development fits many of the precipitation-producing situations, in the tropics fluctuations in weather appear to result from a wide variety of disturbances which vary in character from area to area. The nature of these disturbances is far from being completely understood, partly because of lack of information, especially over ocean areas, but also because they are often associated with very small, difficult to detect, horizontal pressure gradients. Tropical perturbations (disturbances) are often related to mid-tropospheric events not readily identifiable without a good radio-sonde network. The significance of disturbances is highlighted by Riehl (1965) when he states, 'Disturbances produce more than 90 per cent of the rainfall in the tropics', and hence an understanding of their nature and resulting rainfall characteristics is vital. Much progress has been made but many uncertainties remain.

Here, only a brief coverage of key aspects of tropical meteorology and climatology can be presented, designed largely to indicate some of the complexities of tropical rainfall controls. Overviews of tropical meteorology and climatology are presented by Riehl (1979), Krishnamurti (1979) and Hastenrath (1985). Krishnamurti (1981) provides a discussion of major problems, including the significance of sea surface temperatures, surges in the tropical circulation, hurricanes and numerical weather prediction. Other useful references for tropical material include Barry and Chorley (1982) and Lockwood (1985). Despite an increasing array of research, much remains to be done. Unfortunately, developing tropical countries are handicapped by scarce finance and human resources, and much of their limited effort is diverted into key fields such as aviation forecasting. It is vital that sufficient resources are devoted to aspects directly relevant to the needs of water resources and agriculture.

2.3 BASIC FEATURES OF THE TROPICAL CIRCULATION

The major features to be considered are as follows:

1. The sub-tropical high-pressure cells (STHs).
2. The trade winds.
3. The equatorial trough.
4. The south-east Asian monsoon.

Surface locations and seasonal variations of the features are depicted in Figs 2.2 and 2.3. Krishnamurti (1979) points out that seasonal and monthly average maps of pressure and winds are much more useful in

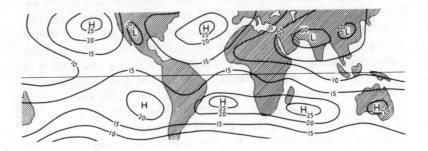

Fig. 2.2 Major features of the inter-tropical circulation in July (after Riehl 1965).

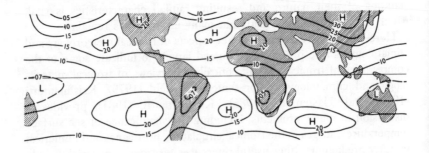

Fig. 2.3 Major features of the inter-tropical circulation in January (after Riehl 1965).

the tropics than in higher latitudes since the major features are also in evidence over short time periods. This is not so true for higher latitudes. These maps suffer from deficiencies because of lack of data, but indicate the major points. They show a series of high-pressure cells between latitudes 20° to 40°N and S with a breadth of some 10°–20° of latitude. Between these two belts of high-pressure cells (STHs) occurs an area of low pressure, generally termed the equatorial trough. Averaged around the globe, the mean annual position of the trough is 5°N, sometimes termed the 'meteorological equator'. Correspondingly, the mean annual position of the STHs in the southern hemisphere is about 5° nearer the geographical equator than that in the northern hemisphere.

Comparison of Figs 2.2 and 2.3 shows marked seasonal variations in the intensity and location of these features. The STHs tend to be centred over the eastern halves of the oceans. During winter, cold air over the land favours the development of high pressure at low levels over the adjacent continents, tending to give the STHs a greater east–west extent. In the southern hemisphere in July (Fig. 2.2) a fairly continuous ridge of high pressure is formed around the globe. In the northern hemisphere in January, elongation of the cells also occurs, with a particularly noticeable area of high pressure extending from the eastern Pacific across the USA and the Atlantic to incorporate part of north-west Africa (Fig. 2.3). Peaks still occur in these general areas, however, especially over oceans, and it must be emphasised that the continental high-pressure areas are very shallow. The shallow, extensive high pressure over the Asian land mass in winter (January) is a noticeable feature.

In the summer hemisphere, the cellular nature of the STHs at the surface is much more in evidence, as is their oceanic location. Figure 2.2, for example, shows two extensive cells in the northern hemisphere, one in the eastern Pacific and one in the Atlantic. Figure 2.2 also shows that in July over Arabia and the general area of the northern Indian sub-continent, a vast low-pressure area is found within approximately the general latitude of the STHs. This will be referred to later in connec-tion with the south-east Asian monsoon. The STHs shift through only about 5° latitude seasonally, moving equatorward in winter and poleward in summer. They tend to be more intense in winter than in summer.

The equatorial trough shows a much greater seasonal latitudinal fluc-tuation than the high-pressure cells. Averaged around the globe, its location is about 5°S in January but 12°–15°N in July. The deeper penetration into the northern hemisphere is evidenced by the mean annual position of 5°N, but this asymmetry results from averaging lo-cations which differ very greatly from place to place. The smaller pen-etration of the trough into the southern hemisphere has been ascribed to the more constant westerly circulation in mid-latitudes there. This constrains the trough nearer the equator in the southern hemisphere than in the northern one (Riehl 1979). It is also apparent from Figs 2.2 and 2.3 that the trough is far from being a regular, continuous feature but is a series of centres tending to concentrate particularly over the summer continents. In January the mean position ranges from 17°S to 8°N, while in July it ranges from 2°N to 27°N. Over most of the western hemisphere (Pacific and Atlantic), the trough shifts by only 5° latitude or less, but in the east, associated with the monsoon regions, the seasonal change is very great, being in excess of 30°. In January, the trough position is easy to locate except in the central South Pacific. However, in July, it is located easily only over Africa, the Atlantic and the Pacific to about 150°E. It is unrealistic to regard the equatorial trough as a finite entity showing seasonal migration. Certainly in some locations a more correct view is of the trough disappearing in one area and at the same time being formed in another. Furthermore, in some areas such as the Indian Ocean and South-East Asia, the structure is extremely complex. These aspects are considered below.

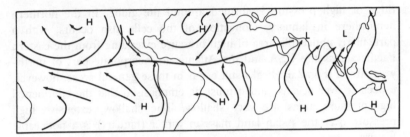

Fig. 2.4   Airflow in July (after Riehl 1965).

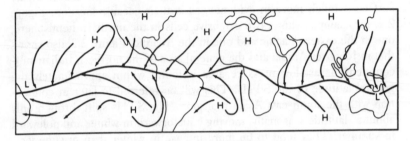

Fig. 2.5   Airflow in January (after Riehl 1965).

While in general the trade winds, originating from the STHs have components towards the equator and westwards, the picture of north-east trades in the northern hemisphere and south-east trades in the southern hemisphere is inadequate. Since the STHs are cells, particularly over the western areas of oceans, airflow may be directly from the east and can have a component away from the equator (Figs 2.4 and 2.5). Therefore, only in the eastern sectors of oceans will convergence of the trade winds tend to be marked. The permanence of the STHs is reflected in the steadiness of the trades and like the former, the latter are strongest in winter. On crossing the equator, there is a tendency for the trades to develop a westerly component. This is particularly so in the case of the south-east trades in July in the south-east Asian monsoon region (Fig. 2.4) and in the Indian Ocean and northern Australia in January (Fig. 2.5). This feature is part of the general circulation and related to the basic tendency for air to conserve angular momentum but enhanced by the monsoon circulation (Riehl 1979). However, the presence of equatorial westerlies cannot simply be ascribed to recurvature of trades as they cross the equator. There is *on average* a westerly component in the Indian Ocean at 2°–3°S in June and July and at 2°–3°N in December and January (i.e. in the winter hemisphere) (Barry and Chorley 1982). Equatorwards of the main root zones of the trades in the eastern Pacific and eastern Atlantic, light variable winds are found which show considerable seasonal variation in extent

(Barry and Chorley 1982). A third major area of light winds occurs in the Indian Ocean and western Pacific.

Early terminology referred to an inter-tropical front where the trades met, but lack of airmass differences in most areas led to this being dropped in favour of the inter-tropical convergence zone (ITCZ – see § 2.6). However, as will be seen later, the ITCZ is by no means a continuous feature in time or space, leading to the use of an alternative term, the inter-tropical confluence (ITC) (Barry and Chorley 1982). Furthermore, areas of convergence are not confined to the equatorial trough.

Love (1985a, b) presents evidence that events in the sub-tropics of the winter hemisphere can have an impact upon the summer hemisphere flow patterns and the development of tropical cyclones in the equatorial trough. Due to the broad ocean area around the Antarctic, westerly winds in the southern oceans develop strongly in both summer and winter. They are much stronger than in the northern hemisphere and as a result, southern weather systems reach the equatorial zone with greater strength and their influence extends beyond the geographical equator (Riehl 1979).

The above simplified description refers to 'average' conditions, and while the basic features are perhaps more permanent than those of higher latitudes there are nevertheless considerable day-to-day variations. The general location and seasonal fluctuations of these features can, in very broad terms, be related to rainfall regimes, but there are marked exceptions. To assume simple links between them and rainfall is incorrect. For a better understanding we must consider the features in more detail and, in particular, look at their vertical structure as well as their location.

In general, rainfall in the tropics is associated with ascent of air, leading to adiabatic cooling. Conversely, subsidence (sinking) of air, with resulting adiabatic warming leads to dry conditions. Hence, detection of areas likely to be wet or dry is essentially concerned with the identification of locations of ascent and descent, respectively. Ascent and descent are associated with horizontal convergence and divergence of air which in simple terms may be thought of as net horizontal inflow and outflow, respectively. Simple models of the relationships are indicated in Fig. 2.6a, b, but often the situation is more complex (Fig. 2.6c). Figure 2.6c indicates the need to consider air movement at a number of layers in the atmosphere to understand weather. Convergence and divergence originate in a number of ways considered below. Lockwood (1974) points out that comparatively weak temperature gradients in the tropics may cause small pressure gradients responsible for large geostrophic winds which are comparable with those in temperate latitudes. Thus, very slight pressure differences, difficult to detect, can result in

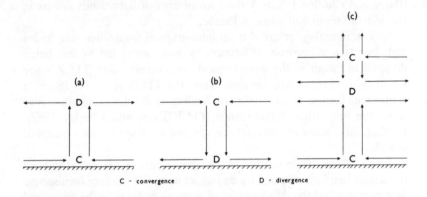

Fig. 2.6 Models of convergence and divergence.

considerable horizontal air movement and perhaps ascent or descent associated with convergence and divergence. Two particularly important aspects in the tropics are high temperatures and humidities in the air. Thus, when condensation occurs, very large amounts of latent heat energy are released and, as will be seen, this is very important to the atmospheric circulation in the tropics and in the generation of disturbances. Furthermore, after condensation, the high liquid-water content of clouds is important in initiating precipitation and in promoting heavy rainfall.

2.4 THE SUB-TROPICAL HIGHS

The formation of these features need not concern us, it being sufficient to point out that they are essentially dynamic in origin although thermal factors and the distribution of land and sea affect their location, shape and intensity. Their characteristics exert a great influence on rainfall – or rather lack of it.

The outstanding characteristic is that they are regions of general subsidence, particularly on their eastern sides. This subsidence from higher levels, which may be associated with the sub-tropical jet stream, is responsible for the general aridity associated with these systems. The major desert areas are found over the western parts of continents under the influence of marked subsidence within the eastern limbs of the STHs. The subsiding air does not penetrate right to the surface, being separated from it by a layer of relatively cool air. Since the main centres of the STHs are located over the oceans, this lower layer is able to accumulate moisture. Dietrich and Kalle (1957) show that sea surface temperatures tend to be below average in the east of the oceans and above average in the west. Thus, relatively low sea temperatures in the general locality of the STHs will limit evaporation both directly and also by increasing atmospheric stability in the lower layers. The descending, warming air above, creating a marked inversion will also affect stability, damping down any tendency for convection to develop which could lead to cloud and rain. Nevertheless, the latent heat energy stored in the lower layers by evaporation in these latitudes is extremely important to the tropical circulation and provides a source of energy which can be utilised by disturbances or when local heating or orographic effects trigger off vertical movement. Meisner and Arkin (1985) point out that forcing of large-scale tropical and extra-tropical circulation by condensational heating is a focus of current research.

Over western continental areas under the influence of the STHs, lack of surface moisture means that despite high net radiant energy, evaporation is low. Since windspeeds within these systems are low, there is little advection of moisture from the oceanic areas and hence moisture content of the atmosphere is very low. Low evaporation means that the radiant energy is available to heat the surface and lower layers of air. Shallow low-pressure areas may therefore develop over the land areas in summer, but any low-level convergence associated with them will be effectively damped down by subsidence at higher levels. Even if a disturbance did produce marked ascent, the low moisture content would inhibit cloud and rain formation and decrease the likelihood of instability developing. Since subsidence is greatest on the eastern side of the STHs, it is here that the inversion is closest to the ground.

2.5 THE TRADE WINDS

The trade-wind inversion is an important characteristic derived from subsidence associated with the STHs (see above). As the air moves equatorwards, it leaves the area of marked subsidence. Consideration of the potential vorticity equation suggests, however, that air moving towards the equator will tend to subside and diverge (Riehl 1969, 1979). While divergence and subsidence continue as the trade-wind air moves equatorwards and westwards (Fig. 2.7a), the height of the inversion increases owing to the convection of energy and water vapour to higher levels (Fig. 2.7b). The inversion is not therefore a material surface. However, the increase in height is not a gradual, even, process. There are considerable variations, even on a day-to-day basis, due to disturbances producing local convergence and ascent with resultant upward transport of energy and lifting of the inversion. Riehl (1979) and Krishnamurti (1979) provide details of the structure, variations and weakening of the trade-wind inversion. The inversion is strongest where its height is lowest, the intensity weakening as the height increases.

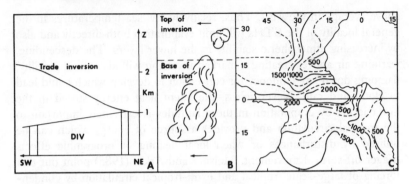

Fig. 2.7   (a) North-east trades: rise of the trade-wind inversion and subsidence of the air column due to divergence as air moves towards the equator (after Riehl 1969). (b) Mechanism of incorporating air in the trade inversion into the lower moist air layer (after Riehl 1969). (c) Height of the base of the trade-wind inversion (metres) in the Atlantic (after Riehl 1954).

Below the trade-wind inversion considerable cloud can develop, but the former inhibits vertical movement, the formation of deep clouds and precipitation. The arid conditions of the sub-tropical high latitudes therefore extend towards the equator over the eastern oceans and western parts of the continents. As the height of the inversion increases however (see Fig. 2.7c), disturbances become increasingly likely to create the necessary conditions for rainfall, the greater depth of moist air and latent heat supply favouring the development of these systems. Air nearing the equator will have characteristics more favourable for the development of disturbances, extensive cloud and precipitation.

Over the western halves of oceans, not only is subsidence associated with the STHs less than in the east, but also air moving into this area will have been considerably modified since leaving the STHs in the east. The warmer sea temperatures over the western side of the oceans will facilitate moisture accumulation and increase the tendency for vertical air movement by increasing instability. In the western ocean areas therefore, the trade-wind inversion will have been lifted to considerable heights and indeed may disappear completely. Conditions are therefore favourable for the development of disturbances, cloud and rainfall. This will also affect the eastern side of the continents, particularly since the airflow will be onshore.

Since the equatorial trough in some areas shows a far greater seasonal shift than the STHs, it follows that the trade winds vary in width both seasonally and longitudinally. They will be broader in the winter hemisphere, penetration from the latter into the summer hemisphere being particularly noticeable in some longitudes (Figs 2.4, 2.5). Penetration into the opposite hemisphere is most noticeable in the northern summer (Fig. 2.4), the greatest development being linked to the south-west monsoon of the Indian Ocean and South-East Asia. In January (Fig. 2.5), penetration is not so marked into the southern hemisphere except over central and eastern Africa and Australia. Deep intrusions of the trough into central Australia do occur in a few years (Riehl 1979). Over the Atlantic and eastern Pacific, the north-east trades do not cross the equator.

After crossing the equator, not only are the trades moving polewards but they also tend to develop a component from the west (Fig 2.4, 2.5). It can be shown that winds from the west have a tendency for ascent while those with an easterly component tend to subside (Flohn 1964). Flohn found that the average frequency of rain associated with tropical easterlies was 8.1 per cent but 25.1 per cent with tropical westerlies. As was suggested earlier, air moving equatorwards tends to subside, but air moving away from the equator has the reverse tendency. Flohn found that tropical air moving equatorwards has a rainfall frequency of 8.1 per cent, but air moving away from the equator a frequency of 12.0 per cent. Air which has crossed the equator will in any case be far from the subsiding influence of the STHs. All these points suggest that when the trades penetrate into the opposite hemisphere rainfall is more likely, but there are exceptions.

2.6 THE EQUATORIAL TROUGH

Recent studies using satellite data show the true complexity of this feature. While to some extent it is difficult to ignore the association between location of the trough and distribution of land and sea, the position, which in any case cannot always be traced, is not simply related to areas of maximum heating. Horizontal variations in surface temperature are comparatively slight over much of the area between the sub-tropics, and hence a simple concept of warm air rising within an equatorial trough is untenable. Instead, the energy for marked ascent originates from latent heat released in cumulus cloud development. As has already been indicated, this energy accumulates from evaporation in the sub-tropical maritime areas and increases within the deepening layer of air below the trade-wind inversion as the air moves equatorwards and westwards. The significance of this energy for the tropical circulation has already been mentioned in § 2.4 and is stressed for example by Petrossiants (1981). In the vicinity of the trough and to the west the inversion may have disappeared or be raised well above the surface. The energy will only be released if some mechanism initiates ascent of air leading to condensation. Heating of the air over land during the day or orographic influences may provide the necessary trigger action. General convergence of the trade winds from the two hemispheres can also initiate ascent.

Important as the above mechanisms are in certain places and at certain times, much of the initial trigger action is due to disturbances creating areas of local convergence and ascent. Following ascent, cloud formation may begin and the resultant release of latent heat provides the ongoing driving force for the major upward movement and energy needed by the disturbances. It is in such areas that heavy rain is likely.

The equatorial trough is not, therefore, a region of continuous ascent with associated cloud and rain. Convergence is intermittent in time and space and maximum convergence and ascent often is found several degrees equatorward of the trough. Even in this area, the inversion layer restricts ascent in some locations. Cold ocean water at the equator in the South Pacific from the west coast of South America to 85°W longitude seems to prevent ascent, the trade-wind air flowing westward to the warm west Pacific where heating and increase in moisture promote upward movement (Bjerknes 1969). This aspect is discussed in § 2.8. Recognition of the significance of disturbances in the development of deep cloud and heavy rainfall overcomes earlier problems when attempts were made to reconcile the association of cloud and rain with general areas of convergence and ascent in an inter-tropical convergence zone (ITCZ) associated with the equatorial trough. In § 2.3, it was pointed out that an alternative term, the inter-tropical confluence (ITC) has been suggested because the ITCZ is not a continuous feature and also convergence is not limited to the equatorial trough. Satellite photographs show that only rarely can ITCZs be identified as long, continuous cloudy areas. Areas of convergence grow and decay either *in situ* or associated with moving disturbances and the position and intensity of the ITCZ changes greatly on a short-term basis. Areas of cloud and rain show very varied structures – sometimes linear, sometimes oval or circular. Clusters of clouds separated by large areas of relatively clear skies occur (Holton *et al.* 1971). The movement of cloud and rain areas is often irregular and clouds and rain sometimes appear to jump between locations without affecting intervening areas and often run counter to seasonal trends. Such irregular space–time variation did not suit earlier concepts of widespread ascent. It was the persistence with attempts to make observed facts fit these concepts that retarded advances in the subject. Many years ago Thompson (1957a) pointed out that a simple association of rainfall with movements of the ITCZ had proved of no value for forecasting. Often the ITCZ cannot be traced and even when it can, rainfall is not always associated with it. In discussing theories about the ITCZ, Krishnamurti (1979) points out that there are marked regional differences in its character.

Chang (1970) and Reed (1970) present evidence that cloud clusters are associated with westward-moving disturbances. Ruprecht and Bucher (1970) point out the high degree of mesoscale organisation of cloudiness and that mesoscale motion is a significant link in energy transport from surface to atmosphere, especially over oceans. Fijita *et al.* (1969) and Williams (1970) have shown that cloud clusters represent concentrations of cyclonic vorticity at low levels and anticyclonic vorticity at high levels with associated strong ascent. Between the clusters, the situation is reversed, with strong subsidence. All this evidence indicates the significance of disturbances to the processes of energy transfer, circulation patterns and the formation of cloud and rain.

Ramage (1968) suggested that where convergence, clouds, rain and disturbances do occur in the equatorial trough – and often they do not – surface heating is interrupted. Accepting a degree of thermal control in the formation of the trough, Ramage suggested that this feature will re-form towards the sun's zenithal latitude. In winter, this will be equatorward of the rain belt and in summer on the poleward side of the rain belt. Sadler (1970) suggested that the ITCZ, viewed as a zone of maximum cloudiness, does not migrate. Clouds build up and decay *in situ*. Bates (1970) theoretically showed that the ITCZ is not zonally symmetric but is a series of disturbances of wavelength 2000 km, having strong centres of rising motion propagating westwards at about 13 kn. He suggests that these arise because of dynamic instabilities in the large-scale airflow without any influence of continent – ocean asymmetries or extra-tropical perturbations being taken into account.

Riehl (1979) indicates that there are two distinct belts of convergence in the Pacific and Indian oceans in July and to a lesser extent in the Atlantic. Between the two zones, marked divergence is found on the equator. In the Atlantic, Krishnamurti (1979) refers to the possible existence of two different disturbance zones. The more northerly one, at about 12°N, is associated with easterly waves and the other, at about 7°N, being associated with ITCZ waves. Satellite observations suggest that the ITCZ waves are modulations of a quasi-stationary ITCZ by the westward passage of the easterly waves further north. Petrossiants (1981) refers to the constant propagation of wave disturbances and vortices along the ITCZ which sometimes develop into tropical cyclones.

In reviewing evidence provided by GATE (Global Atmospheric Research Program (GARP) Atlantic Tropical Experiment), Petrossiants (1981) stresses the complexity of ITCZ structure and the variety of cloud forms associated with it. Averaging of cloud patterns in the ITCZ produces a continuous belt over the Atlantic and Pacific, but although there are days when ITCZ cloud stretches in an almost unbroken band, at other times it breaks up into separate cloud clusters, a series of disjointed strips, or suddenly disappears completely. At least three types of cloud clusters have been identified: clusters associated with squall lines; ribbon-type cloud clusters; and clusters with closed circulations (Petrossiants 1981; Hastenrath 1985). McQuate and Hayden (1984) used cloud patterns to define the mean position of the ITCZ in each of the months February–April over north-east Brazil. They found that rainfall in each month was significantly correlated with the position of the ITCZ.

The above discussion of only a small section of the literature indicates the complexity and uncertainty of the situation. Areas of general convergence of trade winds do exist, depending on the location and intensity of the STHs. Seasonal shifts in windfield convergence do respond to the latter but disturbance activity on a short-period basis masks a simple relationship.

Undoubtedly, what are often identified as areas of convergence are the result of disturbances which may be irregular in space and time. Certainly there may be preferred areas of development of these systems which, when averaged out, suggest a general area of low pressure and an ITCZ. General convergence of the trade winds is not apparent in some areas, the airflow often being very slight and in some cases divergent in the vicinity of the trough. At certain times, westerly winds may be observed in the vicinity of the trough and these are considered semipermanent in some areas. The tendency for such westerlies to ascend has already been indicated.

2.7 THE SOUTH-EAST ASIAN MONSOON

The underlying mechanisms of this system are extremely complex and since much has been written on the subject, the reader is referred to discussions presented by Krishnamurti (1979), Riehl (1979), Ramage (1971), Barry and Chorley (1982), Subbraramayya, Bhanu Kumar and Babu (1981), Pasch (1984), Lockwood (1985). Hastenrath (1985) and a range of papers in Lighthill and Pearce (1981) and American Meteorological Society (1984).

Krishnamurti (1979) and Druyan (1981) stress the importance of general circulation models in gaining a deeper understanding of the monsoon. Examples of the application of models include Druyan (1982a, b) and Fang and Rao (1985). Here the discussion will be limited to indicating that, again, rainfall controls are not so simple as is commonly thought.

The area is characterised by a seasonal reversal of winds at lower levels (Fig. 2.8), most clearly seen over the Indian sub-continent where much research has been undertaken. Lockwood (1985) divides the year as follows: (1) *The north-east monsoon season*: (a) January–February – winter season, (b) March–May – hot weather season; (2) *The south-west monsoon season*: (a) June–September – the season of general rains, (b) October–December – post-monsoon season. During these periods, however, conditions vary considerably in terms of gradual seasonal

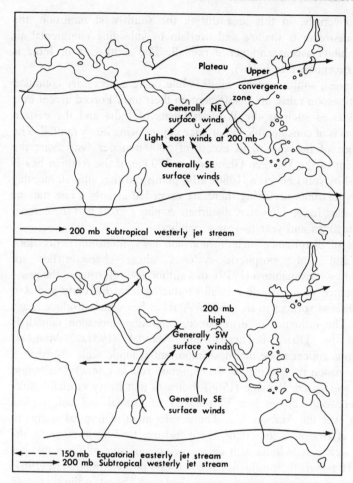

Fig. 2.8 Monsoon circulation: northern winter (upper); northern summer (lower) (after Lockwood 1974).

progressions. There are also short-term fluctuations in the intensity, advance and retreat of the systems. Rainfall conditions associated with the two seasons differ considerably throughout the region and show fluctuations from year to year. Disturbances are again fundamental to rainfall. Over much of the region, the period of the north-east monsoon is extremely arid with, on average, plentiful rain during the south-west monsoon. However, exceptions indicate the complexity. In the north-west of the Indian sub-continent during the north-east monsoon, extra-tropical depressions, usually non-frontal, often originating over the eastern Mediterranean, move across the Middle East to provide rainfall which is especially important since evaporation is at a minimum at this time. Conversely, in this area during the south-west monsoon, the surface airstream is shallow and overlain by subsiding continental air which inhibits ascent and hence rainfall. The south-east of India is another exception (see below).

The moist, south-westerly winds, while being a necessary condition for the monsoon rains, are insufficient on their own. Forced ascent over highland areas such as the Western Ghats of India and the eastern Himalayas is of considerable importance in producing heavy rainfall. The occurrence of heavy rain, in excess of 5000 mm per year along the escarpment of the Western Ghats within 100 km of the Arabian Sea is discussed by von Leugerke (1980), who points out that annual, monthly and daily amounts occurring there are exceeded at only a few stations in north-east India. However, disturbances play a great part in seasonal rainfall regimes and year-to-year variability.

Types of disturbance include monsoon lows, monsoon depressions (§ 2.9) and mid-tropospheric cyclones whose characteristics are discussed by Krishnamurti (1979), this author also referring to the possible existence of mesoscale rainfall-producing features. Fang and Rao (1985) discuss vortices in the eastern Arabian Sea which produce heavy rainfall. The influence of disturbances on Indian monsoon rainfall is reviewed by Dhar, Rakhecha and Mandal (1981a). Monsoon depressions appear to be the most important synoptic scale disturbance in the monsoon trough, playing an important role in rainfall distribution in time and space. Ramage (1968) indicates that heavy rainfall results from 'monsoon' depressions in the Bay of Bengal and sub-tropical cyclones over the Arabian Sea. The development of tropical storms in the Bay of Bengal is important, these features tending to move north-west or west across India with an average occurrence of about two per month. However, their influence is not always positive. Raghaven (1967) found that in particular months when they were absent, while there was less than average rainfall in some areas, especially in the vicinity of the Bay of Bengal, an increase in rainfall occurred elsewhere. Dhar et al. (1981a) found that although there was a relationship between rainfall and frequency of tropical disturbances for individual months (June–September), the season as a whole did not exhibit any significant relation. Dhar et al. (1981a) also cite a number of other aspects of disturbances and rainfall based on earlier work. In the Ganga and Godavari basins, disturbances seem to be capable of producing about 8 per cent and 12 per cent of the respective mean annual totals. Also, when India is free of disturbances in the months of July and August, the net rainfall deficit in the plain areas of the country north of 20°N is of the order of 14–19 per cent of the mean. In the south-east of India, the peak rainfall occurs during the north-east monsoon. Much of this appears to be associated with the generation of disturbances in the Bay of Bengal rather than simply with onshore airflow. The varying import-

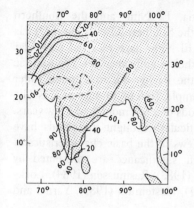

Fig. 2.9 The percentage contribution of monsoon rainfall (June to September) to the annual total (after Ananthakrishnan and Rajagopalachari 1964).

ance of the south-west monsoon period to rainfall is illustrated in Fig. 2.9.

During the south-west monsoon in India, the monsoon trough is normally along the Gangetic Valley. However, the trough sometimes moves further north, close to the Himalayan foothills. When this happens, rainfall declines over most of the country apart from the southern slopes of the Himalayas, adjacent plains and the south-east of the peninsula. This situation is called 'monsoon break' and prolonged and frequent occurrences can cause country-wide drought. Circulation changes associated with the situation are discussed by Subbaramayya, et al. (1981). Increases in rain associated with the 'break' in the foothills and adjoining plains of the Himalayas may be of the order of 100–300 per cent in some parts, but even in this area (e.g. west of 80°E) there are decreases (Dhar, Soman and Mulye 1984). The latter authors point out that increased rainfall in the headwaters of Himalayan streams can cause downstream flooding in areas experiencing almost drought conditions.

McBride (1984) cites similarities between the Australian monsoon and that of South-East Asia, including a monsoon equatorial trough, low-latitude westerlies, onset circulation patterns and marked changes in upper tropospheric flow. Holland, Keenan and Guymer (1984) refer to it as the Australian component of the Asian monsoon. McBride (1984) and Holland et al. (1984) provide reviews of various aspects of the Australian monsoon, including its structure at different seasons, onset of the summer situation, rainfall, tropical cyclones and monsoon depressions.

2.8 EL NINO AND THE SOUTHERN OSCILLATION

In § 2.6, reference was made to the existence of cold surface water at the equator in the South Pacific and its influence on air above it. This introduces two associated phenomena, El Nino and the Southern Oscillation. Although the existence of the Southern Oscillation (SO) has been known since the 1920s, it and El Nino have received particular attention in recent years because of their significance for atmospheric circulation and rainfall, not only in the Pacific but in a much wider context. Rasmusson (1984), for example, refers to the SO as the dominant global climate signal for periods of a few months to a few years. Possible links with extreme events, particularly droughts and floods, have been the focus of much interest. Reviews of the basic characteristics of El Nino and SO (ENSO) and their significance are presented by Rasmusson and Wallace (1983), Cane (1983), Rasmusson (1984), Lockwood (1984), Ramage and Hori (1981), Hastenrath (1985), Yarnal and Kiladis (1985) and Wright (1985).

In the 1920s, attention was drawn to an oscillation in atmospheric pressure between the east and west sides of the Pacific which was termed the SO. Bjerknes (1969) found this surface pressure oscillation to be related to variations in rainfall and sea surface temperatures (SSTs) in the equatorial eastern Pacific, particularly anomalous warm temperatures off the Peruvian coast termed El Nino. As indicated in § 2.6, Bjerknes (1969) felt that the extensive cold water at the equator in the Pacific seems to prevent ascent of trade-wind air which therefore flows westward to the warm west Pacific. It then accumulates moisture and is heated, taking part in large-scale ascent. In the Indonesian area this ascent leads to substantial rainfall and the latent energy released exceeds net radiation at the surface (Lockwood 1984). This circulation between the east and west Pacific is often termed the Walker circulation. In the Atlantic, SST contrasts between east and west are much smaller than in the Pacific, but in January, a thermally driven Walker circulation may develop (Lockwood 1984).

In the eastern Pacific, every year a warm current moves south off Ecuador, displacing the cold surface waters off the South American coast. This feature is known as El Nino. From time to time, El Nino is much more intense and persistent than normal. SST rises along the Peruvian coast and the eastern equatorial Pacific may persist for more than a year and very heavy rain may occur over western South America. Recent occurrences include 1972–73, 1976 and 1982–83. Philander (1985) refers to the complementary phase of El Nino, when low SSTs are experienced near the equator, as La Nina.

Early ideas that rate of upwelling of cold water controlled El Nino have been rejected in favour of determination by fluctuations in the strength of trade winds (Ramage and Hori 1981). As the central Pacific trades decrease in strength, so El Nino intensifies until the trades increase again.

The relationship between the SO and El Nino has been the subject of much study in recent years, Rasmusson (1984) pointing out that a theoretical understanding of the links having only then begun to emerge. There is persuasive evidence that the key element of ENSO episodes lies in large-scale tropical sea–air interaction. The strength of the Walker circulation varies with SST in the Pacific.

Various Southern Oscillation Indices (SOIs) have been adopted in analysis (Lockwood 1984; Wright 1984), one simple one being the Tahiti minus Darwin mean sea level pressure. A negative phase of SOIs when pressure tends to be below average at Tahiti and/or above average at Darwin is associated with abnormally warm SST in the equatorial eastern Pacific.

Prediction of El Nino events has been more difficult than anticipated (Ramage and Hori 1981). Baker-Blocker and Bouwer (1984) examined occurrence over the period 1872–1976 using spectral analysis. Numerous periods between 2.3 and 37.1 years were found but it must be stressed that interpretation of results of such analysis is difficult. They suggested that large random pulses, perhaps of volcanic or solar origin, may act as triggers to shift the atmosphere between quasi-stable states.

Correlations between the SO, El Nino and a range of atmospheric circulation features, including hurricane occurrence and rainfall conditions in a variety of areas have been examined by many workers in recent years. While statistically significant correlations have been established, links are far from perfect. Nor does correlation necessarily imply causality. However, as was indicated at the start of this chapter, the atmosphere is an entity, events in different areas being linked. Certainly even where significant correlations have been established, prediction is still very uncertain since much of the variance remains unaccounted for.

Rasmusson (1984) states that the surface pressure oscillations reflect fluctuations in low-level wind systems linked to major shifts in the large-scale precipitation regime of the Pacific and Indian Oceans. In general terms, when pressure is high in the Pacific, it tends to be low in the Indian Ocean from Africa to Australia and, in both areas, rainfall varies inversely with pressure (Lockwood 1985). Links between the SO and the middle latitudes of both hemispheres also occur. Nobre and Renno (1985) point out that the very intense ENSO event in 1982–83 appeared to be linked to floods in many areas of South America but droughts in others. They point out that the possible links with global climate anomalies of this event are not completely clear. Pittock (1984), in an analysis of southern hemisphere teleconnections concluded that: (1) there is strong evidence of a connection between ENSO and Australian rainfall; (2) other circulation mechanisms also play a major part; (3) ENSO events may be influenced by southern hemisphere high-latitude circulations; (4) at least some of the patterns and sequences of correlation may not remain constant in time. McBride and Nicholls (1983) provide another example of analysis of links between Australian rainfall and the SO.

Links between Indian rainfall conditions, the SO and/or El Nino have received considerable attention (e.g. Mooley and Parthasarathy 1983; Rasmusson and Carpenter 1983; Bhalme, Mooley and Jadhav 1983; Shukla and Paolino 1983; Bhalme and Jadhav 1984b; Mooley, Parthasarathy and Sontakke 1985; Parthasarathy and Pant 1985). Hawaiian rainfall links are considered by Lyons (1982) and Taylor (1984). The nature of some of these and other links are considered in § 3.7.

Relationships between El Nino and hurricanes in the Atlantic are considered by Gray (1984a, b) and in the Pacific during the 1982–83 event by Sadler (1984). A variety of other links between the SO and/or El Nino and other atmospheric features are considered by Stoeckenius (1981), Chen (1983), Allan (1983), Nicholls (1984b), Newell, Selkirk and Ebisuzaki (1982), Selkirk (1984), Pazan and Meyers (1982), Hastenrath (1985), Yarnal and Kiladis (1985).

2.9 DISTURBANCES AND RAINFALL

The importance of disturbances, not only for rainfall but also for energy transport, is evident from the discussion in § 2.7. They are most frequent in areas where vertical ascent is favoured by the absence, weakness or great height of the trade-wind inversion. The main regions are the 'disturbed' trade-wind area over the western oceans and adjacent land areas, certain parts of the equatorial trough and the south-east Asian monsoon system. There are many types and classification is difficult, partly because of lack of understanding but also because types can be interrelated. Some types appear to develop from others or to be composite in character. There are also problems because of the use of different terminology in different regions.

Riehl (1979) points out that since other than on the sub-tropical margins, all systems depend to a considerable extent on conversion of latent heat energy to potential and then kinetic energy, it is unlikely that there are as many physical mechanisms involved as there are superficial aspects of the storms. Barry and Chorley (1982) suggest that five categories of tropical weather systems can be identified, according to their space and time scales. They range from individual cumulus clouds of short duration, through mesoscale systems, 'cloud clusters' which are sub-synoptic in scale lasting for 1–3 days, synoptic scale disturbances and finally planetary waves. Synoptic scale systems are of particular importance since they determine much of the 'disturbed' weather of the tropics. Riehl (1979) points out that it is possible to identify features shared by most synoptic systems, other than tropical storms, which he summarises under the name 'Tropical rainstorms'. Riehl stresses that rain is concentrated in small areas of a synoptic system, often only about 10 per cent of the total area. During passage of a disturbance, in most cases an individual location will receive only one heavy shower. In an area, rainfall varies considerably, some points perhaps having no rain. Krishnamurti (1979) discusses the characteristics of a wide variety of disturbances in different regions. Examples of discussion of rain-producing events in different areas are de Oliveira and Nobre (1985) – the Amazon; Sumner (1981, 1984) – coastal East Africa; and Harrison (1984) – South Africa. To indicate something of the variety and complexity of disturbances, some types are discussed below.

1. Wave disturbances

Within the trade winds and equatorial easterlies are found several types of westward moving wave disturbances whose vertical structure varies between regions. They may be difficult to detect in the surface pressure field. Satellite photographs suggest that they are not so common as was once thought. In the Caribbean, they are characterised by a wave-like

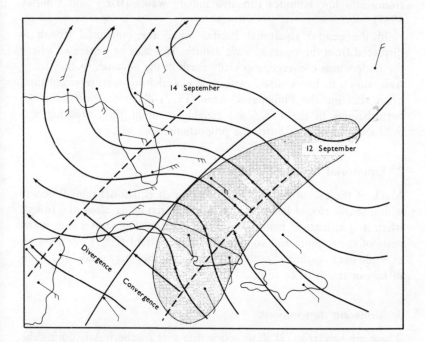

Fig. 2.10 Streamlines for the layer 850–700 mbar for the western tropical Atlantic on 13 September 1957: solid line gives the position of the wave trough in the easterlies; dashed lines give positions one day earlier and later; shaded – main precipitation area (after Riehl 1965).

undulation in the airflow, forming a trough of low pressure. The wave moves in a westerly direction with the easterly airflow blowing through it. In terms of rainfall the key points are that ahead of the trough there is divergence in the lower layers and convergence above, leading to general subsidence. Behind the trough the reverse is true, leading to ascent of air with associated development of cloud and moderate to heavy rainfall (Fig. 2.10). However, when the easterly airflow is slower than the speed of the wave, the reverse pattern of low-level convergence ahead and divergence behind the trough occurs. As would be expected, they tend to occur when the trade-wind inversion is weak or absent. They are of greatest significance over the western ocean areas, particularly during and after the high sun period. During the 'cool' season, stronger subsidence, reflecting the influence of the sub-tropical highs as well as lower sea-surface temperatures, inhibits development.

In § 2.6, reference was made to the possible existence of two sets of linked waves in the Atlantic. Krishnamurti (1979) discusses waves propagating from Africa across the Atlantic which may move over the central American highlands into the ITCZ disturbance belt of the eastern Pacific. Some may be transformed into hurricanes. Penetration of cold fronts into low latitudes can also initiate waves (Barry and Chorley 1982).

In the central equatorial Pacific, when the equatorial trough is displaced from the equator, wave disturbances may be generated where the trade winds converge, especially north of the equator. As they move west, they can break down to form a closed low-pressure circulation, often affecting the Philippines. Agee (1972) illustrates how wave-like perturbations in the ITCZ are capable of amplifying, breaking and producing vortices of hurricane proportions.

2. Equatorial trough disturbances

Much of the ascent of air giving rise to high equatorial rainfall occurs within small, closed low-pressure systems originating along the trough when it is at least 5° latitude from the equator. They travel over many parts of the oceanic and continental tropics. Riehl (1965) discusses an example over northern South America where they may yield 25–75 mm of rain or much more (Fig. 2.11).

3. Monsoon depressions

These are similar to (2) above, occurring over southern Asia during the south-west monsoon. They are important at the head of the Bay of Bengal and along the Ganges Valley, moving west or north-west across the north of the sub-continent with an average frequency of two per

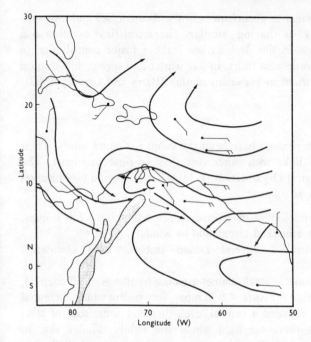

Fig. 2.11 Streamlines at 850 mbar over northern South America on 3–4 July 1952: solid line – equatorial trough; C – centre of disturbance; shaded area – land over 2000 m (after Riehl 1965).

month during July and August. They last for about 4 to 5 days and rainfall would be controlled dominantly by relief if they do not occur. They produce heavy rain, often in excess of 100 mm in 24 hours, over a 400 km-wide strip to the left of the depression track (Dhar, Rakhecha and Mandal, 1981a). The strip extends from about 500 km ahead to 500 km behind the centre. The structure and origins of monsoon depressions are discussed by Riehl (1979). As has been suggested earlier, their occurrence seems to inhibit rain outside their general sphere of activity. Raghaven (1967) compared rain in months without such storms to average rainfall. Results suggest that in July, tropical storms deprive nearly one-third of India of about 10 per cent of its share of monsoon rain, while an almost equal area receives a similar increase. In August, only a small part of India benefits from these storms while a major part is deprived of monsoon rainfall. Much of India receives above-average rainfall from other disturbances when monsoon depressions are absent from areas where they have a beneficial effect. Falls (1970) discusses a monsoon disturbance model over northern Australia closely resembling those of the Bay of Bengal. He also stresses that the rainfall situation is not so simple in this area as would be expected from low-level inflow of moist tropical air into a thermal trough.

Riehl (1979) refers to disturbances in the Arabian Sea and also sub-tropical cyclones as having similar characteristics to monsoon depressions. Those in the Arabian sea make a major contribution to summer rain in north-west India. In late winter and spring, sub-tropical cyclones are important to Hawaiian rainfall (Barry and Chorley 1982).

4. Hurricanes

These violent disturbances have been the subject of much study. Their development and links with other disturbances pose problems. The World Meteorological Organisation (WMO) has adopted a classification of tropical storms as follows:

(a) 'Tropical depression' – low pressure enclosed within a few isobars, either lacking a marked circulation or winds < 17 m/sec.
(b) 'Tropical storm' – several closed isobars, wind circulation 17–32 m/sec.
(c) 'Tropical cyclone' – small diameter (some hundreds of kilometres), minimum surface pressure < 990 mbar, very violent winds, torrential rain. Usually contains a central 'eye', diameter some tens of kilo-metres where there are light winds and lightly clouded sky. In hurricanes, the central pressure at sea-level is commonly 950 mbar or less and hurricane winds are arbitrarily defined as 33 m/sec or more.

Differentiation between these various tropical storms is often difficult, particularly since there is a tendency for a weak system to strengthen, perhaps eventually to full hurricane strength (category (c)), as illustrated by Agee (1972). Burpee (1972) describes easterly waves related to the easterly jet stream found to the south of the Sahara which propagate across the Atlantic, sometimes reaching the eastern Pacific. While few intensify after reaching the Atlantic, they account for about half the tropical cyclones formed there. As already mentioned, waves crossing the Atlantic and Central America into the eastern Pacific serve as the initial disturbance from which many tropical cyclones form (WMO 1979). Closed systems associated with the equatorial trough fall within the above classification, beginning as weak tropical depressions, but some-times developing further. Closed systems are not limited to the equatorial trough and they are not usually found within 5° of the equator. While weak tropical lows are relatively frequent, few develop into full hurri-canes, there being less than fifty in the northern hemisphere during an active year.

There is much evidence that hurricanes form from existing disturb-ances but the reasons why few of the latter do reach the extreme stage are not completely understood. There is also some confusion resulting from the regional differences in terminology applied to systems. For example, under the 'tropical storm' category (17–32 m/sec windspeeds) above, the terms tropical cyclone, cyclone, tropical depression are also used in different areas (WMO 1979). Similarly the 'tropical cyclone' category (> 33 m/sec windspeeds) is also referred to as typhoons, severe cyclones, and hurricanes.

General characteristics of hurricanes are described in texts such as Riehl (1979), Barry and Chorley (1982), Simpson and Riehl (1981), Anthes (1982). Many papers in the *Monthly Weather Review* deal with them, this journal including annual summaries of their activity in the eastern north Pacific and the Atlantic. Lourensz (1981) reviews tropical cyclones in the Australian region. American Meteorological Society Conferences on Hurricanes and Tropical Meteorology also contain many papers. Hurricanes vary considerably in size, both seasonally and regionally. Merrill (1984) has examined the climatology, structure and possible reasons for the size variation. He points out, for example, that those in the western north Pacific are characteristically twice the size of their Atlantic counterparts. Size appears to be only weakly correlated with intensity (Merrill 1984).

Based upon computed and actual values, Riehl (1979) suggests hurricanes produce an average of 400 mm of rain at a point on the track over a 2-day period. However, as Anthes (1982) indicates, the actual rainfall at a point depends on many factors besides the maximum rainfall intensity. These include location with respect to storm track, distribution of rain around the storm, speed of movement, and local effects such as relief and coast orientation. High winds clearly create measurement problems. Anthes (1982) suggests average areal rates of 100 mm/day inside a radius of 220 km and 30–40 mm/day between 220 and 440 km. Simpson and Riehl (1981) discuss the marked differences in rainfall distribution on windward and leeward slopes of mountains. They also point out that heavy rain often occurs long after hurricane characteristics have disappeared from the surface.

Mechanisms need not concern us, but a necessary source of latent heat energy limits development to warm ocean areas, probably at least 27 °C. The significance of the warm ocean has been illustrated by Pope (1971) for a cyclone crossing Madagascar. While crossing the land, the eye disappeared but then re-formed when it regained its source of energy over the warm ocean. Hurricanes are also limited to areas beyond about 5° from the equator. A majority (65%) form in a zone 10°–20° from the equator (Anthes 1982). They tend to develop over warmer western ocean areas away from the subsiding influence of the STHs. While many storms show a connection with the equatorial trough, others originate deep in the trades.

No hurricanes occur in the South Atlantic where the STH is

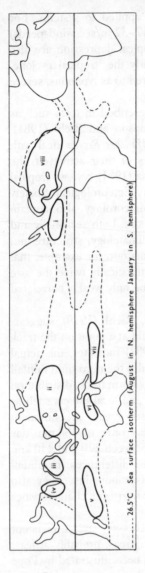

Fig. 2.12 Areas of tropical storm development (after Gray 1968).

persistent and the equatorial trough is never south of 5°S (Fig. 2.3). Much of the South Pacific is largely free of storms except for areas such as a narrow zone off Queensland over relatively warm sea. Like the South Atlantic, the south-east Pacific is an area of general subsidence, the equatorial trough remaining north of the equator (Fig. 2.3). In the southern hemisphere, the main season is from December to March. In the northern hemisphere, the main period of activity is from July to October in the western Atlantic and western Pacific. There is a marked

Table 2.1   Frequency of tropical storms* (after Gray 1968 from Lockwood 1974)

Area number (see Fig. 2.12)	Location	Average number per year	Per cent of worldwide total
I	North-east Pacific	10	16
II	North-west Pacific	22	36
III	Bay of Bengal	6	10
IV	Arabian Sea	2	3
V	South Indian Ocean	6	10
VI	Off north-west Australian Coast	2	3
VII	South Pacific	7	11
VIII	North-west Atlantic (including western Caribbean and Gulf of Mexico)	7	11

* Defined as a warm core vortex circulation with sustained maximum winds of at least 20 m/sec.

peak in September in the western Atlantic. Occasionally, storms affect both areas as early as May or as late as December. A secondary early summer maximum occurs in the Bay of Bengal. Hurricanes in the eastern Pacific, which have developed from disturbances crossing Central America (see above), are prevented from moving further west by a strong anticyclone. They are probably partly responsible for high rainfall over the west coast of Mexico.

The general areas of development of tropical storms are summarised in Fig. 2.12 and their frequencies in various areas in Table 2.1. While the definition of tropical storms (see Table 2.1) is not specifically limited to hurricanes, Table 2.1 and Fig. 2.12 do indicate the basic distribution and relative frequency of the latter. These areas also possess favourable conditions for other disturbances such as those already discussed. Figure 2.12 indicates how well the location fits inside the 26.5 °C sea surface isotherm. Table 2.2 shows the monthly and annual number of storms for six areas. Carpenter et al. (1972) found that over a 78-year period, about 20 per cent more hurricanes formed near new and full moon than near the quarters, showing a stronger peak at new moon than full moon.

The disturbed trade-wind areas where hurricanes occur have the highest daily rainfalls in the world, intensities often being higher than those nearer the equator. Their significance to seasonal totals is varied. Reference has already been made to their importance on the west coast of Mexico where they provide water for agriculture. Simpson and Riehl (1981) indicate other areas such as south-east USA and South-East Asia where they are important in times of water shortage. Pedgley (1969) found that while tropical cyclones approach or cross the coast of Arabia only one year in three, in this dry area they still account for up to 25

Table 2.2 Monthly and average number of storms per year for each major basin (from Crutcher and Quayle 1974)

Basin and stage	Jan	Feb	Mar	Apr	May	Jun	Jul	Aug	Sep	Oct	Nov	Dec	Annual
North Atlantic													
Tropical storms	*	*	*	*	0.1	0.4	0.3	1.0	1.5	1.2	0.4	*	4.2
Hurricanes	*	*	*	*	*	0.3	0.4	1.5	2.7	1.3	0.3	*	5.2
Tropical • *storms and hurricanes*	*	*	*	*	*0.2*	*0.7*	*0.8*	*2.5*	*4.3*	*2.5*	*0.7*	*0.1*	*9.4*
Eastern North Pacific													
Tropical storms	*	*	*	*	*	1.5	2.8	2.3	2.3	1.2	0.3	*	9.3
Hurricanes	*	*	*	*	0.3	0.6	0.9	2.0	1.8	1.0	*	*	5.8
Tropical storms and hurricanes	*	*	*	*	*0.3*	*2.0*	*3.6*	*4.5*	*4.1*	*2.2*	*0.3*	*	*15.2*
Western North Pacific													
Tropical storms	0.2	0.3	0.3	0.2	0.4	0.5	1.2	1.8	1.5	1.0	0.8	0.6	7.5
Typhoons	0.3	0.2	0.2	0.7	0.9	1.2	2.7	4.0	4.1	3.3	2.1	0.7	17.8
Tropical storms and typhoons	*0.4*	*0.4*	*0.5*	*0.9*	*1.3*	*1.8*	*3.9*	*5.8*	*5.6*	*4.3*	*2.9*	*1.3*	*25.3*
South-west Pacific and Australian area													
Tropical storms	2.7	2.8	2.4	1.3	0.3	0.2	*	*	*	0.1	0.4	1.5	10.9
Hurricanes	0.7	1.1	1.3	0.3	*	*	0.1	0.1	*	*	0.3	0.5	3.8
Tropical storms and hurricanes	*3.4*	*4.1*	*3.7*	*1.7*	*0.3*	*0.2*	*0.1*	*0.1*	*	*0.1*	*0.7*	*2.0*	*14.8*
South-west Indian Ocean													
Tropical storms	2.0	2.2	1.7	0.6	0.2	*	*	*	*	0.3	0.3	0.8	7.4
Hurricanes	1.3	1.1	0.8	0.4	*	*	*	*	*	*	*	0.5	3.8
Tropical storms and hurricanes	*3.2*	*3.3*	*2.5*	*1.1*	*0.2*	*	*	*	*	*0.3*	*0.4*	*1.4*	*11.2*
North Indian Ocean													
Tropical storms	0.1	*	*	0.1	0.3	0.5	0.5	0.4	0.4	0.6	0.5	0.3	3.5
Cyclones[†]	*	*	*	0.1	0.5	0.2	0.1	*	0.1	0.4	0.6	0.2	2.2
Tropical storms and cyclones[†]	*0.1*	*	*0.1*	*0.3*	*0.7*	*0.7*	*0.6*	*0.4*	*0.5*	*1.0*	*1.1*	*0.5*	*5.7*

* Less than 0.05   † Winds ⩾ 48 Kn
Monthly values cannot be combined because single storms overlapping two months were counted once in each month and once in the annual.

per cent of the annual rainfall. Milton (1980) indicates the importance of tropical cyclones for water in tropical Western Australia.

Many parts of South-East Asia rely heavily on hurricanes for water supply. For Hong Kong, 24 per cent of the annual total and 40 per cent of the rain in July, September and October is derived from hurricanes (Watts 1969). For various locations in China, Chang (1958) quotes the following proportions of the annual total rainfall resulting from hurricanes: Canton – 21 per cent; Foochow – 17 per cent; Shanghai – 11 per cent; Nanking – 4 per cent; Peking – 3 per cent; Wuhan – 2 per cent. Lu (1954) gives a general figure of 30–35 per cent for the south China coast. Hydrological aspects of tropical cyclones in the humid tropics are presented in Keller (1983). Where they produce much of the total rainfall, their relative infrequency means that the presence or absence of a small number of hurricanes will create marked variations in amount from year to year. In addition to their significance for water supply in some areas, hurricanes have other beneficial effects. For example, Anthes (1982) indicates how they can cause upwelling of nutrient-rich water which is important for fishing. Weaver (1968) illustrates how their destructive effects can even bring about beneficial economic change.

Adverse impacts of hurricanes are reviewed by Simpson and Riehl (1981) and Anthes (1982). Moore and Osgood (1985) developed a model to estimate sugar cane crop damage in Hawaii, estimates being within 10 per cent of the apparent loss associated with an actual event.

In view of their significance in some areas, forecasting occurrence and movement is important. Keenan (1986) provides a recent example of approaches to this. In § 2.8, reference was made to links between hurricanes and ENSO. Chan (1985) and Nicholls (1985) are further examples of investigation of such links.

5. Linear disturbances

In various areas, disturbances produce cloud and rainfall which is linear in pattern. The nature of these disturbances varies, nor are they fully understood. West Africa provides one example. This region is peculiar in that it has something approaching a 'frontal surface'. A moist south-west monsoon airstream from the Atlantic forms a wedge which deepens to the south underneath a hot, dry north-easterly airstream from the Sahara. Towards the northern limits of the south-westerly airstream, its depth is not great enough to promote rainfall under normal circumstances. Some 320 km south of the limit of the monsoon air, however, disturbance lines, several hundreds of kilometres long, travelling west against the surface current produce squalls and thundery showers. Since the northern limit of the monsoon air advances and retreats north and south in spring and autumn, much of the rainfall of West Africa is derived from disturbances at these times. During summer, the monsoon air, certainly in the south, is deep enough itself to allow rainfall, although even then, disturbances of a different kind may be responsible for much of the total. Northern areas, where the monsoon air is still comparatively shallow, still rely on disturbances. Thus, in 1955, 30 per cent of the annual rainfall over coastal Ghana was associated with disturbances, but 90 per cent of the total in the north (Barry and Chorley 1982). Krishnamurti (1979) discusses West African wave disturbances, disturbance lines and squall lines, including their west-ward movement over the Atlantic. The exact nature of these systems is not known, but they are almost certainly related to fluctuations in the upper easterly airstream. Omotosho (1985) illustrates the great import-ance of line squalls in Nigeria.

If the speed of air movement decreases downwind then a 'piling-up' of air results. Such convergence in the lower layers leads to ascent of air and possibly rainfall. Variations in windspeed may be due to an increase in the intensity of the STHs, producing features sometimes called surge lines. Frictional contrasts between land and sea surfaces or the general relief of the land may also result in the slowing down of the lower layers of air. Line squalls (Sumatras) cross Malaysia in the early morning which may be due to night-time convergence of land breezes from the uplands of Malaysia and Sumatra over the Straits of Malacca. Gichiuya (1970) describes disturbances in the south-east monsoon over East Africa which produce an area of maximum rainfall in a band of width 70–120 km. He stresses that these are very different to easterly disturbances of the Caribbean and other parts of the tropics. Westward moving line disturbances originating along the Atlantic coast of Brazil and probably due to the sea breeze front (§ 2.10) are referred to by de Oliveira and Nobre (1985).

6. Cool season disturbances

The poleward margins of the tropics come under the influence of disturbances during the cool season. Cool season weather systems are discussed by Riehl (1979). These take many forms, including cold fronts which move into the tropics from higher latitudes. Cold fronts are rapidly modified, however, and hence their effects are felt only on the margins. Fermor (1971) discusses outbreaks of polar air behind travelling cyclones in the westerlies of higher latitudes which affect Jamaica. Days with these outbreaks of cold northerly winds forming a front are two to three times more likely to be rain days than other days in December–January at Kingston, Jamaica. The average amount of rain on such days is three to six times greater than for other days in these two months. Penetration of mid-latitude frontal systems along the east coast of South America is discussed by de Oliveira and Nobre (1985).

As has been previously indicated, extra-tropical depressions in the westerlies bring important rains to part of north-west India from December to May. By inference, Fig. 2.9 indicates that in this area, the bulk of the rain falls outside the south-west monsoon period.

Cool season disturbances often result from interaction between the upper westerly circulation and the surface easterlies in the sub-tropics. Riehl (1954) discusses cool season cyclones which are usually weak surface features reflecting intense cold low-pressure systems aloft. Interaction between long waves in the upper westerlies and disturbances in the lower easterlies initiates rainfall situations which often contain cold lows above. Such systems are important to water supply in many parts of the sub-tropics. Ali (1953) and Soliman (1953) found that wet periods in Egypt and the Middle East are connected with the development of upper lows above the eastern Mediterranean, the significance of these being shown in Table 2.3.

Table 2.3   Analysis of rainfall at three stations in Egypt during the period December 1951 – November 1952 (after Soliman 1953)

	Alexandria	Port Said	Cairo
Rain due to cold fronts (mm)	9.8	trace	trace
Rain due to instability in absence of upper troughs or lows (mm)	8.4	1.7	trace
Rain due to upper cold troughs or lows (mm)	173.2	105.2	69
Number of days of cold-front rain	3	3	2
Number of days of mere instability rain	6	5	3
Number of days of rain associated with upper cold troughs or lows	70	33	23

2.10 OTHER INFLUENCES ON RAINFALL

Apart from the influences previously discussed, a further set of factors affect rainfall, often on a very local scale. These include the height and general configuration of land and the juxtaposition of land and sea surfaces creating thermal and frictional contrasts.

Ascent of air over high ground produces cooling which can lead to condensation and precipitation as well as perhaps triggering off instability in the air mass. Often, however, highland areas act simply to increase rainfall already resulting from other causes. Such effects will depend on the size and alignment of the barrier with respect to airflow and also on airflow characteristics. Rainfall variation with height is influenced by factors such as water vapour available for condensation and rate of ascent of air. The rate of ascent is critical, and while an increase in rainfall with height is commonly experienced this does not necessarily continue to the summit of the highlands. In many tropical areas, a rainfall maximum occurs at 1000–1400 m. This has been explained by the fact that highest windspeeds, with resultant rates of upward motions and moisture transport, are often found in the low troposphere (Riehl 1979). In temperate latitudes, the maximum is often found at greater heights since windspeeds continue to increase with height. However, as Riehl (1979) points out, even in the tropics there are exceptions to the above generalisation, a rainfall maximum sometimes occurring at greater heights than 1400 m. The presence of the trade-wind inversion may limit ascent above a certain altitude with resultant decrease in rainfall above it.

Thermal contrasts between land and water surfaces can initiate local circulations of which the land–sea breeze effect is well known. Sea breeze mechanisms and their resultant effects on rainfall, including diurnal variations, are discussed by Riehl (1979), Krishnamurti (1979) and Barry and Chorley (1982). The influence of other factors such as prevailing wind direction, atmospheric stability and relief will be superimposed on these effects. Over islands, the development of a sea breeze suggests an afternoon rainfall peak and this does occur in some cases. However, a nocturnal peak has been widely reported. Various reasons have been advanced for the latter, including different radiative and stability conditions between cloudy and cloud-free areas, semi-diurnal pressure oscillations and diurnal variations in air–sea temperature difference. Factors involved in diurnal rainfall variations are still the subject of considerable debate, further comments being presented in § 4.3. Increased friction over land may create convergence in onshore winds which are slowed down (see above, Linear disturbances, § 2.9) or divergence in offshore winds which accelerate.

The impact of such local influences and a list of examples is indicated by Flohn (1970). Coutts (1969) illustrates the effect of Mount Kilimanjaro (5899 m) in Tanzania on rainfall distribution which has great influence on agriculture and water supply. During the 'long rains', south-easterly air is the main rain-bringer, producing a rainfall maximum on the south and south-east slopes (Fig. 2.13). The north-easterly airstream during the 'short rains' produces a north-east slope maximum (Fig. 2.14). From April to September an inversion between 3965 m and 4575 m limits the cloud top, but thick orographic cloud forms overnight of sufficient depth to produce precipitation. From November to March, the inversion often breaks down, convection clouds build up in late morning and afternoon leading to showers and storms. These storms move downwind from the mountain and are rejuvenated on the western lee side by convergence of air diverted round the mountain and by anabatic air generated

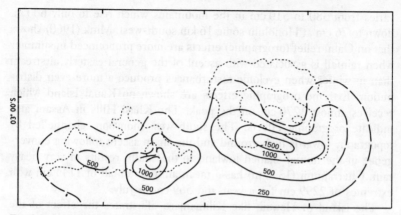

Fig. 2.13 Kilimanjaro: rainfall in the 'long rains' (March–May), mm (dotted lines 5000, 10 000, 15 000 ft contours after Coutts 1969).

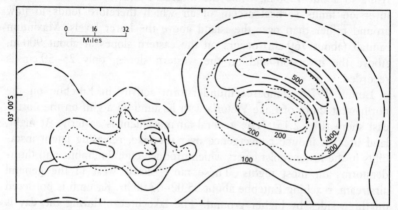

Fig. 2.14 Kilimanjaro: rainfall in the 'short rains' (November–December), mm (dotted lines 5000, 10 000, 15 000 ft contours after Coutts 1969).

by heating on the southern and western slopes during the afternoon. This effect is shown at Lyamungu (1250 m) on the southern slopes where 40 per cent of the rainfall during November–March occurs between 12.00 and 21.00 hr, compared with only 25 per cent from April to September when convection is limited by the inversion. Riehl (1979) provides another example of seasonal contrasts in rainfall regime between north and south facing slopes on the north coast of Venezuela. Very complex rainfall regimes are also found on the islands of the Philippines and Indonesia.

The Hawaiian Islands illustrate complex rainfall patterns resulting from relief and airflow characteristics (Riehl 1954, 1965, 1979; Barry and Chorley 1982). Marked contrasts occur between the high-rainfall windward eastern slopes and the dry western sides. On Oahu, rainfall varies from 380 to 510 cm in the mountains which rise to only 600 m, down to 76 cm at Honolulu some 16 km south-west. Mink (1960) shows that on Oahu relief (orographic) effects are more pronounced in summer when rainfall is associated with ascent of the general easterly airstream than in winter when cyclonic disturbances produce a more even distribution. Extreme orographic effects are shown on Kauai Island which receives almost 1270 cm on the peak. The Khasi Hills in Assam also indicate orographic effects. The south-west monsoon, channelled by topography towards high ground and the sharp ascent following convergence in the funnel-shaped lowland to the south, results in very heavy rain. Cherrapunji (1340 m) has a mean annual rainfall of 1144 cm with extremes of 2299 cm for a year and 569 cm in July.

The island of Hawaii, like Kilimanjaro, illustrates the effect of the trade-wind inversion. Two broad volcanoes occupy much of the island, rising to about 4200 m, while the inversion is at about 2000 m. The inversion limits vertical ascent of air which therefore tends to flow around rather than over the island above the lower levels. Maximum rainfall (about 760 cm) occurs on the eastern slopes at about 900 m. Above the inversion and on the western slopes, only 25–50 cm is recorded.

Lumb (1970) compares thunderstorm activity in Entebbe on the north-west shore of Lake Victoria with Kisumu on a gulf on the north-east shore (Fig. 2.15). The general air movement is easterly. At night, land breezes induce convergence over the lake, releasing latent instability in the lower moist layers which produce cumulonimbus and thunderstorms on most nights. These move slowly west in the general airstream, reaching Entebbe about 05.00–07.00 hr. Kisumu is bordered on three sides by higher ground. The lake breeze during the day is funnelled up the gulf and reinforced by an anabatic wind to create a strong lake breeze front during the afternoon over high ground at the head of the gulf. The front meets the general easterly airstream and this convergence, together with thermal convection, results in a strong afternoon thunderstorm maximum. Rainfall is highest over the Lake itself, with the maximum displaced towards the western side, presumably because of the influence of the general easterly airflow. Although other factors may play a part, this lake maximum presumably represents the influence of the lake breeze mechanism (Riehl 1979). The possible effects of the creation of large reservoirs are discussed in § 9.2.

The lee side (west) of the island of Hawaii is arid except for a narrow belt near the west coast. Here, a strong sea breeze moves inland against the trades in summer, producing high rainfall near the coast. The resulting summer rainfall maximum is the reverse of the winter maximum in the east. A variety of sea-breeze–trade-wind interactions in the Hawaiian Islands is shown in Fig. 2.16.

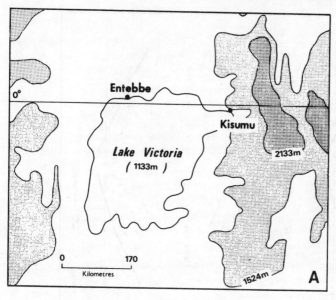

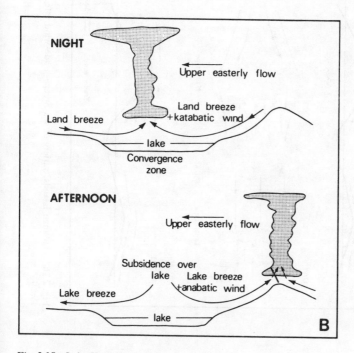

Fig. 2.15   Lake Victoria: (a) Location of Entebbe and Kisumu. (b) Weather over northern lake area (after Lumb 1970).

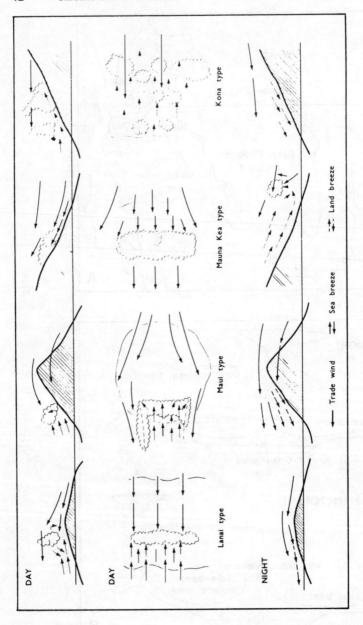

Fig. 2.16 Hawaiian Islands: variety of sea-breeze–trade-wind effects (after Leopold 1949).

Convergence and divergence can be set up in winds blowing parallel to the shore because of differential friction. This can influence rainfall amounts. Lahey (1958) suggests that such divergence partly accounts for the dry northern coast of Venezuela where it runs closely parallel to the east-north-easterly trade. Where the Venezuelan coast runs north–south friction induces convergence and an increase in rainfall. However, other effects including relief, the presence of cold water and general airflow characteristics also have an impact. Thus separation of the various influences in such cases is very difficult.

The complex interaction of broad-scale and local influences on the spatial, seasonal and diurnal variations of rainfall is illustrated by Chia Lin Sien (1968) over Selangor, Malaya. He suggests that they result not only from general convergence between major airstreams but also sea-breeze convergence during periods of easterly winds, convection storms, orographic influences and coastal rain in the early morning associated with 'Sumatras' discussed earlier.

2.11 CONCLUSIONS

The controls and origins of tropical rainfall are very complex. The general circulation of the atmosphere, disturbances and local factors combine to produce spatial and temporal rainfall variations which are significant for agriculture and water resources. In Chapters 3 and 4 these rainfall characteristics responsible for much of the weather diversity in the tropics are analysed.

CHAPTER 3

RAINFALL SEASONALITY AND VARIABILITY

3.1 INTRODUCTION

From the varied origins of rainfall discussed in Chapter 2, it is apparent that characteristics will vary considerably in time and space. Despite this, certain generalisations can be made to differentiate conditions from those in higher latitudes. The character of rainfall will vary not only from place to place but with time at a particular place. The nature of the rainfall can vary during the growing season and this will have major implications. In this chapter seasonal rainfall regimes and variability are discussed. Other rainfall characteristics are examined in Chapter 4.

The implications of the character of rainfall are discussed in this chapter and in Chapter 4, but many are amplified in later chapters, especially those related to agriculture. The effects of rainfall conditions are not constant, varying not only with other physical conditions such as soil type and relief but also with economic, technological, social and political conditions. Moreover, the human factors vary with time and it is easy to mistake their influence for a change in rainfall characteristics. Traditional peasant activities were well adapted to the environmental conditions but changes, often induced from outside, can lead to a breakdown in the system or make it more difficult for communities to adapt to harsh conditions such as a series of dry years, particularly in semi-arid areas. Many examples are found in a symposium on drought in Africa (Dalby and Harrison Church 1973). Fosbrooke (1973) indicates how overpopulation resulting from, for example, disease control, leads to a decrease in the fallow period of the bush fallow system with resultant drop in fertility and yields. As the soil loses its humus content it becomes less moisture retentive, and hence rainfall is less effective and nutrients become less available to plants. In fairly low rainfall areas the farmer thinks that rainfall is decreasing, whereas often it is simply that the same rainfall has become less effective. Swift (1973) shows how political factors such as restriction on movement of Tuaregs makes the population less able to cope with drought than in the past. Provision of water supply such as bore-holes in semi-arid areas can lead to local overstocking, destruction of vegetation and erosion. In dry years, this produces disasters perhaps more serious than in the past. These aspects are considered later.

Oguntoyinbo and Odingo (1979) indicate how in East and West Africa, traditional systems in harmony with the physical environment were broken down by inappropriate land use, socio-economic and technological changes. They point out that impacts tend to be greatest in more marginal areas, making communities particularly vulnerable to rainfall variability and worsening the effects of drought. Ogallo (1984) points out that although rainfall variability and change has important impacts on agricultural production, many human factors play a part. He contrasts the situation in some tropical Latin American countries and parts of Asia such as India, where sound planning and increased food production resulting from technological innovations have greatly reduced famines and deaths due to climatic variability, with that in other countries of these regions and also particularly countries of sub-Saharan Africa. In the latter, unstable economic, social and political systems, as well as constraints such as lack of skilled local human resources, fluctuations in markets and prices and a range of socio-economic factors have in many cases played a major role in declining agricultural production. This has highlighted the effects of climatic variability, particularly drought occurrence in drier regions.

Although mistakes have been made in the past, appropriate technological changes present opportunities for improving the situation. For example, means of combating desertification, salinity and waterlogging problems and possibilities of land reclamation and irrigation in arid areas are presented in Bishay and McGinnies (1979). Michaels (1982) used a modelling approach to assess climatic sensitivity of high yielding variety 'green revolution' wheat in India and Mexico. Increased adoption of the high yielding 'package' led to significant increase in yield sensitivity to important climatic factors including rainfall. Therefore, without any change in climatic variability, production will become more variable. The above aspects will receive more attention in later sections. However, even from this brief discussion, it should be apparent that assessing the impacts of rainfall characteristics is very complex. Furthermore, the effectiveness of rainfall must take into account a variety of characteristics, including storm size, intensity, frequency and time of occurrence as well as soil characteristics, vegetation, slope and evaporation rates.

3.2 SEASONAL RAINFALL REGIMES

From Chapter 2 it is clear that a picture of a seasonal north–south movement of a rain belt associated with a migrating equatorial trough is a great oversimplification. However, it is useful to start with this simple model and then modify it. Working from the equator polewards, the following zones are present:

(A) At the equator a double rainfall maximum but with no real dry season.
(B) A double rainfall maximum separated by a more pronounced dry season during the low sun period, but only a slight dip in the high sun period since the equatorial trough would not be far from the area.
(C) A single rainy season and a single dry season.
(D) A semi-arid region on the equatorward margin of the STHs with a short rainy season in the high sun period and a long dry season.
(E) An arid area associated with the STHs.
(F) A semi-arid region on the poleward margin of the STHs with a short wet season in the low sun period under the influence of extra-tropical disturbances.

This simple pattern is modified in a number of ways. The location, structure and shape of the STHs and the varying characteristics of the trade winds create contrasts between longitudes. Over the eastern parts of oceans and adjacent land, marked subsidence promotes arid/semi-arid conditions. Over the western sectors of oceans and adjacent land, the decreased effects of the STHs and 'disturbed' character of the trades promote wetter conditions, the onshore airstreams advecting moisture over the eastern parts of continents. Over western continental areas, therefore, drier conditions of the sub-tropics will extend equatorwards, while on the eastern continental areas, wetter conditions of low latitudes might be expected to extend polewards.

Seasonal movements of the equatorial trough vary considerably around the globe (Ch. 2). Even where the trough can be recognised, its character varies greatly and is not always conducive to rainfall. The great importance of disturbances to rainfall, both within the vicinity of the trough and away from it, were highlighted in Chapter 2. Since disturbances vary greatly in character and occurrence a simple associ-ation of rainfall with an equatorial trough is not found.

The shape and distribution of land and sea exert a considerable influence, as do local factors such as those indicated in Chapter 2. Figures 3.1–3.4 indicate the complexity apparent even from monthly average values. In January (Fig. 3.1), a belt of heavier rain is found south of the equator over South America and Africa. In the longitudes of Asia, the belt of rainfall is very wide, covering both sides of the equator, including northern Australia. This is a region of very widespread convergence (Lockwood 1974). By April (Fig. 3.2) it is not difficult to recognise a general northward progression of a rain belt, especially over South-East Asia.

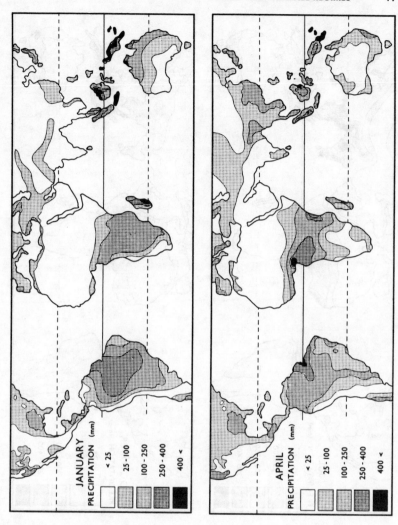

Fig. 3.1 Average rainfall: January.   **Fig. 3.2** Average rainfall: April.

In July (Fig. 3.3), the northward shift has progressed further, the association with the south-east Asian monsoon being particularly noticeable. Comparison of areas within the latitudinal belt 15°–30 °N indicates the lack of a uniform zonal belt of rain, e.g. compare the Caribbean, North Africa and South-East Asia. Figure 3.4 shows that in October a general swing south of the belt of heavier rain is in progress.

Figures 3.1–3.4 show that a simple zonal picture is subject to considerable exceptions. Also, certainly in some areas, the concept of a migrating rain belt is incorrect. In Chapter 2 it was suggested that,

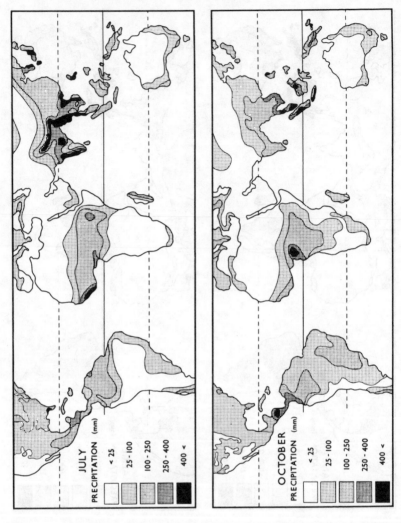

Fig. 3.3 Average rainfall: July.   **Fig. 3.4** Average rainfall: October.

in some regions, as an equatorial trough disappears in one area another develops somewhere else, and this same sequence can occur in terms of a rain belt. The rainfall distribution on individual days or months will often depart considerably from the average pattern, showing sudden jumps often counter to the seasonal trend and being patchy in character, developing and decaying *in situ* (Hammer 1972; Johnson 1962).

Examination of rainfall at individual stations on a continental basis (Fig. 3.5) illustrates some general points.

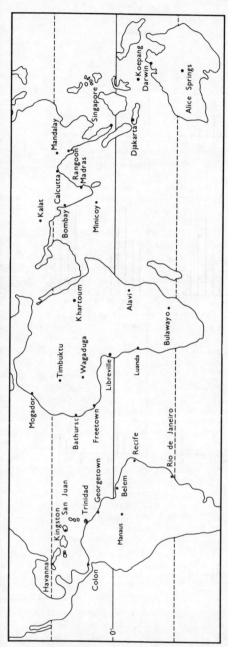

Fig. 3.5 Location of rainfall stations.

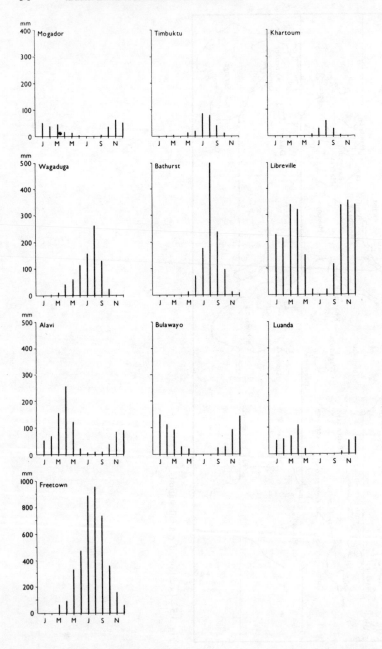

Fig. 3.6   Rainfall graphs: Africa.

Africa (Fig. 3.6)

Mogador (Morocco) illustrates the 'winter' maximum on the poleward side of the STH in contrast to the high sun rainfall on the equatorward side of Timbuktu and Khartoum. Further south, at Wagaduga and Bathurst, the 'summer' rainfall period is longer and still more so at Freetown. The control on rainfall exerted by the depth of the moist south-westerly airstream over West Africa was discussed in Chapter 2.

The asymmetrical movement of the equatorial trough over West Africa is illustrated by Libreville (1°N). While this station shows a double rainfall peak, during the northern high sun period, the equatorial trough is so far north that here, despite being near the equator, dry conditions prevail. During the southern hemisphere high sun period, however, the trough does not move so far south and hence Libreville still has rainfall in December and January. The greater penetration of the trough over central and eastern Africa is illustrated by Alavi (7°S) having a regime like that of Libreville, and the contrast between Bulawayo and Luanda. Luanda (9°S), with only a short summer rainfall, is the equivalent of northern hemisphere stations such as Timbuktu (17°N). Riehl (1979) reports that the low rainfall at Luanda appears to occur when the equatorial trough adopts a complex structure with two branches, such as happens in the Indian Ocean in July. Bulawayo (20°S) in central Africa, with a fair amount of rain from summer disturbances, has more rain than northern hemisphere stations such as Timbuktu (17°N) and Khartoum (16°N).

South-East Asia–northern Australia (Fig. 3.7)

The broad belt of rainfall on both sides of the equator during the southern hemisphere high sun period has already been referred to. The rainy period over northern Australia is illustrated by an arid station (Alice Springs) and Darwin. Various aspects of the structure, associated rainfall and variability of the Australian monsoon situation are discussed by Nichols, McBride and Ormerod (1982), McBride (1984), Kininmonth (1983), Holland, Keenan and Guymer (1984). McBride (1984) stresses many similarities with the northern hemisphere or Indian monsoon. A monsoon equatorial trough is accompanied by low latitude westerly winds which during December to March extend from 50 to 180 °E. Nichols, McBride and Ormerod (1982) however point out that about 30 per cent of the Darwin wet season rainfall occurs before the large scale rearrangement of the tropical circulation takes place. They present evidence that the pre-monsoon part of the wet season is related to the SO (§ 2.8). During the northern hemisphere high sun period dry conditions are experienced at these two Australian stations and also at Koepan (Timor)

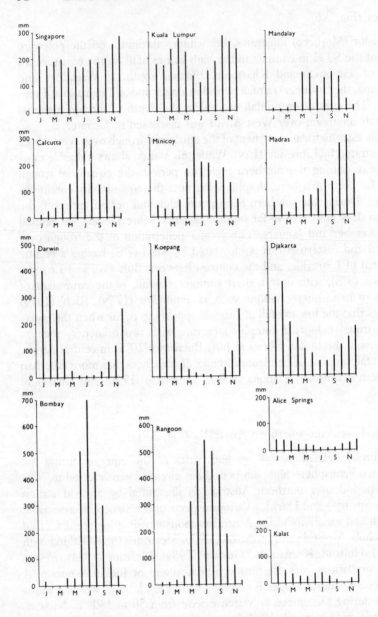

Fig. 3.7 Rainfall graphs: northern Australia–South-East Asia.

at 10 °S and Djakarta (6 °S). At this time the south-easterly airstreams are subsident and dry which affects southern hemisphere stations. At Singapore, the rainfall peak from November to January is in agreement with the broad convergence zone on both sides of the equator at that time. During the northern hemisphere high sun period, however, despite the fact that the general convergence area and rain have moved well to the north (Lockwood 1966, 1974), Singapore still has a fair amount of rain. Kuala Lumpur (3°N) illustrates the double peak in April and October which might be expected in this latitude. Rangoon (17°N) has a single peak in July but Mandalay, further north (22°N), has a double peak in May and September. The latter feature is thought to be related to jet-stream fluctuations and the nature of the south-east Asian monsoon.

In the north-west of the Indian sub-continent, Kalat illustrates the importance of winter rainfall associated with extra-tropical depressions (Ch. 2). The sudden 'burst' of the monsoon on the west coast of India is illustrated by comparing May and June figures for Bombay. The more gradual increase in the north-east is shown by Calcutta where early rainfall is associated with pre-monsoon thunderstorms and disturbances generated in the Bay of Bengal. Stations in the south-west of India such as Minicoy experience a double rainfall peak, amounts decreasing from June to September when the rain area is well to the north but rising again in October as the equatorial trough and associated disturbances leave the area. Madras illustrates the rainfall peak during the north-east monsoon in south-east India (Ch. 2).

South America and the Caribbean (Fig. 3.8)

Comparison of graphs for Manaus, Belem, Recife and Rio de Janeiro indicates the very great influence of the South Atlantic STH which is nearer the equator and the western side of the ocean than its northern hemisphere counterpart. The effect is amplified by the projection of the wedge-shaped area of north-east Brazil towards the high. In the southern hemisphere 'winter', Manaus (3°S) and Belem (1°S) have little rain, indicating that the influence of the subsident, divergent air from the STH extends right to the Amazon Basin on the equator when the equatorial trough is well to the north. At Manaus, rainfall begins to increase in October with the southward movement of the trough, but since the latter does not penetrate very far south over north-east Brazil because of the influence of the Atlantic STH, there is no dry season corresponding to that in the northern hemisphere high sun period and rainfall is heavy from December to May. Belem, further east, is even more under the influence of the STH and rainfall does not really begin

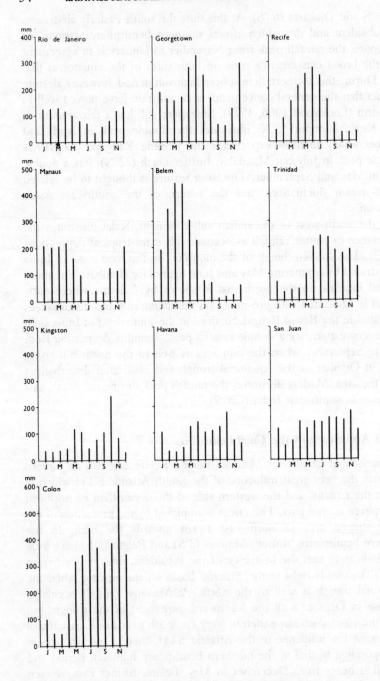

Fig. 3.8 Rainfall graphs: South America–Caribbean.

to increase until January, much of it occurring from February to May in which month the trough once again retreats northward. The continued influence of the STH over the north-east hump of Brazil even in 'summer' is illustrated by the fact that Recife (8°S) has little rainfall at that time. This is true of the whole coastal area from 5°S to 13°S. Recife does have rain mainly from March to June. There are some doubts as to its origin, but weak disturbances moving north from higher latitudes are probably important. A preliminary analysis of the penetration of mid-latitude frontal systems into this general area is presented by de Oliveira and Nobre (1985). The mountains inland from this coast, rising to 1000 m, effectively stop the inland penetration of the moist surface air layer below the trade-wind inversion. Thus, a short distance inland is an arid area which receives no rainfall of the type of Recife and has very little summer rainfall due to the continued influence of the STH.

Various aspects of the rainfall climatology of north-east Brazil are presented by Nobre, Nobre and Moura (1985), particularly in relation to large-scale circulation anomalies and drought prediction. The significance of condensational heating over the Amazon and its fluctuations in determining tropical and extra-tropical weather is pointed out by de Oliviera and Nobre (1985). They indicate that understanding such fluctuations needs consideration of three principal rain producing mechanisms, including those associated with mid-latitude fronts referred to above.

Rio de Janeiro (23°S) despite being well south of Recife is not so much under the subsiding influence of the STH. It therefore illustrates features which might be expected of an area to the west of the STH cells with a summer rainfall maximum which is not, however, very marked, some rainfall occurring in each month.

Georgetown (7°N) demonstrates the passage of the equatorial trough with rainfall peaks in June and December. However, over the Caribbean area in general the picture is far from simple (Gramzow and Henry 1972), and it is often impossible to relate rainfall regimes to movement of the equatorial trough and ITCZ. While double rainfall peaks are found at a number of stations such as Trinidad, Colon and Kingston, they appear to be related not simply to the general movement of an equatorial trough but to the occurrence of disturbances (Riehl 1954). Satellite photographs show that not only do weather systems along the trough frequently reach the South American coast near Trinidad but that systems within the trade winds also influence the island and those to the north. Havana, as well as two peaks in summer also has a minor winter peak associated with cool season disturbances. The double summer peak at this station is related to movements of an upper air trough (Riehl 1954). San Juan (Puerto Rico), like Rio de Janeiro, may be regarded as more typical of areas west of the STH cells with a broad summer peak resulting from disturbances in the trade-wind belt. However, orographic factors often create contrasting rainfall regimes in such areas. Riehl (1979) points out that the number of hurricanes actually hitting Puerto Rico is small and their influence on rainfall less than often assumed. Heavy rains, up to three times the monthly average, have occurred in 24 hours at the tail end of hurricanes however.

The above brief account of rainfall at a selection of locations hopefully indicates the complexity of and departures from a simple seasonal pattern. Further illustrations of the complexity, for example over the Pacific and Indian Oceans, are presented by Riehl (1979).

Regional distribution of rainfall regimes

Despite the complexity of the situation, certain broad generalisations can be made. In Chapter 1 it was pointed out that it is the relation between rainfall and evaporative demand which is important. This relationship and the topic of climatic classification systems are examined in later chapters. Nevertheless, the pattern of seasonal rainfall regimes in many cases can be related to agricultural systems and problems of water supply. Therefore, a regional sub-division of regimes is important. Any sub-division will be somewhat arbitrary, there being a range within each class, and boundaries represent transitional zones rather than sharp features. On a world scale, considerable generalisations are necessary, the often marked variations due to local factors, particularly relief, not being shown. In some areas, such as Malaysia, it may not be possible to divide the year into wet and dry seasons and heavy rain may occur at any time. However, even in such areas, short dry spells occur although not of sufficient length or regularity to be called dry seasons (Dale 1959).

The drawing of boundaries is hampered by lack of data, but in view of their transitional nature and considerable variations from year to year this may not be too great a problem. The problem of variability is highlighted by Meher-Homji (1974) for India who compared the rainfall regime found using average values with the proportion of years conforming to this regime. He found, for example, that while New Delhi has average values suggesting a 'tropical' regime (summer rainfall), only 37 per cent of years had this classification while 53 per cent had a 'bixeric' regime (two dry and two rainy seasons). Taylor and Tulloch (1985) classified years in terms of amount and distribution of rainfall at Darwin over the period 1870–1983. They found that about 77 per cent of years had significant departures from the two most frequent rainfall patterns.

The character of rainfall in a particular month (e.g. intensity, duration, frequency) and the elements of the atmosphere–soil–plant system determining the effectiveness of rainfall in relation to evaporative demand make it impossible to define strictly a 'wet' month. A period of five months, the middle one having, say, 25 mm of rain but the other four having over 100 mm, might be regarded as a single wet season. A month with 25 mm but with months having less than 10 mm before and after could not in many cases be regarded as wet. The time of year during which rain falls is important. North of the Sahara, the 'winter' rainfall is more effective than the summer rain to the south because of varying evaporative demand. Thus Harrison Church (1973) suggests 400 mm as the limit of the dry zone north of the Sahara but 500 mm south of the desert. Some problems of the definition of wet and dry seasons are discussed by Bowden (1964) with illustrations of previous work using varying monthly rainfall amounts as well as attempts to take other factors into account such as the frequency of rain days.

The use of monthly values of any kind to define seasonal regimes is suspect, not only because rainfall conditions during short time periods are critical for agriculture but also because the onset and end of the wet season – either on average or for individual years – does not co-incide with calendar months. This has led to the use of pentades (5-day periods) to define rain seasons. Examples of this approach are the work of Gramzow and Henry (1972) for Central America, Ilesanmi (1972) for Nigeria and Ananthakrishnan and Rajagopalachari (1964) for India. Torrance (1967) used pentades to examine the progress of the mean rainy season, occurrence of rainy spells and dry spells in the rainy season and variations in the length of the dry season in central Africa. Ananthakrishnan, Pathan and Aralikatti (1981) used pentades to study the northward advance of the ITCZ and onset of the summer monsoon rainfall for island stations in the Bay of Bengal. They found onset dates to be earlier than those indicated on existing diagrams. Subbaramayya, Babu and Rao (1984) used daily rainfall as well as surface pressure and wind charts to determine the northern limit of the Indian monsoon for the period 1 May to 15 July over the period 1956–80 at 200 stations. The mean and standard deviation of the dates of onset were calculated. Correlations between the onset dates at different stations were assessed to identify regional relationships. For West Africa, daily rainfall models have been used by Stern, Dennett and Garbutt (1981) to derive a probability distribution of the starting date of the rains. The use of pentade and daily data is discussed in more detail in § 3.5. Virmani (1975) defines the growing season in India as being made up of weeks with a 70 per cent probability of receiving at least 10 mm per week and 'sowing rains' as more than 20 mm rain on not more than two consecutive days in such a week. Oldeman and Frere (1982) use backward and forward accumulation of 10-day totals to determine the onset and termination of

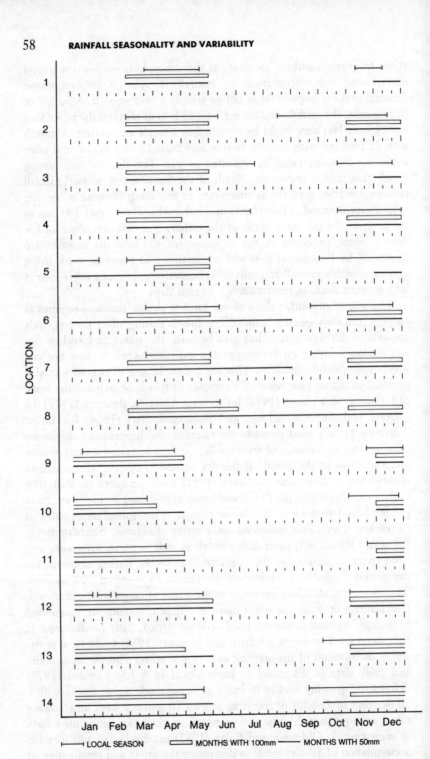

Fig. 3.9 Tanzania – comparison of traditional seasons with those based on monthly totals of 50 mm and 100 mm (after Jackson 1982).

the wet season. This is done for each year and then probabilities are assessed. Other approaches to seasonal regime definition include the use of spectral or harmonic analysis of mean monthly rainfall, e.g. Keen and Tyson (1973), Fitzpatrick, Hart and Brookfield (1966) and Potts (1971). While the latter approach provides a quantitative description of the seasonal regime, its practical utility to water resources and agriculture is doubtful, especially bearing in mind some of the problems discussed above. Definitions of 'wet' and 'dry' will vary both with crop type and agricultural system, e.g. peasant subsistence as distinct from plantation agriculture.

Jackson (1982) compared estimates of seasons based on monthly totals of 50 mm and 100 mm with 'traditional' views (i.e. of the local people) in Tanzania (Fig. 3.9). Although some of the smaller discrepancies could be accounted for by the fact that the traditional estimates used months divided into three parts, in a number of cases the differences were quite considerable. These results indicate how seasons defined by meteorological parameters such as rainfall may differ considerably from those of local farmers.

A rain gauge may be in an atypical area. In a semi-arid region, for example, a settlement – and hence probably the rain gauges – may be located in relatively wetter areas. Rainfall measurement and representativeness of rain gauges is an extremely complex field, as illustrated, for example, by WMO (1981), O'Connell (1982), Sevruk (1982). However, in tropical areas where raindrop size is large and windspeeds often low, some of the problems may not be so great as in higher latitudes (Jackson 1971a). Problems of rainfall estimation over oceans and islands are highlighted by discussions of Reed (1980, 1982) and Dorman (1982).

Bearing in mind the above problems, the following general classification of rainfall regimes is presented. The distribution of the types is shown in Fig. 3.10.

I. Humid tropics

Ia. While a seasonal variation may occur, in general terms there is no dry season. Annual totals usually >2000 mm and all months at least 100 mm.

Ib. No pronounced dry season, annual totals < 2000 mm but a drier period with a few months having < 100 mm. This is a transitional category between 'humid' and 'wet and dry' areas. It occurs on the poleward margins of (Ia) but also on the windward (east) coasts of continents and islands where it may extend over a considerable width of latitude and is associated with the 'disturbed' trade winds. Because advection of moisture onshore is involved orographic effects often create marked local variations.

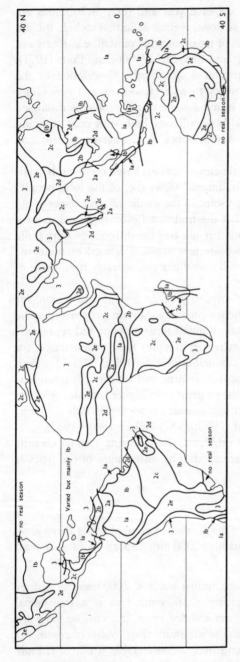

Fig. 3.10 Distribution of seasonal rainfall types (for explanation see text).

II. Wet and dry tropics

Any division is fairly arbitrary, each type in most cases showing a gradual transition to an adjoining type.

IIa. Annual totals 1000–2000 mm. Two rainy seasons with short dry seasons or months with lower rainfall, usually a few months < 50 mm.

IIb. Annual totals 650–1500 mm. Two short rainy seasons separated by a pronounced dry season (a few months < 25 mm) and a short drier season.

IIc. Annual totals 650–1500 mm. One fairly long rainy season (normally 3–5 months each with > 75 mm) and one long dry season.

IId. Annual totals > 1500 mm. One season of exceptionally heavy rain and one long dry season. Typically illustrated by some south-east Asian monsoon situations.

IIe. Annual totals 250–650 mm. One short rainy season (3–4 months each with >50 mm) and one long dry season.

III. Dry climates

Annual totals <250 mm, little rain at any time, but can be concentrated in a very short 'wet' season perhaps of only a few weeks.

In § 8.3 these regimes are related to agricultural systems. While arrangement in latitudinal zones is apparent in some areas (Fig. 3.10), there are considerable departures from this simple picture. The latitudinal extent of the humid climates, particular (Ib) in South-East Asia and the east coast of the Americas and Caribbean has no counterpart over eastern Africa where no 'humid' areas occur. The reasons for this are discussed by Jackson (1971c), Griffiths (1972) and Trewartha (1961), the latter describing it as 'the most impressive climatic anomaly in all of Africa'. The concentration of rainfall in short season(s) in many places, especially Africa and monsoon Asia is apparent. Griffiths (1972) indicates that more than two-thirds of Africa has over half its annual rainfall in a period of 3 months. In the semi-arid zone of northern Nigeria, between 70 and 90 per cent of the rainfall occurs in the months of July, August and September (Griffiths 1972). In Queensland the proportion of the rainfall occurring in the summer half-year varies from 60 per cent in the extreme south to over 95 per cent in the Gulf country and Cape York Peninsula (Dick 1958). The bulk of these high percentages falls in the four wettest months. This concentration is a major problem in many areas. Since soil-moisture storage is limited, plants are under stress in the dry season and an important need is to store surplus runoff and conserve water for use in the dry season. This will be discussed in Chapters 8 and 9.

3.3 VARIABILITY

Rainfall variability has been the subject of much comment, but in reality comparatively little quantitative analysis of this factor has been undertaken in tropical areas. Lack of such analyses has led to a number of generalisations being made from often insufficient evidence. Frequently, analyses end with a statistical statement of the degree of variability without consideration of the implications, which can only be understood when related to the physical, social, economic and political environment. However, there have been some useful reviews including consideration of implications, such as those by Fukui (1979) for the humid tropics (which includes 'wet and dry' regions) and Mattei (1979) for the semi-arid tropics. Fukui (1979) analyses variability, including climatic change, and its implications for agriculture. Different crop types (e.g. annuals, tree crops) and their adaptability to suit variability are considered. Mattei (1979) considers impacts of variability on agriculture and how land use and management practices (see § 3.4), crop types and technology should be adapted to overcome problems. For Africa, Oguntoyinbo and Odingo (1979) provide a similar coverage to the above, with particular attention being paid to implications of land use and socioeconomic changes as far as environmental deterioration and desertification are concerned. Swift (1973) discusses strategies adopted by Tuareg tribesmen as protection against variability (especially drought years) as well as the impact of economic, social and political changes beyond their control. Further comments on these aspects are presented in this section and the following one as well as in later chapters.

Mooley *et al.* (1981) examined impacts of marked rainfall deficit and excess on the All India Indices of food grain prices and food grain production. While the impact of a marked deficit was usually clear, that of a marked excess was apparent in only some cases. The latter is understandable since an excess may or may not result in floods with associated crop damage. Mooley *et al.* (1981) also considered remedial measures which could make the Indian economy less dependent on rainfall variability. For Queensland, Australia, Dick (1958) highlighted the importance of economic and technological factors to the impact of rainfall variability, providing an interesting comparison with the developing world. Variability may become more critical since new high-yielding crop varieties can be less tolerant of fluctuations in water supply. This was illustrated by the modelling experiment of Michaels (1982) already referred to in § 3.1.

A frequently asserted claim is that rainfall in the tropics is more variable than in higher latitudes. Caution must be exercised in considering such a sweeping generalisation. Conditions vary greatly throughout the tropics. In Chapter 1 the critical nature of water resources in much of the tropics was emphasised and, understandably, fluctuations in the supply (i.e. rainfall) are bound to have greater significance and hence receive more attention than in some other regions. In addition, the economic situation in many tropical areas makes it difficult to cope with variability. Widespread droughts or floods in, say, parts of the USA or Australia, serious as they are, do not have the disastrous results akin to those in North Africa in recent years. This alone tends to give an impression of greater variability in the latter area. To some extent, therefore, the assumption of greater variability in the tropics could be a question of perception.

Not only can a wide range of social and economic factors influence the impact of rainfall variability: they are often sufficient in themselves to produce food shortages and famines commonly attributed to climatic influence. Mascarenhas (1968) illustrates how food shortages cannot be simply correlated with low rainfall and cites a wide range of economic and social factors which can be responsible, including the introduction of cash crops.

Variability receives particular attention in semi-arid areas where droughts cause havoc. Especially in such areas, rainfall is characterised by the occurrence of a few extremely high values and a skew distribution (see below). Superficial examination of these extreme values supports a concept of high variability. Examples of large extremes for a variety of African stations are shown in Table 3.1. In low-rainfall areas, extremes have considerable impact upon averages and hence the latter are of little value. At the Red Sea station of Hurghada, with a mean annual rainfall of only 3 mm, 41 mm fell on 8 November 1939 (Griffiths 1972).

Table 3.1 Rainfall range (mm) for African stations (from Griffiths 1972)

		Jan	Feb	Mar	Apr	May	June	July	Aug	Sept	Oct	Nov	Dec	Year
(a) Mogadiscio	Max.	9	3	257	245	324	349	240	182	239	192	179	76	997
(Somali)	Ave.	1	0	9	58	56	82	58	40	23	27	36	9	399
	Min.	0	0	1	5	7	13	13	10	5	4	4	1	57
(b) Harbell	Max.						1203							
(Liberia)	Ave.						461							
	Min.						123							
(c) Meru (Kenya)							October range: 1386–15							
(d) Quelimane (Mozambique)							Annual range: 530–2501							
(e) Maintirano (Madagascar)							Annual range: 45–240% of mean							

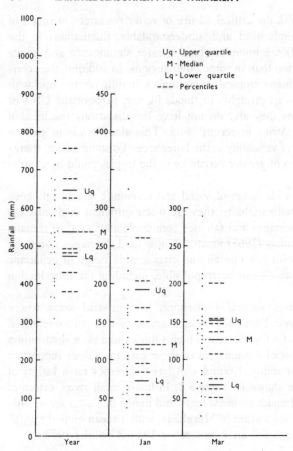

Fig. 3.11 Dodoma: annual, January and March rainfall.

The skew distribution in semi-arid areas means that commonly used analytical techniques assuming a normal frequency distribution are inappropriate. A discussion of how such techniques create an impression of great variability is given by Gregory (1969). The principle can be illustrated by data for Dodoma, Tanzania (Fig. 3.11). The highest values (1083 mm for the year, 417 mm for January) suggest great variability, but if these and perhaps one or two other large totals are ignored the range is not so striking. Considered as an amount of rainfall, such high values are not of great importance, although their characteristics, especially high intensity, have an impact – perhaps creating flooding and erosion problems. The low-rainfall values in semi-arid areas, while not so 'extreme' as the high values, are of far greater practical importance since they signify drought. It is, however, not these which make casual inspection or inappropriate analysis suggest great variability but the few very high values. In higher rainfall areas which perhaps receive less attention because variability does not have such a great influence, Gregory (1969) suggests that the latter is no greater than in higher latitudes.

The above discussion is not meant to imply that tropical rainfall does not show considerable variability but to suggest that generalisation is dangerous. In areas where water supply is marginal, any variability has great significance and hence is perceived to be high. Furthermore, in such areas, superficial examination and inappropriate techniques heighten an impression of large variability. There are areas where variability is high but there are also regions where it is not excessive, certainly no greater than in higher latitudes with similar average rainfall. The situation is complex, Huke (1966) for example, suggesting that variability is markedly lower in Burma than Oahu (Hawaii), indicating perhaps that rainfall derived chiefly from trade winds is less reliable than that derived from the south-east Asian monsoon. At one location variability may show seasonal differences. For example, in a mountainous region of Costa Rica, during December–April, orographic rain predominates while in the other rainy period (May–November), convective showers are more important. Chacon and Fernandez (1985) found that the coefficient of variation (see below) was large in the former period and small in May–November. Altitudinal differences in the coefficient were also found.

In dry areas, a few years of low rainfall can lead to a deterioration in vegetation which is not always reversed in succeeding years of high rainfall. A decrease in protective vegetation cover in dry years makes an area more susceptible to flooding and erosion when wetter conditions appear, and the deterioration may be accelerated. In a wetter area, the same absolute magnitude of variation might not have a significant effect on vegetation and land use. These aspects are considered in more detail below and in later sections (e.g. §§ 4.1, 6.3, 7.1, 9.3). As increased demands for agricultural production may lead to expansion in marginal rainfall areas, conditions in the latter will be even more critical than in the past.

In an analysis of variability, the time period considered is important. Expressions of annual rainfall variability, reliability and probability (see below) suffer from the same kind of limitations as annual averages in that they give no indication of shorter period conditions. At certain times in the growth season of a crop, presence or absence of water is critical and hence indications of variability over short time periods are of great importance. Figure 3.12 illustrates how, within a month, even mean weekly totals show considerable variation.

In areas with a marked seasonal regime, variability at the start and finish of the rains is particularly important. An indication of the variation in length of wet and dry seasons is shown in Table 3.2 for the Congo

Fig. 3.12 Harbell: mean weekly rainfall (after Griffiths 1972).

Table 3.2 Date of start of dry and wet seasons and duration of dry season (days) (10th and 90th percentiles and extremes) 1930–59 (from Griffiths 1972)

	Start of dry season		Start of wet season		Length of dry season		Start of dry season (extremes)	Start of wet season (extremes)	Length of dry season (extremes)
	P_{10}	P_{90}	P_{10}	P_{90}	P_{10}	P_{90}			
Lower Congo (Kinshasa)	15/5	2/6	5/9	9/10	103	139	9/5–5/6	26/8–12/10	90–147
Ruzizi Valley (Bujumbura)	9/5	8/6	26/8	3/10	88	138	1/5–21/6	17/8–15/10	79–153

Note: P_{10} date – 10% of years will have an earlier date, P_{90} date – 90% of years will have an earlier date, i.e. in 80% of years the date will lie between P_{10} and P_{90}. Similarly for Kinshasa, in 80% of years, the length of the dry season will be between 103 and 139 days.

and Burundi. Oliver (1969) illustrates variation in the length of the wet season for Abu Deleiq, in the Sudan, where records exist since 1905. The shortest rainy season lasted from 28 August to 15 September (1917), and the longest from 18 March to 28 October (1928). For a station in the western Llanos region of Venezuela, Fig. 3.13 indicates the erratic timing of the start of the rainy season (April) as well as variations in amount (Benacchio *et al.* 1983). After a prolonged dry season, soil moisture will be virtually nil and the ground perhaps too hard to work. The peasant farmer must wait for the rains to be able to work the ground and have moisture available for crops. A delay of even 1 or 2 weeks can destroy all hopes of a normal harvest. Swift (1973), in discussing droughts in the West African Sahel region points out that a delay of even a week or two in the onset of the rains can mean serious animal losses, but indicates that it is not unusual for there to be no consistent rain and no new pasture until a month later than average.

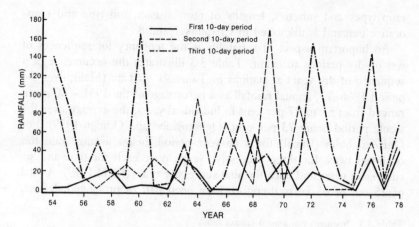

Fig. 3.13 Rainfall in the first, second and third ten-day periods of April at Acarigua, Venezuela (after Benacchio *et al.* 1984).

Oguntoyinbo and Odingo (1979) point out that in Nigeria in 1973, the rains seemed to start early, encouraging early planting but then they dried up, resulting in crop loss. Livestock losses were also considerable because of an early move north in anticipation of the rain which suddenly stopped. The greater variability of rain at the start and finish of the rains than in mid-season is also pointed out by Odumodu (1983) for the Plateau State, Nigeria.

In the middle of the rainy season, not only might rainfall occurrence be expected to be more certain but soil moisture reserves will be able to cope with absence of rain for short periods. Towards the end of the season when the meteorological situation is changing, a greater degree of variability can be expected. There are certain periods in the growth season when lack of water is critical. In the *Symposium on Drought in Africa* (Dalby and Harrison Church 1973), it was pointed out that in northern Nigeria even 'drought-resistant' crops would not have produced any yield in 1972 because the 3 or 4 weeks of drought occurred at a time when the crops were forming their yields. It is therefore important to consider the time of occurrence of drought – whether rains are late or a dry period occurs in early, mid or late season. Here, not only crop type but variety will be important. If the farmer has sown a variety which performs best when the season is early and the rains are a few weeks late, only poor results will be obtained as compared with a variety performing well when the season is late. Unfortunately the peasant farmer probably does not have access to alternative varieties and thus, as well as advice on timing of planting, making available several varieties of a crop could be worth while. This complex issue involving crop types and varieties, lengths of rainy season, soil type and evaporative demand is discussed in Chapters 7 and 8.

An important aspect of variability is the tendency for sequences of wet or dry periods to occur. Table 3.3 illustrates the occurrence of a sequence of dry years at stations in Tanzania. At Gao (Mali), over the period 1966–73, annual rainfall as a percentage of the 1931 – 60 mean ranged from 55 to 97 per cent in individual years, the average over the whole period being 73 per cent (Oguntoyinbo and Odingo 1979). For Quixeramobim (Brazil) for a 65-year period, below average rain has occurred twice for three successive years, once each for four and six successive years and once for nine successive years (Riehl 1979). A total of 25 out of 65 years therefore have been part of prolonged deficits.

Table 3.3 Sequences of annual rainfall (mm)

Narok Forest (average 1811)

Year	1936	1937	1938	1939	1940	1941	
Total	1389	1202	1536	1429	1539	1428	

Mvumi Mission (average 544)

Year	1948	1949	1950	1951	1952	1953	1954
Total	360	374	391	576	377	459	340

Similiar characteristics have been noted by Swift (1973), Huke (1966), Morth (1970), Grove (1973) and Grundy (1963). Such conditions mean that short-period records can be very misleading and have very serious effects on agriculture and water supply. It is the tendency for such sequences to occur, perhaps more than the magnitude of departure from average conditions, which has serious implications. One exceptionally dry year or a wet year creating problems of flooding and erosion can be survived, although even here it is more difficult for the peasant subsistence farmer to do so than, say, a plantation system or the highly capitalised system of temperate areas. The real problems arise when such conditions recur over a series of years, the peasant subsistence farmer again being especially hard hit.

The Sahelian drought situation of 1973 is more correctly seen as the culmination of a number of poor years rather than as a single disastrous year. Thomas and de Bouvrie (1973) indicate that in the Mali, Niger and Upper Volta Sahelian regions the drought started in 1969 and as early as 1968 in some areas. In these regions annual precipitation was consistently not more than half the long-term average. For example, at Agades in northern Niger, the 30-year average is 164 mm, but only 40 mm and 74 mm fell in 1970 and 1972, respectively. Several places in northern Mali, Chad and the Central Niger Republic received only 15 per cent of their normal totals (Oguntoyinbo and Odingo 1979). The

Table 3.4 Senegal – mean annual rainfall and annual rainfall 1966–72 (mm) (from Burdon, Drouhin and Dijon 1973)

	Period of mean	Mean	1966	1967	1968	1969	1970	1971	1972
Podore	1940–69	335	243	271	184	432	254	136	80
Linguere	1937–69	506	504	556	279	679	295	328	219
Diourbel	1931–60	700	709	811	352	874	592	663	372

cumulative effects on the environment of these consecutive dry years is far greater than those due to a single dry year following a succession of wet years. In Senegal, Burdon, Drouhin and Dijon (1973) indicate that the drought producing the disaster in the first half of 1973 can be traced back to low rainfall in 1968, 1970, 1971 and 1972 (Table 3.4). The good rainfall in 1969 was not sufficient to reverse the dry trend and the drought had been intensifying over some 5 years. Daniels (1980) reports that the UN Conference on Desertification held in Nairobi in 1977 considered that the drought of 1968–72 was the culmination of desiccation dating back to the late 1950s or early 1960s in most parts of the Sahel. Though prolonged, severe and widespread it was not unprecedented, similar occurrences appearing 30, 50 and 60 years earlier (Daniel 1980). As has already been indicated, the effects of these droughts may be amplified by economic, social and political factors. Grove (1973) points out that the risks are greatest in marginal areas when a succession of 'good' years are followed by a long drought. Stocks of animals are built up in the former leading to overgrazing in the latter. Cultivation of land during good years is especially dangerous and perhaps the main cause of desertification, against which it is necessary to take preventive action. In the following dry years, ploughed soil or areas where natural vegetation has been removed are at the mercy of wind erosion. They are also liable to water erosion since flash floods still can occur in dry years. These aspects are considered in more detail in later sections such as § 3.4. Studies of the perception of pastoral farmers in western Queensland over the last 100 years indicate their inclination to view wetter years as 'normal' and the drier years as 'bad'. They tend to gear their activities to the better spells of years (Oliver pers.comm.).

If sequences of dry years coincide with attempts to introduce new seeds or crops and new techniques such as earlier planting or settlement schemes, then the result may be complete failure and unwillingness of the people to attempt further changes.

Bearing all the above points in mind, it is appropriate to consider some methods of analysing rainfall variability and their practical application, using illustrations from a variety of areas. Even as an indicator

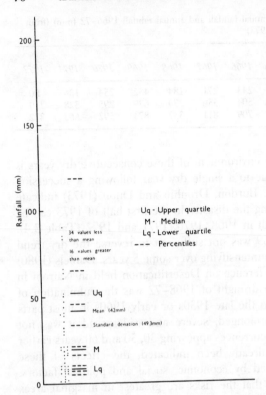

Fig. 3.14   Thursday Island: May rainfall (1890–1939).

of 'average' conditions, the commonly used arithmetic average (mean) has faults where data are skew. Figure 3.14 shows the May rainfall over a 50-year period for Thursday Island. The average is 42 mm but thirty-four values are less than this (68 per cent of the years). As an alternative, the median or middle value may be used where half the years had less than the median and half more. The median for the data in Fig. 3.14 is 20 mm, only half the average, and in many ways it provides a more meaningful picture.

Maximum and minimum values (Table 3.1) give a crude indication of variability although as has been pointed out, particularly in semi-arid areas, the occurrence of a few extremely high values can be misleading. What is needed is a measure which takes all values into account, not only to obtain an expression of the degree of variability but also for reliability and probability assessments. The mean deviation and standard deviation are commonly used to express variation, but as in the case of the arithmetic average they suffer from disadvantages with a skew distribution.

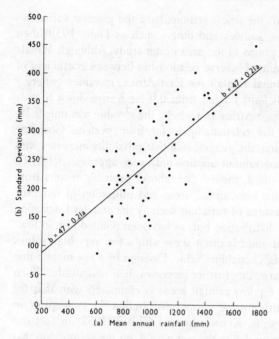

Fig. 3.15 Mozambique: mean rainfall against standard deviation (after Gregory 1969).

It is not unexpected to find a general relationship between mean values and absolute values of variation, a characteristic which some have used to predict variation if the mean is known (e.g. Griffiths 1961). It is unrealistic to expect such a simple relationship to be universal, bearing in mind the varied origins of tropical rainfall. For example, if much of the rainfall is derived from infrequent heavy falls such as those associated with hurricanes, then a small change in their number from year to year can result in much higher variability than in an area where the rainfall originates from varied, less heavy storms. For Mozambique, Gregory (1969) has shown that a relation between mean rainfall and variability is far from perfect (Fig. 3.15) and clearly other factors influence the degree of variation.

To allow comparison in relative terms, variation can be expressed as a percentage of the mean. Useful as this technique is in climatological studies, such data are not as helpful in practical terms as absolute measures of variation such as mean deviation and standard deviation. Studies utilising a measure of relative variability (relative variability, coefficient of variation) often refer to an inverse relation between this measure and mean rainfall. Dick (1958), for example, states 'this is a normal relationship', Huke (1966), '*All other things being equal, theory dictates* that stations with the greatest total volume of rainfall have the least annual percentage variation and the driest stations have the greatest variation.' Interestingly, both these authors and others such as Dale (1959) then go on to point out exceptions in the area under study. Although Woodhead (1982) found a general inverse relationship between coefficient of variation and mean annual rainfall for East Africa, r^2 values (where r is the correlation coefficient) ranged from 0.9 for Kenya down to only 0.2 for Tanzania. For East Africa as a whole, the r^2 value was only 0.39, again illustrating that the relationship is far from perfect. Cochemé (1966) is correct in saying 'the general rule that variability increases with decreasing mean annual rainfall applies only very approximately'. To some extent the often-cited 'theory' (see above) probably results from the inclusion of data for low-rainfall areas and utilisation of inappropriate techniques. Measures of variation such as the standard deviation require a near-normal distribution but, as has been pointed out, in low-rainfall areas the distribution is often skew with a few very high values tending to produce a high variability value. Division by a low mean value to produce a relative variability further increases their magnitude. When the relative variability for low-rainfall areas is compared with that for higher rainfall areas with a more normal distribution the concept of an inverse relationship results. Kenworthy and Glover (1958) in fact use the percentage variation to define the wet period, on the assumption that a wet month will have a distribution more normal and hence a more moderate variability. They included a month with a variability of < 50 per cent (in many cases < 30 per cent) in the rainy season.

If data are approximately normal, Gregory (1969) suggests that the relative variation not only is not closely related to the mean value but also that, contrary to general opinion, it is no greater in the tropics than in higher latitudes. Gregory (1969) cites variability values of 10–20 per cent in Britain, similar to those of northern Nigeria and Sierra Leone. In Ghana, 25 per cent is the highest value, most being less than 20 per cent. In Mozambique, annual variability is somewhat higher, but exceeds 35 per cent only in very limited dry areas. Data and maps of relative variation are found in a variety of papers, e.g. Dale (1959), Dick (1958), Gregory (1969), Lockwood (1974), Griffiths (1972), Brookfield and Hart (1966) and Meher-Homji (1974). Figure 3.16 shows relative variability for the world which indicates the large variability of the semi-arid areas resulting from the techniques of analysis.

For skew distributions, just as the median can replace the arithmetic average, so measures such as mean deviation and standard deviation can be replaced by the quartile deviation and percentiles, the latter being particularly important in probability calculations. These values are shown on Figs 3.11 and 3.14, 50 per cent of the values lying between the upper and lower quartiles and 10 per cent between adjacent percentiles. The 10th and 90th percentiles and extremes for Kinshasa are shown in Table

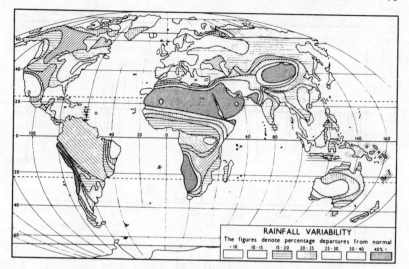

Fig. 3.16 Annual rainfall variability (after Riehl 1954).

Table 3.5 Lower Congo (Kinshasa) monthly and annual rainfall (1930–59) 10th and 90th percentiles and extremes (mm) (after Griffiths 1972)

	Min	P_{10}	P_{90}	Max
Jan	1.7	32.9	227.1	320.6
Feb	48.6	67.0	234.2	329.8
Mar	58.0	79.6	317.3	428.9
Apr	58.9	112.4	326.9	378.6
May	22.2	42.7	217.8	280.0
June	0.0	0.0	22.7	37.6
July	0.0	0.0	1.1	34.0
Aug	0.0	0.0	17.8	24.4
Sept	1.6	5.0	76.3	100.4
Oct	19.5	52.6	206.8	281.6
Nov	84.3	135.0	334.8	347.7
Dec	47.7	101.3	275.8	326.6
Year	1824.0	1170.0	1655.0	1824.0

Note: P_{10} – 10% of years less than this value, P_{90} – 90% of years less than this value. Thus for January, 80% of years will have between 32.9 and 227.1 mm.

3.5. For non-normal distributions, they provide a more realistic picture of the variability of rainfall. The quartile deviation:

$$\frac{\text{Upper quartile} - \text{Lower quartile}}{2}$$

can be expressed in relative terms by dividing by the median.

Nicholson (1981, 1983) and Hulme (1984) in examining African rainfall variations used normalised scores which are rainfalls standardised

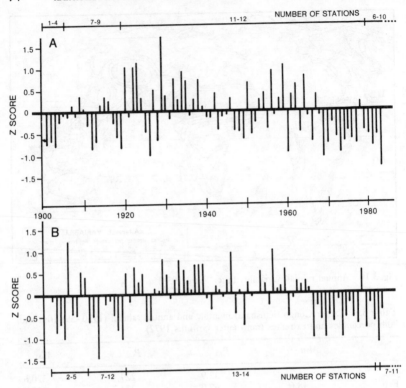

Fig. 3.17 Normalised series of annual rainfalls in central Sudan (based on 1951–80 means and standard deviations). A: north-central Sudan 14–16°N, B: south-central Sudan 12–14°N (after Hulme 1984).

in terms of means and standard deviations. An example for stations in Sudan is presented in Fig. 3.17. Mooley *et al.* (1981), using a network of 306 gauges over India, examined the total volume of rainwater for each of the years 1871–1978. The mean, standard deviation and coefficient of variation were calculated and the series was found to be homogeneous, random and fitting a Gaussian distribution. Years with deficit and excess greater than 20 per cent were examined. Years where 40 per cent or more of the country had a deficit or excess were found to have a random occurrence. Mooley, Parthasarathy and Sontakke (1982), in examining summer monsoon (June–September) rainfall for India during 1871–1978, devised a monsoon excess index (MEI) based on the percentage area with a specific percentage rainfall excess. The MEI series was homogeneous, random and fitted a gamma probability model. Years with large MEI values had a random distribution. Among a large number of other studies of Indian rainfall variability are those of Dhar,

Rakhecha and Mandal (1981b), Dhar *et al.* (1982), Mooley and Partha-sarathy (1983), Parthasarathy (1984).

An extremely interesting analysis of rainfall variability over India is presented by Meher-Homji (1974) who examined the stability of four factors – number of rainy days, dry months, precipitation and seasonal incidence. Such an approach acknowledges the fact that variability of more than one rainfall characteristic must be analysed and that the factors do not vary in the same way. He defines a 'probable year' as one corresponding in precipitation, rainy days and dry months to the mean ± 1 standard deviation and having a seasonal regime of the most common type. The proportion of years conforming to the probable year varied from only 8 per cent to 65 per cent at different stations. Hastenrath (1985) discusses variability in a variety of tropical areas including Indonesia, India, Brazil, Central America and the Caribbean and sub-Saharan Africa.

3.4 DROUGHT

In the previous section, drought was referred to but some further comments are in order here. Perhaps understandably, a simplistic attitude to drought is common. Occurrence is regarded as simply lack of rain and the 'problem' ended with a return to wetter conditions. Reality is more complex. Drought must be considered as a water supply–demand situation. Demand is influenced by evaporative demand (Ch. 5) but also by human activities in an area which create particular water requirements. As far as supply is concerned, rainfall is merely an initial input to the atmosphere–soil–plant water system (§ 6.3). Relationships between rainfall and water eventually becoming available to meet a demand are complex, being influenced by short-period rainfall characteristics for example (§ 4.1), the complexities of the system (§ 6.3) and rainfall–runoff relationships (§ 4.4).

How people use land is important and here it is appropriate to distinguish between land use and land management. Land use can be thought of as *what* an area is used for (e.g forestry, pastoralism, agri-culture, urbanisation) whereas management is *how* the activity is carried on (e.g. soil and water conservation practices, cultivation techniques, livestock densities, irrigation and drainage practices). Often it is the management which is crucial in terms of hydrology and environmental effects such as soil erosion. Unfortunately, land use is sometimes blamed for adverse environmental effects when the real culprit is the adoption of inappropriate management practices. Land use and management have considerable impact on drought occurrence, severity and environmental deterioration, as has already been suggested, for example, due to overgrazing (§ 3.3). These aspects are considered later, particularly in Chapters 8 and 9.

Although drought involves a shortage of water in relation to a particular need, the problem is to decide on the degree of shortage necessary for a drought situation to be recognised. This involves three elements – the magnitude of the deficit, its duration and its areal extent. In § 3.3, it was suggested that it is often the occurrence of a series of dry years/seasons which is crucial rather than only a large deficit in a single year. Since the effect is often cumulative, it is frequently difficult to say when a drought really began. Certainly in terms of drought effects, it may also be difficult to say when the situation ends. A deficit over a small area may have limited effects whereas the same magnitude of deficit over a large area could warrant classification as a major drought. However, the scale of human activity, whether peasant smallholder or large-scale plantation agriculture will have a bearing here (§ 4.2). Socio-economic and political factors as well as scientific and technological developments must be considered (§§ 3.1 and 3.3).

The above points indicate problems of defining drought other than in very general terms of a water deficit and difficulties in deriving quantitative drought indices. Drought impacts are difficult to assess, either in terms of effect on plants, for example (§§ 6.1, 7.3) or in economic terms. Measures to ameliorate drought impacts, by appropriate land use and management as well as by various socio-economic means are complex. These are discussed by Mattei (1979), Oguntoyinbo and Odingo (1979) and Ofori-Sarpong (1983).

Unfortunately, certainly until recently, simplistic attitudes to drought led to emphasis on short-term 'solutions' (e.g. emergency relief) without addressing long-term aspects such as appropriate land use and particularly management practices as well as socio-economic and political factors. This is not meant to imply that major rainfall deficits such as those of the 1970s and 1980s in the Sahel will not always create major problems which require short-term action. However, at the very least, their effects can be minimised and less extreme rainfall deficits correspondingly reduced in their impact by longer term approaches. Aspects are considered in Chapters 8 and 9. One final problem is the tendency to regard large rainfall deficits as freak occurrences and wetter years as the 'norm' when the former are really a normal part of environmental variability which must be adapted to.

Drought in sub-Saharan Africa has received much attention and reference to some studies illustrates not only the severity of the situation but also its complexity and some of the above points. Exact delimitation of the spatial extent and period is difficult. Ogallo (1984) states that after drought conditions from 1968 to 1973, in 1974–75 there were substantial rainfall increases in the drought areas of sub-Saharan Africa but that the drought returned in 1976 and continued in some parts to date (i.e. 1984). Motha *et al.* (1980) speak of drought over eleven West African countries from 1968 to 1975, with 1973 being the worst year when the 300–400 mm zone of annual rain was at least 200 km further south than normal. Daniels (1980) refers to drought from 1968 to 1973 in West Africa with a return of rain in some areas in 1974–75 but persistence of drought in others. In 1977, drought re-invaded West Africa (Daniels 1980). De Boer (1984) refers to the Sahel drought from 1969 to 1974 and Ofori-Sarpong (1983) talks of drought from 1970 to 1973, 1975 to 1977 in Burkino-Fasso. Nicholson (1983) speaks of drought in sub-Saharan Africa as persisting until at least 1980. Not only do these differences in period relate to definition problems (see above), they also indicate that despite its widespread nature (Nicholson 1980, 1981, 1983; Nicholson and Chervin 1983), there were regional contrasts. Ofori-Sarpong (1983), for example, indicates regional variations in Burkino Fasso, some stations having above average rainfall in some years over the period 1970–77. Hulme (1984) points out that in the generally dry year of 1983 in central Sudan, some stations had well above average rainfall and demonstrates the extreme localisation of annual totals. Spatial variations in rain are discussed in § 4.2.

Several authors refer to earlier equivalent droughts. Daniel (1980) suggests similar ones occurred 30, 50, 60 years earlier. Motha *et al.* (1980) write of a prolonged drought in the 1940s in northern Nigeria and de Boer (1984) cites three previous periods in the Sahelian–Sudan zone in this century when annual rainfall was 30–50 per cent below average. Nicholson (1980) examined correlations between annual rainfall departure maps for the period 1901–75. Frequently occurring anomaly types representing several Sahel drought patterns were identified. She concluded that rainfall fluctuations are more commonly characterised by contraction or expansion of the desert belt than by its north–south displacement. Motha *et al.* (1980) refer to a strong relationship between rainfall and ITCZ position (§ 2.6) for the Sahel zone of Nigeria. Further comments on links between rainfall anomalies and atmospheric features are contained in § 3.7.

Evaluation of drought impacts is difficult. For Burkino-Fasso, Ofori-Sarpong (1983) suggests that impacts on livestock were especially severe, the 1972–73 loss being estimated at 35 per cent. Production of millet, sorghum, maize, rice and groundnuts fell by about 40 per cent. Large-scale rural to urban and north to south migration occurred in 1970–73.

De Boer (1984) suggests that due to greater population and animal numbers, the impact of the 1969–74 drought was greater than with earlier rainfall deficits of similar magnitude. In Mali, sheep and goats numbered 8 million in 1961, 11 million in 1970, dropping to 7.8 million in 1972–75 but increasing again to 11.5 million in 1979 (de Boer 1984).

Cattle showed similar fluctuations. Oguntoyinbo and Odingo (1979) present a detailed discussion of drought impacts, such as the fall in groundnut production in Nigeria. Consequences for population and economic development in West Africa are discussed by Ojo (1983).

Focus on the African situation is not to imply that severe droughts have not occurred elsewhere. Ogallo (1984) refers to the Latin American situation, especially north-east Brazil, where, for example, Riehl (1979) indicates wholesale population migration from the Quixeramobim area when low rainfall persists for more than one season. Hastenrath (1985) discusses drought in various tropical regions and in § 3.7, reference is made to droughts in a variety of areas.

3.5 RAINFALL PROBABILITY AND RELIABILITY

Interesting as the analyses discussed in § 3.3 are in climatological terms from the points of view of agriculture and water resources, rainfall probability and reliability data are of more practical concern. Needs will vary with crop type, evaporative demand and soil type. In particular, the period over which calculation is needed will vary as will the critical values of rainfall in these periods. The problem of skew data in such calculations increases the shorter the time period and the lower the rainfall. Gregory (1969) suggests that for annual falls of more than 750 mm normality is a reasonable assumption and this is often true for wet season falls (Kenworthy and Glover 1958). For falls less than this, the possibility of non-normality is greater and Griffiths (1961) found that for annual falls <635 mm there was a 45 per cent chance of data being non-normal. Mooley and Appa Rao (1971) for India found that while the normal distribution gave a good fit to seasonal and annual rainfall at stations in some parts of India, it did not give a good fit over the major part of the country where instead a gamma distribution was more suitable.

Particularly for shorter periods, transformation of data to produce normality may be necessary. Manning (1956) and Griffiths (1961) found that a square-root transformation was best but Gregory (1969) shows that for Zaria neither this nor a logarithmic transformation is necessarily a better fit (Table 3.6). Even within an individual area, the shape of the distribution curves may vary, working against the adoption of a single transformation.

Much work on probability and reliability has been carried out in East Africa. Woodhead (1982) provides a comprehensive review of probability studies in East Africa and their value for agriculture and hydrology. He points out that not only did a revised mean annual rainfall map published in 1971 by the East African Meteorological Department differ significantly from one drawn in 1959, it indicated anomalies and inconsist-

Table 3.6 April rainfall, Zaria, 1904–59: the proportion of occurrences falling within given limits in untransformed (x) and transformed ($\sqrt{x}$ and log 10x) data, as compared with the normal frequency distribution (after Gregory 1969)

Form of data	x	$\sqrt{x}$	Log (10x)	Normal frequency distribution
Average	1.72	1.18	1.0837	
Standard deviation	1.26	0.58	0.4575	
±1 standard deviation	66%	71.4%	75%	68.3%
	(37 items)	(40 items)	(42 items)	(38 items)
±2 standard deviations	96.4%	89%	89%	95.5%
	(54 items)	(50 items)	(50 items)	(53 items)

encies in an 80 per cent probability map published by the East African Meteorological Department (1961). This indicates the need to update such maps as more information becomes available. Woodhead (1982) calculated ratios W_{10}, W_{20} of the 10 per cent and 20 per cent probable annual rainfall to mean annual rainfall at 34 East African locations and investigated their relationships with location and mean rainfall. It should be noted that the 20 per cent probable rainfall was that expected to be exceeded in 80 per cent of years. For Tanzania and Uganda, single ratios appeared appropriate. However, for Kenya there appeared to be a strong dependence on mean annual rainfall and hence estimating probable rainfalls involved the product of the mean value and corresponding W values. Woodhead (1982) also found that the East African relation of probable to mean annual rainfall conformed to a worldwide pattern. He also found that for seasons of 3,4,5,6 or 8 months' duration, the 20 per cent probable seasonal rainfall for Tanzania could be estimated by multiplying seasonal mean values by 0.83.

Gregory (1969) has combined data from various sources to produce a map for a large part of eastern Africa showing the annual rainfall likely to be reached or exceeded in 80 per cent of years. All such calculations are, strictly speaking, related to an infinite number of years (or at least to a very large number). If there is an 80 per cent probability of receiving a certain amount of rainfall then this means that 80 per cent of a long series of years have this amount. It is incorrect to say that this amount will occur in four out of five years. While probability estimates applicable to long time periods are useful, the agriculturalist is not so concerned about how things will average out over a 100-year period. The peasant subsistence farmer, in particular, is concerned with the next few years. Poor rainfall three years in a row may mean starvation. In any case, long-term estimates must assume no essential 'change' in climate.

For short-period assessments, the binomial frequency distribution may be used (e.g. Jackson 1969c). Table 3.7, for a number of Tanzanian stations, compares the long-term probability of receiving 800 mm per year

Table 3.7 Rainfall probability – Tanzania (from Jackson 1969 c.)

Station	Per cent probability of receiving 800 mm	Per cent probability of receiving 800 mm in 4 years out of 5	Per cent probability of two successive years receiving <800 mm
Kingolwira SE	61	36	16
Lugoba Mission	83	79	3
Marios SE	52	23	23
Kikeo Mission	94	96	<1
Morogoro	68	48	10
Kikondeni	76	65	6
Tungi SE	49	18	26
Melela	32	3	46
Maneromango	65	44	12
Utondwe Salt Works	59	32	16
Kisiju	96	98	<1

with the probability of receiving 800 mm in at least four out of five years and the probability of two successive years receiving < 800 mm, the latter two calculations using the binomial frequency distribution.

Table 3.8 shows that it is important to remember that calculations are based on sample data. Even average values vary considerably from period to period although they all fall within the years 1942–66. It is important, therefore, to assess standard errors of probability and reliability figures. Sampling errors may indicate that both probability and variability differences between two stations are not statistically significant. Methods of assessing standard errors and significance of differences are discussed by Gregory (1969) who shows, for example, that differences in coefficients of variation for Gambia and Britain are not statistically significant.

Table 3.8 Average annual rainfall, standard deviation and probabilities for various periods (Tanzanian stations) (from Jackson 1969 c.)

Period	Ruvu Sisal Estate				Kingolwira Prison Farm				Bagamoyo Agricultural Office				Marios Sisal Estate			
	1	2	3	4	1	2	3	4	1	2	3	4	1	2	3	4
1942–66	990	279	75	64	838	212	57	28	1062	308	80	74	807	173	52	21
1942–61	926	245	70	53	850	270	57	28	1086	319	82	78	779	176	45	13
1942–56	906	187	71	55	824	221	54	24	1090	244	88	89	772	157	43	11
1942–51	940	190	77	68	874	246	62	37	1165	259	92	94	811	157	53	23
1950–64	987	290	74	61	850	257	58	30	1051	326	78	69	785	188	47	15

Column 1: average annual rainfall (mm).
Column 2: standard deviation (mm).
Column 3: per cent probability of receiving 800 mm in a year.
Column 4: per cent probability of receiving 800 mm in at least 4 out of 5 years.

Statistical significance is relevant to the construction of rainfall probability and reliability maps. Isolines should be for values which are significantly different, and, particularly in the middle ranges, a wide interval may be necessary (Gregory 1969). Since extremes are likely to be most important agriculturally and are also most likely to produce a map of statistical significance, Dow (1955) recommends mapping only upper and lower extremes.

The tendency for sequences of wet or dry conditions to occur (§ 3.3) casts doubt on the validity of short-term probability estimates. It indicates the need to examine the possible existence of short-term cycles and fluctuations in rainfall, although as will be seen in § 3.8 this involves considerable problems. Mooley and Appa Rao (1971) and Mooley (1971) point out the need to examine sequences of rainfall for independence before undertaking probability assessments. For India they found that persistence of conditions was important for some time periods, consecutive daily, 5-day and 10-day periods not being independent. They found that consecutive 15-, 20- and 25-day periods were independent and monthly periods pairwise independent, but not tripletwise or quadrupletwise. Rainfall in the first half of the summer monsoon was found to be independent of that in the second half. Parthasarathy (1984) found no persistence in Indian summer monsoon rainfall for 306 stations over the period 1871–1978. The question of independence of monthly rainfall has also been considered by Glover, Robinson and Taylor (1955), and in the next section, shorter period independence is considered in discussing rainfall modelling.

Since the water needs of a crop vary within the growing season it is not always sufficient to have a probability estimate for the whole season or even a particular part of it. It may be important to have a probability estimate for the occurrence of a particular sequence of amounts throughout the season (e.g. 50 mm in the first month, 100 mm in the second, 150 mm in the third and so on). Estimates of this kind have been carried out by Robinson and Glover (1954) assuming monthly totals to be independent and by Glover et al. (1955) if they are not independent.

As an alternative to calculations using the normal distribution, percentiles may be used. Thus, 80 per cent of values lie above the 2nd percentile, i.e. it can be said that there is an 80 per cent probability of receiving at least the second percentile value. In Fig. 3.14 it can be seen that there is an 80 per cent probability of receiving at least 8 mm but only a 20 per cent probability of receiving 65 mm. This method has the advantage that no real calculation is required and it is quick, but in particular it overcomes the problem of a skew distribution. Again, however, results are based only on a sample and apply to a long series of years.

Oldeman and Frere (1982) calculated monthly rainfall probability for south-east Asian stations using a ranking method adopted more normally in calculating return periods (§ 4.1). Straight line plots of rainfall against probability on a log-normal scale were obtained. For 19 locations a straight line plot of mean monthly rainfall and the 75 per cent probable rainfall was also obtained. Other examples of the calculation and use of rainfall probability data include Odumodu (1983) for Plateau State, Nigeria, and Virmani (1975) for India.

3.6 DAILY AND PENTADE VARIABILITY

The significance of variability of rainfall over short time periods has already been referred to in § 3.2 and the use of 5-day periods (pentades) for analysis pointed out. Griffiths (1959) considered the totals for three consecutive pentades and designated the middle one as 'rainy' if

1. The total rainfall for the three pentades together amounted to at least 76 mm.
2. The rainfall amounted to 7.6 mm or more in each of at least two of the three pentades.

Figure 3.18 for two East African stations, Nairobi and Gulu, shows the distribution of rainy pentades for individual years. The figures illustrate the degree of variability from year to year, particularly in terms of defining the 'start' and 'finish' of the rains. As has already been suggested, variability in the start and finish of the rains may be especially critical. Lineham (1967) utilised the rainy pentade technique of Griffiths to define not only the average start and finish of the rainy season, but also variability and probability estimates of both. In addition, the 'quality' of the rainy season was assessed using the mean number of rainy periods as well as their variability and probability of occurrence. Other studies of seasonal rainfall utilising pentades have already been indicated in § 3.2.

Evans (1955) for southern Tanzania considered the question of planting dates for crops in relation to the start of the rains. Other East African studies of various aspects of probability assessments for periods less than a month and start of rains are those of Manning (1956), Woodhead (1970) and Woodhead, Waweru and Lawes (1970).

Examination of daily rainfalls further illustrates the magnitude of variability in the tropics and indicates the importance of day-to-day changes in the occurrence of disturbances. Even in 'wet' areas with no real dry season, considerable day-to-day variations occur and these can be illustrated by examples from Malaysia (Dale 1959) and Singapore (Nieuwolt 1968).

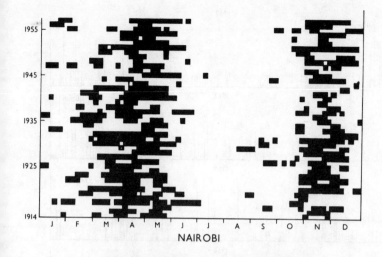

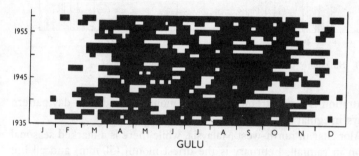

Fig. 3.18   Nairobi and Gulu: rainy pentades (after Griffiths 1959).

For Singapore, daily rainfalls over the period 1962–66 (Fig. 3.19) illustrate how irregular is the pattern. Even during the wet months, dry spells occur and wet spells are found in the relatively 'dry' months. For the period 1951–60, wet spells longer than 6 days (each day with > 0.25 mm) occurred in every month of the year and dry spells of more than 6 days (< 0.25 mm) occurred in every month except November, even though the climate is wet overall. The longest wet spell lasted 14 days and the longest dry spell 20 days. Even in a wet location such as Singapore, much of the rain is concentrated in short periods of a few days of heavy rain. In the relatively dry months, a large part of the total may fall in a few days. The proportion received in the three wettest days of the month reached 78 per cent in August 1963, 81 per cent in September 1962, 90 per cent in February 1962 and March 1963 and 100 per cent in April 1963. In very dry months, the wettest day may bring much of the rain, e.g. 63 per cent in March 1963. While the

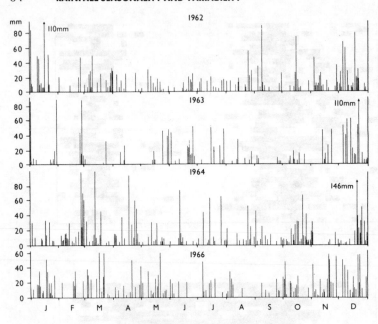

Fig. 3.19   Singapore: daily rainfall 1962–66 (after Nieuwolt 1968).

concentration is less extreme in the wet months, the day-to-day pattern is always irregular.

At Alor Star in north-west Malaysia, while there is a marked seasonal variation in rainfall, February is the driest month (50 mm) and all but two months have more than 100 mm. Even in such a relatively wet location, long dry spells occur. No rain occurred for 49 days during January–February 1940 and periods of 21 days without rain occur relatively frequently (16 in 23 years). In 23 years, there were forty-six periods of 14 days with no rain. The number of days with rain also shows great variation from year to year. At Penang for example, the number has varied between 119 and 229 and Dale (1959) states that in Malaysia, the number is liable to vary by ± 40 per cent of the annual average. Thus, even in wet locations and wet seasons, it does not rain every day, much of the rain occurring in only a limited number of rain days and the number varying considerably from year to year. This is clear evidence of the significance of disturbances rather than the occurrence of rain associated with a general, uniform area of convergence.

Riehl (1979) provides examples from Venezuela of how a few events produce much of the rain. For example, of 14 rainstorms in a 100-day season in 1969, two episodes produced 50 per cent of the rain and seven produced 75 per cent. For one month in June in north central Venezuela with 200 mm, 4 rainstorm days produced 150 mm. If the two largest storms with 87 mm had not occurred, the rainfall for the month would have been almost average but these two storms raised the total to almost twice the average.

Hutchinson and Sam (1984) illustrate the patchy nature of daily rainfall in Gambia, particularly at the start of the season in June. They point out that after the first rain (days with $\geqslant 12.7$ mm being used as a threshold), the second heavy rain may be expected 10 days later. Each succeeding rain occurs at ever decreasing intervals, finally stabilizing for any rainfall amount at about 3 days which matches the average frequency of easterly waves (§ 2.9).

In drier areas, in particular, falls are sporadic. At Alice Springs, Australia, for example, of the annual total of 265 mm, in the average year sixteen groups of 1–3 wet days supply 171 mm, and two groups exceeding 3 days account for the remaining 96 mm (Winkworth 1970).

This variability on a day-to-day basis and for periods less than a month, even in wet locations is critical to plant growth, particularly in the early part of the rainy season before soil-moisture reserves have been built up. It limits the practical utility of variability and probability studies over periods of, say, a month or a season. The distribution of rain days through a month is important. The effectiveness of 9 rain days evenly spread over the month could be different from three groups of 3 rain days. This aspect is considered in later chapters.

For South and South-East Asia, Morris and Rumbaoa (1980) review a number of studies involving daily rainfall analysis and discuss the importance of such work for rice-based cropping systems research. They consider a rainfall total of 50 mm in a 3-day period as a lower criterion providing enough moisture for germination of dry seeded rice and about 10 days of initial growth. Return periods (§ 4.1) of maximum 1-, 2-, 3-day rainfalls in the dry-to-wet and wet-to-dry transition periods were determined to assist in rice cropping pattern adaptations.

The effect of rainfall variability will vary depending on factors such as crop type, soil type and the length of the period considered, as well as economic and social factors. Peasant subsistence farmers are more likely to be hard hit than those in more wealthy areas such as tropical Australia. Not only are they less likely to have supplementary irrigation to cope with variations of rainfall, but their general economic circumstances make them less capable of withstanding 'poor' years. Furthermore, the great importance of agriculture in most tropical areas means the fluctuating yields because of varying rainfall will have a tremendous impact throughout the whole economy.

The importance of short-period variability has led to a number of approaches to statistical modelling of conditions, including the question of independence (§ 3.5). A discussion of the principles of such modelling is outside the scope of this book but it is useful to indicate some studies dealing with tropical conditions. For stations in East and West Africa and South-East Asia, Jackson (1981) examined the dependence of the occurrence of wet and dry days on the situation on previous days. Dependence was found to vary between stations and seasons, with at least the previous two days often needing consideration (i.e. a higher than first order Markov model being suggested). For dry spell characteristics in semi-arid stations in India, Singh *et al.* (1984) found a second order Markov model to be a better fit than a first order one. Composite charts of the probability of occurrence of rain around depressions were used to examine the role of the latter in alleviating droughts. Singh *et al.* (1981) fitted a range of models (e.g. first- and second-order Markov, logarithmic) to runs of wet and dry days at 22 Indian stations during the summer monsoon. Persistence in daily and 5-day amounts was examined using autocorrelation coefficients of various lags and persistence in 5-day amounts was also examined using contingency tables. Singh and Kripalani (1982) also examined dependence of daily rainfall in India using Markov, logarithmic, autoregressive and autoregressive moving average models. They also found that rainfall intensity was more on spells of two or three days than on isolated days.

Stern (1980a) used first- and second-order Markov models to calculate the probability distribution of number of rain days, total rainfall and lengths of dry spells for Nigerian and Indian stations. Stern (1980b) also used a two-part model where occurrences were modelled as a Markov chain and amounts of rain on rain days fitted by a gamma distribution, one aim being to illustrate the advantages of such a model over analysis of 10-day and monthly totals. The basic two-part model approach was used by Garbutt *et al.* (1981) to compare rainfall at 11 locations on a north-south transect in West Africa. Mean rainfall per rain day was found to be relatively constant and hence variation in probability of rain was the most important aspect in seasonal and locational differences. This rainfall model was also used to assess the probability distribution of date of start of the rains in West Africa by Stern, Dennett and Garbutt (1981) as already indicated in section 3.2. The use of daily rainfall models in agricultural planning is discussed by Stern and Coe (1982). Analysis where the probability of an event (e.g. dry spell, start, end, and length of growing season) is estimated directly from its relative frequency of occurrence or from a particular frequency distribution is considered by Stern, Dennett and Dale (1982a) and then results compared with those using a daily rainfall modelling approach by Stern, Dennett and Dale (1982b) for Indian and Nigerian data. One further example of daily rainfall analysis in West Africa found no dependence between rainfall at different times in the rainy season (Dennett, Rogers and Stern 1983). Differences in rainfall pattern between seasons with early, medium and late starting dates were found to be of no practical importance.

3.7 RAINFALL VARIABILITY LINKAGES

Prompted partly by interest in droughts, in recent years much interest has been shown in linking rainfall variations to features discussed in Chapter 2. In most cases, concern has been with establishing statistical relationships. While these do not prove causality, their establishment can lead to greater understanding. They may have predictive value, although without a causal link such prediction must be treated with caution.

Relationships have been established between rainfalls at different times at the same or different places as well as between rainfall and atmospheric circulation features and related elements such as sea surface temperatures (SSTs) (§ 2.8). For example, links between ENSO events and tropical rainfall are discussed by Yarnal and Kiladis (1985). The situation in various tropical regions is discussed by Hastenrath (1985). Examples from the tropical Americas, Africa, the Pacific, Australia and India illustrate the variety of studies.

Ogallo (1984) refers to a range of studies showing how in some years anomalous climatic patterns have covered large areas ranging from regional to inter-continental to inter-hemispherical. These studies examine climatic teleconnections between conditions in north-east Brazil and West Africa and between rain in the tropical Americas and changes in general atmospheric circulation over other tropical areas, mid- and high latitudes. They also examine teleconnections between rainfall anomalies in sub-Saharan Africa and circulation over other parts of the globe. These studies, together with the fact that rainfall anomalies often affect large areas (§ 3.4) suggest that causes of anomalies tend to be global rather than regional (Ogallo 1984).

Stoeckenius (1981) found that seasonal and annual precipitation anomalies for various regions between 30 °N and S were related to variations in intensity of zonally orientated circulations in the tropics which were in phase with variations in the SO (§ 2.8). Hastenrath (1984) examined general circulation features linked to both the annual cycle and inter-annual variability of rain over various regions of the tropical Americas and Africa. In most cases it appeared that rainfall anomalies tended to be linked to departures in the large-scale atmospheric and ocean fields corresponding to pattern changes in the annual alternation of wet and dry seasons. This suggests that inter-annual variability is largely an enhancement and reduction of the annual cycle.

A number of studies have focused on Latin America, especially Brazil. Chung (1982) found links between surface wind and SST for the tropical Atlantic coast and precipitation at Quixeramobim (north-east Brazil), particularly the occurrence of abnormally dry conditions. Relationships between atmospheric circulation and rainfall anomalies in two large regions of north-east Brazil were examined by Pao-Shin Chu (1983). For example, excess rainfall during the peak rainy season in the south was marked by negative pressure anomalies over the South Atlantic, weak onshore south-east trades and below average SST along the southern Brazilian coast. These features appeared to be related to southern hemisphere frontal systems (§§ 2.9, 3.2). During dry years, departures were almost the reverse. Rainfall anomalies over Brazil in February–March 1981 have been related to atmospheric circulation features by Kousky (1985). Nobre, Nobre and Moura (1985) found that phase changes of a large-scale teleconnection pattern linking the North Atlantic to East Asia might have value in predicting extreme precipitation anomalies in north-east Brazil although the generating mechanism and causal links are not known. In a preliminary study, Nobre and Renno (1985) examined links between abnormal drought over most of the Amazon and north-east Brazil, extensive flooding in south Brazil, Paraguay and northern Argentina in 1983 and the 1982–83 ENSO episode (§ 2.8).

Reference has already been made in § 3.4 to studies of variability and drought in sub-Saharan Africa. Nicholson (1981) suggests that a north-ward displacement of the ITCZ may account for wetter years but that a weakened 'intensity' of the rainy season, independent of ITCZ position, is the most likely cause of drought in the region. Nicholson and Chervin (1983) found that major rainfall fluctuations in sub-Saharan and semi-arid zones of southern Africa seemed to be primarily linked to factors modifying the intensity and frequency of disturbances. During most drought years the location of the ITCZ seemed to be 'normal'. They also found synchronous occurrence of droughts in the sub-tropics of both hemispheres, extreme spatial coherence of rainfall anomalies and persistence of anomalies. Ofori-Sarpong (1983) discusses studies suggesting that droughts in West Africa are related to features such as southward movement of mid-latitude pressure belts with consequent shift of sub-tropical highs and changes in latitudinal temperature gradients creating changes in global circulation patterns. For southern Africa, Tyson (1981, 1984) relates year-to-year rainfall variations to atmospheric circulation features at various levels.

Lyons (1982) found that most El Nino winters in Hawaii are dry and Taylor (1984) found significant differences between winter months in relationships between rainfall and the SO (§ 2.8). Reiter (1983) indicates teleconnections between rainfall variations in the equatorial Pacific and SST anomalies in the northern and southern hemispheres and with the SO. He also suggests extra-tropical connections with tropical rainfall regimes. Links between the SO and Pacific rainfall anomalies are also discussed by Sadler (1984). Douglas and Englehart (1981) found a statistical relationship between autumn rain in the central equatorial Pacific and subsequent winter rain in southern USA. Wet winters in south-central Florida appeared to be linked to warm water events along the equator and dry winters more commonly associated with cold water events in the eastern tropical Pacific.

Nicholls, McBride and Ormerod (1982) found that an index of the date of onset of the north Australian wet season could be predicted several months in advance using Darwin winter pressure. They found only a weak relationship between wet season amounts and date of onset. Amounts in mid- and late parts of the season were not related to either onset date or amount early in the season. Changes in large-scale flow associated with persistent dry spells in the high rainfall northern area of Australia as well as the meteorological situation related to high rainfall events in the arid interior are examined by Kininmonth (1983). Correlations (simultaneous and lagged) between areal rain for 107 Australian rainfall districts and SO indices (§ 2.8) showed seasonal and regional variations (McBride and Nicholls 1983). Differences were also found in correlations for the two periods 1932–53 and 1954–74. McBride (1984) found that Australian monsoon onset was accompanied by marked changes in upper tropospheric flow, tropical easterlies increasing and the sub-tropical jet stream shifting abruptly polewards. Nicholls (1984a) built on earlier work of Nicholls, McBride and Ormerod (1982) (see above) to produce a method of predicting the probability that the wet season will commence late.

Shukla and Paolino (1983) found that the tendency of the Darwin pressure anomaly, used as a SO index (SOI) was a good indicator of Indian summer monsoon rain anomaly. Bhalme, Mooley and Jadhav (1983) found an inverse relationship between a SOI and a drought area index and a positive relationship with a flood area index for India. Bhalme and Jadhav (1984b) also found a significant correlation between summer monsoon rain and seasonal SO indices. A SOI for different months and seasons showed opposite tendencies during deficit and excess years of all-India summer monsoon rain (Parthasarathy and Pant 1985).

Relationships between Indian summer monsoon rain and both a SOI and SST anomalies in the eastern Pacific have been examined by Mooley, Parthasarathy and Sontakke (1985). Mooley and Parthasarathy (1983) found that during 22 El Nino years (§ 2.8), Indian summer monsoon rain was mostly below average in most areas, a similar result being found for Sri Lankan and Indian rainfall by Rasmusson and Carpenter (1983).

Campbell, Blechman and Bryson (1983) hypothesise that tidal effects modulate the advance of the Indian summer monsoon front. A significant negative correlation was found between south-west (summer) and north-east monsoon rainfalls for Tamil Nadu, India by Dhar and Rakhecha (1983). Relationships between Indian summer monsoon rain, differences in depression track behaviour and monsoon trough location have been examined by Mooley and Parthasarathy (1983). Diehl (1984) has also examined links between monsoon rain and a range of atmospheric variables.

Possible human factors modifying atmospheric composition and in turn circulation are discussed in § 9.2. It must be remembered that although studies such as those indicated above may establish significant correlations, the unexplained variance is often large and hence predictive value may be limited. Nevertheless, such studies, many of them preliminary, hopefully point the way towards increased understanding and possible prediction.

3.8 FLUCTUATIONS AND TRENDS

In § 3.3 the occurrence of series of wet or dry years and their great importance was noted. This raises the question as to whether identifiable fluctuations or trends in rainfall exist. This aspect has received considerable attention in recent years with the occurrence of drought conditions in the Sahel and other tropical areas. Such fluctuations have implications for water resources and agriculture, and if they exist it is dangerous to rely on short-period rainfall records in planning water-resource development. Here, we need not be concerned with any changes on a geological time scale but with those over, say, 100–150 years at most. One problem is the very wide range of time scales to be considered, ranging from several days (e.g. Singh and Prasad 1976) to fluctuations with a periodicity of 50 years or more.

A review of the literature presents an extremely confusing and conflicting picture. The problems of analysis and interpretation of the practical significance of possible fluctuations must be considered. Lack of data and lack of consistency in data between areas is a major problem. Some authors (e.g. Morth 1970; Riehl 1954) point out the dangers of utilising single-station data, a point to be discussed in § 4.2. Analyses have been concerned with amounts (for months, seasons, years), but as will be apparent from Chapter 4 other characteristics such as intensity, duration, frequency and time are also of great importance. Fluctuations in these characteristics can be more significant than those in amount. Analyses which are concerned with conditions over a month or longer ignore the very great importance of rainfall over short time periods.

A number of factors act to create fluctuations and the result will be a composite of their influence. There are variations from year to year which tend to mask fluctuations over longer periods. Because of the complexity of the pattern Veryard (1963) stressed the need to test the reality of any alleged fluctuation, pointing out that many workers fail to do this. Many 'cycles' are identified purely on the basis of inspection and this is very dangerous, even experienced workers tending to see cycles in any data, even in a random series. Curry (1962) discusses climatic change as a random series, concluding that the latter 'appears to be able to produce changes of the required magnitude and periodicity'.

Techniques of analysis may produce apparent fluctuations not existing in reality. Discussions of a range of techniques of time series analysis can be found in WMO (1971), Box and Jenkins (1976), Chatfield (1975), Salas *et al.* (1980), Murphy and Katz (1985). Running means are commonly used, the limitations of this technique being discussed by WMO (1971), Kraus (1955) and Veryard (1963). Indirect evidence for climatic change is particularly difficult to assess, especially in the case of desert advance. Daniel (1980) states that most analyses imply that the droughts of the 1970s in Africa (§ 3.4), although severe, are normal to the climate, having happened before and will occur again. He concludes that the desertification accompanying the drought was related to heavy stocking and dry land cultivation. Ofori-Sarpong (1983) discusses climatic fluctuations and human disruption of ecological systems as causes of drought in West Africa. Desertification is a complex issue, combining both human elements and climatic variability (§ 3.4).

Even if a trend or fluctuation is identified, extrapolation into the future is dangerous since there may be no justification for assuming its persistence. A trend (e.g. general increase in rainfall with time) may simply be part of a long-term fluctuation. Therefore, identification of physical 'causes' of changes is important for future prediction. Because of the above problems, many dismiss analysis of time series as a waste of effort. Even accepting the dangers of future prediction, studies of past conditions can be useful in, say, utilisation of short-period records at a particular station or the analysis of rainfall–runoff relations. Furthermore, the possible existence of fluctuations has important implications for water resources, for agriculture and for rainfall probability and reliability assessments. As was indicated in § 3.3, it is often the persistence of, say, dry conditions rather than the occurrence of a single exceptionally dry period which has the greatest impact.

The practical importance of a fluctuation or trend, both over the period considered and in the future if it continues, is difficult to assess. The magnitude and rate of change must be considered. Vegetation and human activities may adjust fairly easily to a gradual change over a long period but not to rapid fluctuations. Cooper and Jolly (1969) suggest that except for the occurrence of extreme values in particularly harsh climates, the long-term structure of plant and animal communities is a response to average climatic conditions not isolated events, but this is perhaps an oversimplification. Adjustment relative to average conditions may differ between areas of high and low variability. Curry (1962) points out how the random occurrence or non-occurrence of a single extreme event can have a great effect upon the type of vegetation. The occurrence of a few years of extreme conditions, not only in a harsh environment, may initiate a reaction which is not reversed when conditions return to normal. A period of more favourable conditions can allow the development of a particular vegetation which persists even after conditions have deteriorated because it creates its own environment.

An extreme illustration of the consequences of isolated events was referred to in Chapter 2 when it was noted that a destructive hurricane over the island of Grenada led to a radical change in the whole economy, ultimately for the better. Such 'random' events may be just as important as more gradual long-term changes whose continuation is in any case difficult to forecast. Human adjustment to climatic changes is not necessarily similar to that of plants and animals, involving as it does important questions of perception and response to a large number of social and economic variables. The response of a peasant farmer to a fluctuation will not necessarily be the same as that of a plantation owner. Although mainly concerned with the impacts of changing carbon dioxide levels, Ausubel (1983) indicates something of the wider problems in assessing impacts of climatic change.

If a trend is of sufficient magnitude and continues for long enough then some adjustment will take place in response to the different 'average' conditions. The period over which changes are considered is important and Curry (1962) points out the limitations of the official definition of climatic changes involving differences between non-overlapping 10-year periods. A short-rooted grass would feel that a 'change' had occurred before it was felt by long-rooted grass or trees. To the peasant farmer, three dry years may constitute a climatic 'change' because he does not have the resources to cope.

A variation in rainfall can have an effect proportionately greater or less than its quantitative change. For example, a 10 per cent increase in rainfall may increase streamflow by much more than 10 per cent. Whether or not any change in rainfall is potentially beneficial, since economic systems are geared to existing conditions, adjustments will have costs (Pittock 1972). Investments may be written off before their time and changes in, say, agricultural systems will almost certainly involve some cost. The rate of change is likely to be critical. Over a lengthy period, economic and other costs can be spread out and gradual adjustments made. Rapid changes may have very serious implications.

Climatic changes are liable to be especially critical in marginal areas.

This will become even more important as demands for increased production may lead to developments in marginal areas. High-yielding varieties may be less tolerant of fluctuations in water supply (e.g. Michaels 1982, see § 3.1), and in any case production demands are increasingly focusing attention on the provision of optimum water requirements.

It is against this complex background that studies of climatic fluctuations and trends are examined. Unfortunately, few such analyses concern themselves with the implications for agriculture and water resources other than in very general terms. Reviews by Mattei (1979), Fukui (1979), Oguntoyinbo and Odingo (1979) which discuss climatic variability, agriculture and land use provide useful material. Veryard (1963) has reviewed earlier studies of climatic fluctuations during the period of meteorological record. Past climates in the tropics are discussed by Hastenrath (1985). Except for parts of South-East Asia, Kraus (1955) indicates that rainfall was above the 1881–1940 average in the nineteenth century in the tropics but that there was an abrupt decrease at the turn of the century bringing drought or at least dry conditions to tropical Australia, parts of the Pacific, Colombia, Mexico, north-east Brazil, the Caribbean, Africa, south-west Asia and Sri Lanka (Fig. 3.20). Kraus suggests that the period of comparatively high tropical rainfall began in the 1820s or earlier. He postulates that the sudden decrease resulted from a narrowing of the rain belts and a shortening of the wet season and that the effects were greatest in the sub-tropics at the arid zone boundaries (Kraus 1958).

The decrease at the start of the twentieth century involved changes in the long-term mean of 30 per cent or more (e.g. difference of 84 per cent at Aden before and after 1894). That it is dangerous to consider annual data only is shown by the fact that at Freetown, while the early and late rains decreased after 1897, the particularly heavy July–August rains led to a continuous rise in the annual figures up to 1906. While changes occurred simultaneously in climatologically corresponding locations north and south of the equator, the central Sahara and Amazon, rainfall did not seem to alter a great deal. Kraus (1955) indicates that the dry period at the start of the twentieth century may have ended or at least been temporarily interrupted in the 1930s, most of the curves in Fig. 3.20 showing a tendency to flatten out or rise. Australian towns such as Townsville and Georgetown show the 1940s and 1950s to be much wetter.

Grove (1973) states that the decrease early in the twentieth century did much to strengthen the tendency for people living near deserts to suppose conditions were deteriorating and the deserts advancing. For Africa, Grove says that the somewhat improved conditions in 1950–63 resulted in an increase in population and cattle in semi-arid Africa which

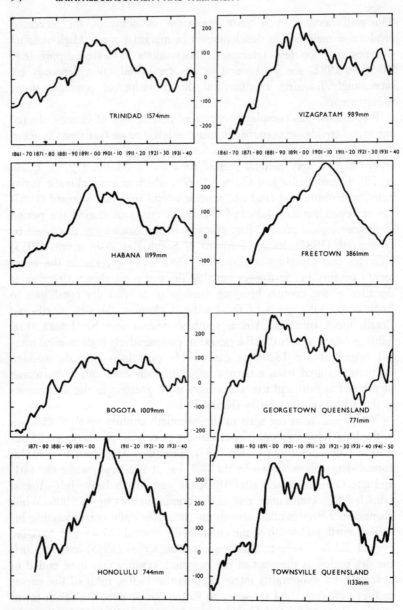

Fig. 3.20 Cumulative percentual deviations from the annual mean (1881–1940) (after Kraus 1955). Use of percentual deviations allows comparison of records from widely different regimes.

increased the risk in drought years compared to 60 or 70 years previously. Lamb (1973) suggests that from 1951 to 1969, the northern sub-tropical high moved nearer the equator and, while weaker, brought a tendency for reduced rainfall along the southern fringe of the desert zone of North Africa. The equatorial rain belt had a reduced latitudinal migration and therefore some equatorial locations had more rainfall in this period. Lamb suggests a similar pattern in the southern hemisphere and that from 1970–72, the high-pressure belt moved to even lower latitudes, this almost certainly causing the intensification of drought along the equatorial fringe of the desert zone.

Jenkinson (1973), in an analysis of eight West African marginal stations, found well-marked maxima centred about 1920, 1931 and 1957, and well-marked minima centred about 1913, 1942 and 1971 with shorter periods of marked deficit in 1925–26 and 1947–49. The longer periods of marked deficit occurred at roughly 30-year intervals. Grove (1973), however, in a study of desertification in Africa concluded that there was no indication of any long-continued upward or downward trend nor any obvious cyclic periodicity, although years of high and low rainfall tended to bunch together in small clusters of 2 or 3 years and in longer periods of 10 to 15 years. Ojo (1983), in an analysis of West African data over a 50–60-year period comments on a tendency for fluctuations of wet and dry years to occur about every 2–3 years. Ojo also refers to spatial differences in variability pattern, a point already referred to in previous sections and which is considered further in § 4.2. Morth (1967) suggested that a 10-year cycle could be fitted to rainfall data of the Lake Victoria region, while Lumb (1966) proposed the interaction of several cycles of different length from 5 to 14 years over East Africa.

Melice and Wendler (1984), in an analysis of 80 years of data for three areas in southern Tunisia, found no statistically significant decrease in precipitation. Dennett, Elston and Rogers (1985) for West Africa refer to an earlier study of theirs in 1976 where they concluded that the drought years of 1968–74 in the Sahel did not differ significantly from a random series and hence did not indicate a climatic change. They updated this study to 1983, finding the relatively dry conditions continuing, mainly because of a decline in falls during August, the wettest month. However, better falls early in the season have meant the drought has been less severe in the late 1970s than in the early 1970s. They suggest that agricultural planning should utilise rainfall data for the last 20 years.

Dyer (1982) for southern Africa data over the period 1921–74 found little evidence suggesting that within a year variation in rainfall was undergoing any kind of upward trend. Hutchinson (1985) found a marked decrease in annual rainfall over the previous 30 years for five Gambian stations. He suggests this results from a reduction in rain producing systems rather than an intensity decrease. A key point seemed to be a marked mid-season rainfall reduction which could have serious agricultural consequences.

Kraus (1955) points out that the south-east Asian monsoon area is more complex than elsewhere with less variation in annual rainfall but with changes in the seasonal pattern. While some stations showed a similar fluctuation in annual values to other parts of the tropics, over much of India, for example, there were none despite an observed tendency for droughts after 1898. The early rains, however, were above average up to the end of the nineteenth century after which they decreased. The main rains, however, were below average in the nineteenth century, but their increase after this cancelled out the change in the early rains giving no trend in the annual values.

Rao (1963) found no statistically significant change in either annual or seasonal rainfall over arid and semi-arid zones of India in the previous 80 years. This confirmed the findings of Pramanik and Jagannathan (1954). The latter suggested a tendency for deficient rainfall near the hills in north and north-east India to become more frequent, and also a tendency for rainfall in excess of the mean to be more frequent in the case of south-west monsoon rainfall except in some places where the reverse was true. In a study of 100 years' data for Karachi, Naqvi (1958) found a rising trend but did not claim it as statistically significant. However, he maintained that the data revealed a 50-year cycle with an amplitude of about 38 mm.

Dhar, Rakhecha and Kulkarni (1982a) found no evidence of long-term trends over the period 1877 to 1976 in Tamil Nadu, south-east India during the north-east monsoon period. Mooley and Parthasarathy (1984) in an analysis of the All-India summer monsoon rainfall over the period 1871–1978 found a continuous rise in the 10-year mean from 1899 to 1953. They distinguished four major rainfall periods in the series.

A number of Indian and Sri Lankan studies have examined data for cycles, using techniques such as power spectrum analysis (see references cited earlier dealing with time series analysis). These include findings of short-period cycles (2–30 days) for summer monsoon rain in different regions (Singh and Prasad, 1976), and for periods of years (Dhar, Rakhecha and Kulkarni 1982a; Mooley and Parthasarathy 1984; Suppiah and Yoshino 1984a, b).

For annual rainfalls, 1910–75 at central equatorial Pacific islands, Meisner (1983) found that amounts were smaller and more variable in the earlier part of the century. However, most adjustments were of the order of 10 per cent, suggesting that even 20–30-year records would provide reasonably accurate averages. Taylor (1970), investigating a Gilbertese tradition of drought every 7 years looked at records of equatorial Pacific island stations within the area 175°E–160°W. He found conspicuous droughts having a duration of 1 to 3 years with a crude period of 5 to 7 years, direct observation confirming the existence of an oscillation for at least the previous 75 years and indirect observation to perhaps 130 years.

Schmidt-Ter Hoopen and Schmidt (1951) for the Indonesian Archipelago found that available data indicated opposed variations in rainfall amounts occurring in the north and south of the area, a minimum at southern stations and a maximum at northern stations occurring in and about 1925. They attributed this to anomalous oscillations in the ITCZ. Both Morth (1970) and Grove (1973) stress the tendency for large anomalies to affect very large areas. Kraus (1955) suggests that more 'local' droughts are comparatively short-lived, but that widespread droughts such as occurred at the turn of the century last longer when the drier conditions continue with some interruptions for at least 15 years. He stated that the same applied to high-rainfall conditions and that this might have some predictive value.

A number of analyses have considered links between rainfall and sunspot activity. Bhalme and Jadhav (1984a) indicate a number of studies either for or against such links. They found an association between occurrence of large-scale flood events in India and the double (Hale) sunspot cycle. A tendency for occurrence of more frequent floods in the positive (major) sunspot cycle than in the negative (minor) sunspot cycle was claimed. Ananthakrishnan and Parthasarathy (1984) suggest a weak positive association between Indian annual rainfall (1871–1978) and sunspot activity, although most of the variance in rainfall is not accounted for by this. For southern Tunisia, Melice and Wendler (1984) refer to some indication of links at some stations. Ward and Russell (1980) refer to regional and seasonal contrasts in association between rainfall and sunspot trends. Over an 80-year period they found a positive correlation between rainfall and trends in sunspot numbers in south-east Australia, but a negative correlation for the south-west and Cape York Peninsula. On the north coast, a negative correlation in summer and a positive correlation in winter were found. However, in Queensland and western NSW, seasonal links were the opposite of these. Ward and Russell (1980) state that it is not clear whether links are accidental or functional.

An interesting point is that some authors such as McNaughton (1971) have found that certain dates tend to favour more rain than others although the cause is open to question.

The above examples illustrate something of the range and conflict in evidence. Berlage (1957) listed all the cycles 'discovered', yet felt that none had been established beyond doubt. Flohn (1960) stated that it had 'not been possible so far to establish truly persistent periodicities

in climatic evolution'. In the *Symposium on Drought in Africa* (Dalby and
Harrison Church 1973), one paper stated that there was no conclusive
evidence of cyclic or permanent change in climate on a continental scale
and that the ability to predict droughts was practically nil. This brief
review indicates that the uncertainties still remain.

CHAPTER 4
OTHER RAINFALL CHARACTERISTICS

4.1 INTENSITY–DURATION–FREQUENCY

The often marked seasonality of rainfall (§ 3.2) exerts an overall control on water availability and agricultural systems while variability (§ 3.3) imposes a degree of uncertainty. Important as these aspects are, it is often the characteristics of individual rainstorms which are especially significant, particularly their intensity, duration and frequency of occurrence. These three aspects are interrelated and hence best considered together.

Figure 4.1 shows that average intensity decreases the longer the period (duration) considered. Similarly, high intensities will occur less

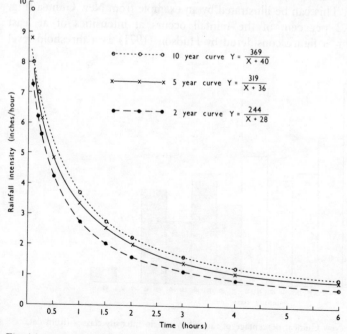

Legend within figure:
- o····o····o 10 year curve $Y = \dfrac{369}{X + 40}$
- x——x——x 5 year curve $Y = \dfrac{319}{X + 36}$
- •– –•– –• 2 year curve $Y = \dfrac{244}{X + 28}$

Vertical axis: Rainfall intensity (inches/hour)

Horizontal axis: Time (hours)

Fig. 4.1 Intensity–duration–frequency relations (after Lim 1969).

frequently than lower. This is shown in Fig. 4.1 by the series of curves for different return periods. The return period is the *average* interval of time within which rainfall of a specified amount/intensity will be equalled or exceeded once. The data in Fig. 4.1 are, however, for a point. Important as this may be, for agriculture, water resources and problems such as urban storm water drainage, it is conditions over an area which are more significant. In the centre of a storm, intensities may be high, but as a larger and larger area is considered, *average* intensities will decrease. The rate of decrease will vary with the nature of the rainfall, a particular point being the often very localised nature of tropical rainfall and marked intensity gradients (§ 4.2). Relationships between point and areal rainfalls have been investigated for different types of rainfall in various areas to allow extrapolation from point (gauge) values. For example, Dhar and Bhattacharya (1977) cited by Dhar, Rakhecha and Mandal (1981b) for severe storms in the north Indian plains derived the following equation:

$$P = P_{m}e^{-KA^n}$$

where P = average depth over area A, P_m = maximum point rainfall at the storm centre, and K, n are constants varying with duration.

While the above relationships are universal, a key point is that in the tropics a high proportion of rainfall occurs in large storms of high intensity. This can be illustrated by an example from New Guinea (Fig. 4.2): 34.2 per cent of the rainfall occurs at intensities of at least 25 mm/hr, a figure considered by Hudson (1971) as a threshold level

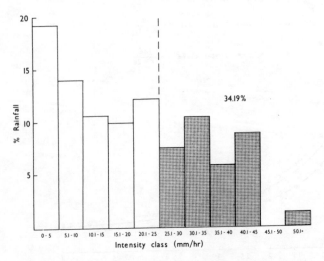

Fig. 4.2 New Guinea: percentage of rainfall in certain intensity classes (from data supplied by N. Turvey).

at which rainfall becomes erosive. This compares well with a figure of 40 per cent for the tropics given by Hudson (1971) and contrasts markedly with the value of 5 per cent he gives for temperate areas. A further 12 per cent of the rainfall occurs with an intensity approaching the erosive value (20.1–25 mm). While in temperate latitudes intensities rarely exceed 75 mm/hr, and then only in summer thunderstorms, in the tropics, intensities of 150 mm/hr occur regularly and a rate of 340 mm/hr for a few minutes has been recorded (Hudson 1971).

Figure 4.3 shows that 47 per cent of the total rainfall occurs in single storms of more than 25 mm and 20.4 per cent in storms of more than 50 mm, illustrating the degree of concentration in comparatively large storms. Furthermore, this concentration is associated with a small proportion of the total storms. Thus, Fig. 4.4 shows that the 47 per cent of total rainfall in storms of more than 25 mm is produced by only 9 per cent of the storms, and storms of more than 50 mm which produce 20.4 per cent of the rainfall are only 2.5 per cent of the total number (six in all). The largest 25.4 per cent of storms account for 77.8 per cent of the rainfall. On the other hand, the smallest 61.3 per cent of storms (0–5 mm) produce only 10.8 per cent of the rainfall. Riehl (1954, 1965) points out that the production of much of the rainfall by a small proportion of storms appears to be a universal feature, not confined to the tropics. He states that 10–15 per cent of rain days account for 50 per

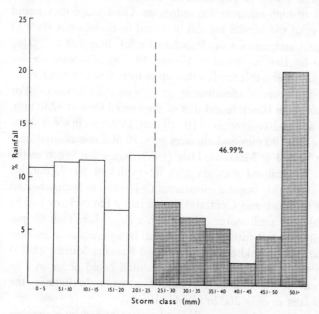

Fig. 4.3 New Guinea: percentage of rainfall in certain sized storms (from data supplied by N. Turvey).

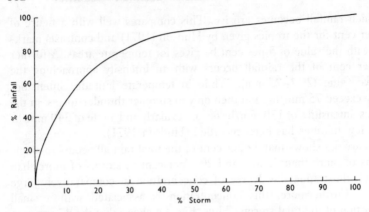

Fig. 4.4 New Guinea: cumulative percentage of rainfall against percentage of storms (from data supplied by N. Turvey).

cent of the rainfall, 25–30 per cent of days account for 75 per cent, and 50 per cent of days with smallest rain amounts produce only 10 per cent of the total. These figures agree well with those for a short period in New Guinea (Fig. 4.4). Riehl (1979) presents other examples which indicate the concentration of rainfall in a limited number of events, and this aspect has already been considered in § 3.6.

Mohr and Van Baren (1959) illustrate the concentration of rainfall in large storms of high intensity for Indonesia. On average they found that 22 per cent of the annual rainfall occurred in cloudbursts (falls of intensity at least 1 mm/min for not less than 5 min) compared to a value of 1.5 per cent for Bavaria found by Geiger. Making allowance for the high totals in the tropics, this implies that up to forty times as much rainfall in Indonesia occurs in cloudbursts as in temperate latitudes. For Bogor, Mohr and Van Baren found that of the annual total of 4230 mm, 900 mm fell in about sixty showers of 10–20 mm, 2600 mm in some sixty-five showers of 20–100 mm and showers of <10 mm contributed only about 700 mm/year. For Malaysia, Dale (1959) states that much of the rainfall occurs in localised showers with intensities of 50–75 mm/hr. Henry (1974) states that tropical rainstorms (>10 mm) over South-East Asia and parts of South and Central America had a life cycle of 1–3 hr and usually only one such storm occurred in a day. More than 60 per cent of the daily rain fell within a 1 hr period. In an analysis of intense storms at Bidar, India, Rakhecha, Mandal and Ramana Murthy (1985) found that up to 56 per cent of the daily rainfall could occur in 1 hr and 73 per cent of the rain in 2 consecutive hours. In Batavia, the annual rainfall falls in only 360 hr (Riehl, 1954).

Rainfall intensities within storms have been examined by a number of authors (e.g. Arenas 1983). For Ibadan, Nigeria, 69 per cent of storms had only one peak, 18.7 per cent had two peaks, 8.5 per cent had three peaks and 3.6 per cent had more than three intensity peaks (Ayoade and Akintola 1982; Oguntoyinbo and Akintola 1983). Most single peak storms in Ibadan attained peak intensities in less than 15 min and those with a second peak attained it at about 30 min from the start. Severe storms at Bidar, India, had peaks either early or just after the middle of the storm (Rakhecha, Mandal and Ramana Murthy, 1985). One study relating short-period intensities to synoptic systems is that of Bonell and Gilmour (1980) for north-east Queensland, Australia.

All the above characteristics are of great importance for both soil erosion and the effectiveness of rainfall for agriculture. Effective rainfall in agricultural terms is that entering the soil and remaining within root range. Rainfall lost as surface runoff or draining beyond root range is ineffective. Large storms of high intensity may result in considerable loss as surface runoff and as drainage beyond root range. More moderate temperate-area rainfall is likely to be proportionately more effective than that of tropical areas. The problem in the tropics is compounded by the significance of a few heavy storms. Not only can much of their total be ineffective, but since they are few in number, there will be considerable intervals when the soil may dry out and plants suffer moisture stress. These aspects will be considered further in Chapter 7.

While small storms produce only a small proportion of the total rainfall they are numerous, and particularly in view of the above disadvantages associated with the small number of large intense storms, it is important to consider their contribution to the water balance. Because of the high evaporative demand in the tropics (Ch. 5), it is often stated that 'light' showers are ineffective and should be ignored in assessing the water balance (Riehl 1954). The question of what constitutes a 'light' shower itself presents a problem, various authors quoting values from 2.5 mm to 5 mm maximum (Riehl 1954). To ignore the contribution of these relatively frequent light showers is a mistake. Glover and Gwynne (1962), for example, illustrate how maize, by concentrating rain at the base of the plants, makes light showers normally neglected when discussing the water balance of considerable importance to plant survival during dry periods. In addition, light showers, whether intercepted by vegetation or barely wetting the surface-soil layers, in being evaporated utilise energy and increase atmospheric humidity to some extent, thereby reducing plant transpiration demands, although a simple equivalence is unlikely. Evaporation of light rainfall from vegetation may also provide beneficial cooling of leaf surfaces during high-temperature conditions.

It is commonly felt that light showers may be more important if they occur at night than during the day when they immediately evaporate. Since they are either intercepted by the vegetation canopy or at best wet only the top layer of the soil, they are still available for evaporation at a reduced rate at night or will evaporate the next morning. Nor will the plant utilise the moisture at night. Hence, it may be doubtful whether time of occurrence of light showers is as important as is generally thought.

The significance of intense tropical rainfall in creating surface runoff (see above) has implications for soil erosion. Comments on the mechanisms of soil erosion are presented in Chapter 7, but here relationships between erosive power and rainfall characteristics are reviewed. El Swaify, Dangler and Armstrong (1982) present a major review of erosion by water in the tropics, including types of erosion, mechanisms, an inventory of erosion in different areas, prediction and erosion control measures. Characteristics of individual drops (number, size, distribution, terminal velocity) are of primary consideration. An example of the experimental investigation of raindrop impact stress of concern for soil erosion using various techniques is presented by Ghadiri and Payne (1981). Analyses of size distribution, terminal velocity and their relationships with a range of parameters such as momentum, kinetic energy, intensity are indicated by El-Swaify *et al.* (1982). Various authors have attempted to relate storm kinetic energy to rainfall intensity (Fig. 4.5) and raindrop size, kinetic energy and intensity (Table 4.1).

Wischmeier, Smith and Uhland (1958) looked at relationships between soil loss and various rainfall parameters from experiments in the USA. They found the best predictor of soil loss was a product of kinetic energy (E) and the maximum 30-minute intensity (EI_{30} index). Due to lack of short-period intensity data in many areas, attempts have been

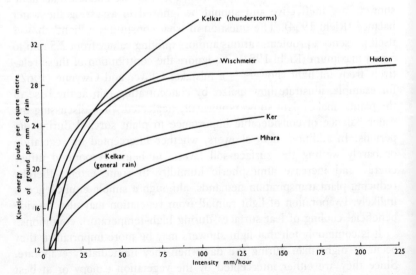

Fig. 4.5 Kinetic energy–rainfall intensity relations (after Hudson 1971).

Table 4.1 Interrelationships between raindrop size, storm kinetic energy, and intensity, as established by various workers (from El Swaify, Dangler and Armstrong 1982)

Location	Climate/type of precipitation	Intensity range studied	D_{50} versus I	E versus I (Units of E)	(Units of I)	Reference
Washington DC	Temperate/predominantly frontal	0.03 to ≈2 in/hr Limited data 2–4.6 in/hr	$D_{50} = 2.23\,I^{0.182}$	—	$\left(\dfrac{\text{in}}{\text{hr}}\right)$	Laws and Parson (1943)
		0.15–4.6 in/hr Extrapolated to 10 in/hr	—	$E = 916 + 331\log_{10} I$ $\left(\dfrac{\text{ft ton}}{\text{ac in}}\right)$	$\left(\dfrac{\text{in}}{\text{hr}}\right)$	Wischmeier and Smith (1958)
Zimbabwe	Subtropical/convective thunderstorms	< 9 in/hr Limited data > 6.5 in/hr	Peak D_{50} of 2.55 mm at I of 3–4 in/hr	$E = 758.52 - \dfrac{127.51}{I}$ $\left(\dfrac{\text{ergs} \times 10^3}{\text{cm}^2}\right)$	$\left(\dfrac{\text{in}}{\text{hr}}\right)$	Hudson (1965)
Miami FL	Temperate/five types	≈0.1 to ≈9.5 in/hr Limited data > 6 in/hr	—	$E = 8.37\,I - 45.9$ $\left(\dfrac{\text{ergs}}{\text{cm}^2\ \text{sec}}\right)$	$\left(\dfrac{\text{mm}}{\text{hr}}\right)$	Kinnell (1973)
South Central USA	Temperate	< 10 in/hr Limited data > 5 in/hr	$D_{50} = 1.63 + 1.33\,I - 0.33\,I^2 + 0.02\,I^3$	$E = 429.2 + 534.0\,I - 122.5\,I^2 + 78\,I^3$ $\dfrac{\text{ft ton}}{\text{ac in}}$	$\dfrac{\text{in}}{\text{hr}}$	Carter et al. (1974)
Western Nigeria	Humid tropical/convective thunderstorms	Approximately 0.5–9.5 in/hr	D_{50} Range: 1.5 to 4.5 mm D_{50} Increases with I	—	—	Aina et al. (1977)

Explanation of symbols: D_{50} is median volume drop diameter; E is kinetic energy per unit volume of rainfall; I is rainfall intensity. Units are reported as published by authors.

Table 4.2 Summary of alternative estimates for the rainfall erosion index and its components (from El-Swaify, Dangler and Armstrong 1982)

Location	Climate (average annual precipitation)	Relationship	Author's units Qty	Author's units Units	Reference
Sefa, Senegal	Tropical (1300 mm)	$EI_{30} = 1.2A - 4$ $E = 2.5A - 1$	A E EI_{30}	mm kg m/m^2 ton/ha	Charreau (1969)
Allokoto, Niger	Dry tropics (495 mm)	$EI_{30} = 0.0158\,A\,I_{30} - 1.2$	A I_{30} EI_{30}	mm mm/hr ton/km^2	Delwaulle (1973)
Western Nigeria	Tropical (1500 mm)	$E = (198 + 84 \log_{10} I_{30})A + 24$	A I_{30} E	cm cm/hr ton m/ha	Wilkinson (1975)
Northern Nigeria	Tropical (1100 mm)	$E = (41.1A - 120.0) \times 10^3$	A E	mm ergs/cm^2	Kowal and Kassam (1976)
Large area of West Africa	Dry to humid tropical (500–2100 mm)	$\dfrac{Ram}{Ham} = 0.50$	Ram Ham	(ft ton/Ac) × 10^2 mm	Roose (1977b)
Madagascar	Humid tropics (1300 mm)	$E = 2.325A - 3.945$	A E	mm ton m/km^2	Bailly et al. (1976)

Explanation of symbols: A is rainfall amount; Ham is average annual rainfall amount; *E* is kinetic energy of rainfall per unit area of ground surface; EI_{30} is rainfall erosivity index after Wischmeier and Smith (1958); I_{30} is maximum intensity of rainfall sustained for 30 min; Ram is total yearly average EI_{30}.

made to relate EI_{30} and other energy values as erosion indices to more widely available rainfall data including totals for longer durations and different return periods (see below). Some examples are presented in Table 4.2.

The use of the EI_{30} index in the tropics has been criticised due to data requirements and the fact that since it was derived in temperate areas it may not be successful in lower latitudes due to different climatic conditions. Low-intensity rain, composed of small drops of low velocity, has low energy. Also, with well-structured tropical soil profiles such rain will produce little surface runoff and transportation of soil will therefore be negligible. Therefore Hudson (1957) suggested a threshold value, above which rain is erosive, of about 25 mm/hr in Africa. Omitting rain of lower intensities, Hudson found an excellent correlation between splash erosion and the total energy of the rest of the rain and felt that this index ($KE > 25$ index) was more appropriate to tropical and sub-tropical rain. However, El Swaify et al. (1982) point out that the EI_{30} index also omitted storms of less than 12.6 mm which were separated by 6 hr or more even in the USA unless the maximum 15 min intensity was > 24 mm/hr.

Lal (1976) found a high correlation and insignificant differences between the $KE > 25$ index, EI_{30}, rainfall amount (A), maximum intensity for a minimum duration of 7.5 minutes (I_m), AI_m, and kinetic energy (E). Arnoldus (1977) found a poor correlation between EI_{30} and soil loss at Benin, Nigeria. However, Lo et al. (1985) found EI_{30} was a suitable index of erosivity of storms in Hawaii. They also established a relationship between maximum 30 min and 60 min intensities to assist estimation of EI_{30} where data for less than hourly periods were not available. EI_{30} was tested for relationships with rainfall totals on a storm, daily, monthly, seasonal and annual basis together with Fournier indices (see below). Average annual rainfall was chosen as the best estimator of average annual EI_{30} index. Stocking and Elwell (1973) found that EI_{15} was best correlated with soil loss in Zimbabwe. Elwell and Stocking (1973a, b) also examined relationships between momentum and energy and between these two parameters and soil loss for intensities above certain threshold levels.

Fournier (1960) related suspended sediment in African rivers (NB not soil erosion as such) to rainfall data as follows:

$$D = K_1 \left(\frac{p^2}{P} \right) - K_2$$

where D = sediment yield, p = mean rainfall for the wettest month, P = mean annual rainfall, K_1 and K_2 are constants depending on climatic type. Arnoldus (1980) proposed a modification of this approach using the term $\sum_1^{12} Pi^2/P$ where P_i is monthly precipitation.

The above brief review indicates the uncertainties with respect to indices of rainfall erosivity in the tropics. Such indices are important in estimating areal and temporal variations in energy available for erosion and hence assessing conservation measures. Naturally, the actual erosion occurring will vary with a number of factors such as soil type (erodibility), slope and vegetation cover (Chs 7, 8). In addition to El Swaify *et al.* (1982), useful material is contained in a variety of texts (e.g. Morgan 1979; Norman, Pearson and Searle 1984; Hudson 1971). Lewis (1985) discusses some of the problems of deriving and using a suitable rainfall index in the tropics in the Universal Soil Loss Equation (USLE) (§ 7.1).

Since in the tropics much of the rainfall occurs in a comparatively small number of intense storms it follows that much of the erosion will occur in a limited time period. Examples of this are given in Chapter 7. Furthermore, as has been indicated, a large proportion of tropical rainfall occurs with high intensities. Taken with the fact that rainfall *totals* are often higher in tropical areas than in temperate latitudes and high average intensities are also more frequent, then the far greater energy available for erosion than in temperate areas is readily apparent. This is why soil erosion is a far greater problem in tropical and sub-tropical areas.

Rainfall characteristics such as intensity are not necessarily directly related to total rainfall amounts over periods of a month or year. Jackson (1986a), for data from a range of tropical locations, found rainfall totals had a better relationship with number of rain days (i.e. frequency of occurrence of rain) than with mean daily intensity. Furthermore, there was a tendency for rain to occur in fewer days with higher mean daily intensities than the regressions predicted for stations with a limited rainy season, especially at the start of the season, as well as dry months sometimes having fairly high mean daily intensities. The implications for hydrology and soil erosion are discussed in § 7.1. Relationships between intensity, duration, return period and rainfall total have been found to exist. An example for Zambia applicable for annual rainfalls up to 1000 mm is

$$I = \frac{26 \times (0.16R - 83) \,(\log d + 2) + 10}{t + 15}$$

where I = intensity (mm/hr); R = annual rainfall (mm); d = return period (yr); t = duration (min) (after Griffiths 1972).

Intensities as well as individual storm totals can be considerable in arid and semi-arid areas. In Mozambique for example, Pafuri with an annual mean rainfall of 380 mm has recorded over 100 mm in 1 day, and Tulear (Madagascar) with a mean annual total of only 380 mm has recorded 102 mm in 1 day (Griffiths 1972). At Biskra in Algeria, rainfall on 27 and 28 September 1969 was over twice the mean annual rainfall of 148 mm (Winstanley 1970). Port Sudan, with an average annual rainfall of 110 mm has recorded 111.5 mm in 1 day (Oliver 1969). Alice Springs, Australia, with an average annual rainfall of about 267 mm has a highest daily recorded fall of 147 mm while the probable maximum daily rainfall (see below) is about 450 mm (Weisner 1970). For the west Rajasthan desert region of India with an average annual rainfall of 310 mm, falls of 250–500 mm per day have occurred (Dhar and Rakhecha 1979). One-day falls of 50–120 mm with a return period of 2 years and probable maximum falls of 300–900 mm are suggested.

In the central and western Sahara area, an intensity of 46 mm/hr has been recorded at Tamanrasset for a period of 63 min, 174 mm/hr at Golea for 3 min and 92.4 mm/hr at Beni Abbas over 25 min (Griffiths 1972). While rain may be infrequent in such areas, when it does rain it can rain hard and with a sparse protective vegetation cover this results in considerable erosion, surface runoff and flooding (Ch. 7). The high falls in September 1969 over Tunisia and Algeria referred to above (Winstanley 1970) caused such devastation that nearly 600 people were killed and 250 000 made homeless. Oliver (1969) points out that while very heavy falls do occur in such areas, they are not so common as is sometimes stated. For the central Sudan, only 5–8 per cent of falls amounted to 10 mm or more in 1 hr.

The question of intensities over particular durations likely to occur with certain return periods (recurrence intervals) is extremely important. Return periods can be calculated in a number of ways, aspects being discussed by Weisner (1970) and Lockwood (1968), the latter using Malaysian examples. Perhaps the most common method is to arrange the data (e.g. largest daily rainfalls in each year) in descending order and assign a rank m to each, where $m = 1$ for the largest value and $m = n$ for the smallest, n being the number of values. Although a number of plotting position formulae exist, one often used is $T = (n + 1)/m$ where T is the return period.

These return periods can be plotted graphically against their corresponding amounts or intensities and a line of best fit computed or inserted by eye, allowing estimation of the return periods of various amounts or intensities and vice versa. Commonly, one or more of the highest values lie well off a straight line formed by the lower values. This is the subject of considerable debate (e.g. Kishihara and Gregory 1982) which cannot be considered here. However, it produces doubt as to the location of the line. Examples from the Uluguru Mountains, Tanzania, are shown in Fig. 4.6. Other examples of analysis of return periods for various durations include Dhar and Ramachandran (1970) for Calcutta, Tavares and Ellis (1980) for north-east Brazil, Morris and Rumbaoa

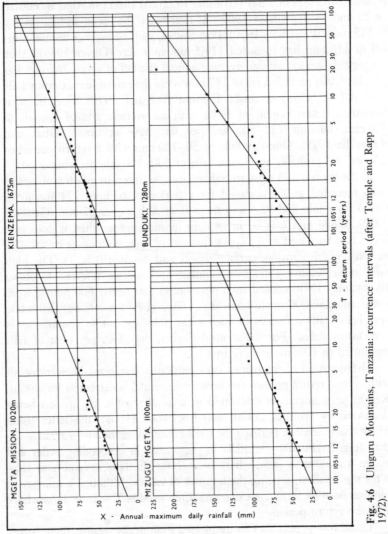

Fig. 4.6 Uluguru Mountains, Tanzania: recurrence intervals (after Temple and Rapp 1972).

(1980) for the Philippines, Giambelluca *et al.* (1984) for Oahu (Hawaii).

Areas registering the highest intensities will vary according to the duration considered, reflecting the varying characteristics of the disturbances creating rainfall. For short time periods, say up to 1 hr, thunderstorms may produce the heaviest falls. Since the intensity of these systems is likely to be greatest where there is the greatest supply of moist air, highest values are likely in the general area of the equatorial trough. This is illustrated by Fig. 4.7 showing 1 hr falls likely to be equalled or exceeded once every 2 years.

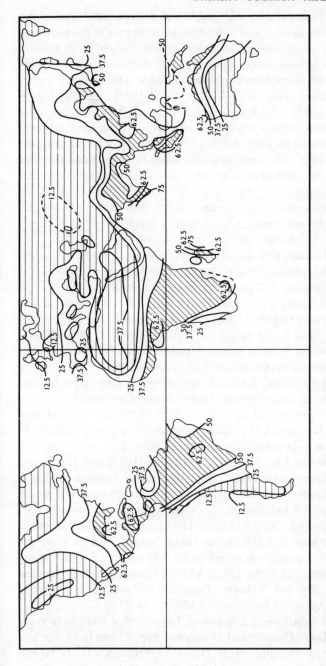

Fig. 4.7 One-hour precipitation (mm) likely to be exceeded once every 2 years (after Reich 1963).

For longer time periods, high totals will be associated with major organised disturbances of which the most extreme case is the hurricane – especially a slow-moving, declining system. Since hurricanes in particular do not usually occur within about 5° of the equator, it follows that maximum totals are found in higher latitudes. This certainly applies to falls of duration more than 6 hr (Lockwood 1974). The duration of most concern will vary with the purpose of study. Urban storm-water drainage problems or the water balance of a very small stream catchment requires information on short-duration conditions. Over catchments, the concentration time – the time taken for water to travel from the most distant part of the catchment to its exit – is an important parameter and for a small stream may be only 15–30 min. For large catchments, if, for example, interest centres on reservoir capacity, then concentration times are considerable and hence rainfall over a much longer period, say 12–36 hr, may be of more interest.

The critical return periods will vary with a range of physical, economic, social and political factors. Danger to life because of flooding, for example, may demand that protective structures should be able to cope with conditions having a very long return period, i.e. the probability of their occurring is negligible. Flooding of a small airfield or a limited area of agricultural land is not so critical and the economic cost of flood protection measures must be balanced against the cost of flood damage. Thus, in the latter case, protection may be limited to conditions with a much lower return period. Economic considerations are of great importance in developing countries with limited financial resources.

Lack of data for tropical areas, particularly in the case of conditions for periods of less than 24 hr make it impossible to give indications of highest falls and intensities experienced. Bahr Dar (Ethiopia) has received 204 mm in 2 hr. Mohr and Van Baren (1959) state that twelve stations in Indonesia have recorded daily values between 500 and 600 mm, six between 600 and 700 mm and one has recorded 702 mm in one day. Figure 4.8 indicates the high daily falls having 10- and 100-year return periods for west Malaysia. The maximum daily fall recorded in Madagascar exceeded 500 mm in Diego-Suarez. Stations along the east coast of Madagascar can record nearly 100 mm in a day during any month of the year (Griffiths 1972). Victoria (Cameroon) has recorded 510 mm in 1 day and Tukuyu (Tanzania) 432 mm/day. Townsville (Queensland, Australia) has recorded 650 mm in 21 hr.

Monrovia (Liberia), for a 2 hr period, has recorded 100 mm in every month from May to October and 18 mm in 5 min, 43 mm in 15 min and 63 mm in 30 min (Griffiths 1972). Dale (1959) reports a fall of 50 mm in 15 min at Kuala Lumpur and 719 mm in 96 hr at Kota Bharu, Malaysia. Arenas (1983) cites 24 hr values of 762 mm and 584 mm for Barbados and 229 mm in 2 hr for Saint Lucia in the Caribbean. Arenas

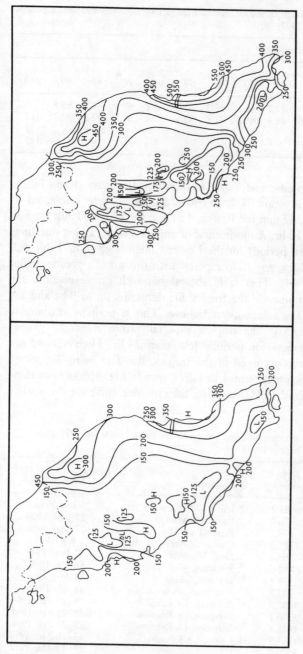

Fig. 4.8 West Malaysia: daily precipitation likely to be equalled or exceeded once every 10 years (left); daily precipitation likely to be equalled or exceeded once every 100 years (right) (after Lockwood 1967).

Table 4.3　Lower Congo (Kinshasa) rainfalls for various durations and return periods (mm) (after Griffiths 1972)

Return period (years)	Duration (min)									
	10	20	30	40	50	60	70	80	90	24 hr
2	23.3	37.5	46.5	56.2	62.0	66.1	67.3	69.5	69.8	
10	30.6	49.3	61.2	74.4	82.4	87.9	89.5	92.5	92.7	117
25										132
50										143

(1983) presents intensities for various durations for heavy storms hitting Cuba from which amounts for various durations were calculated as 29 mm for 5 min, 47 mm for 10 min, 150 mm for 1 hr, 400 mm for 5 hr and 624 mm for 12 hr. An indication of amounts over varying durations with various return periods for the Lower Congo is given in Table 4.3. All the above values may be compared with the world's greatest point rainfalls (Table 4.4). This table shows that with one exception, the record falls occur outside the tropics for durations up to 9 hr and are far in excess of any values quoted above. This is perhaps attributable to the lack of data in the tropics, especially from recording gauges providing information for periods less than 24 hr. High values, not recorded, will have occurred in the tropics, this fact being supported by the way in which maximum 1 hr falls shown in Fig. 4.7 occur in tropical areas. Dhar *et al.* (1982) compare the envelope curve for the world's

Table 4.4　World's greatest observed point rainfalls (from Lockwood 1974 after Palhaus 1965 and Chaggar 1984)

Duration	Depth (mm)	Location	Date
1 min	31.2	Unionville, Maryland	4 July 1956
8 min	126.0	Fussen, Bavaria	25 May 1920
15 min	198.1	Plumb Point, Jamaica	12 May 1916
20 min	205.7	Curtea-de-Arges, Romania	7 July 1889
42 min	304.8	Holt, Missouri	22 June 1947
2 hr 10 min	482.6	Rockport, West Virginia	18 July 1889
2 hr 45 min	558.8	D'Hanis, Texas	31 May 1935
4 hr 30 min	782.3	Smethport, Pennsylvania	18 July 1942
9 hr 00 min	1086.9	Belvouve, La Réunion	28–29 Feb 1964
12 hr 00 min	1340.1	Belvouve, La Réunion	28–29 Feb 1964
18 hr 30 min	1688.8	Belvouve, La Réunion	28–29 Feb 1964
24 hr 00 min	1869.9	Cilaos, La Réunion	15–16 Mar 1952
2 days	2500.0	Cilaos, La Réunion	15–17 Mar 1952
5 days	4301.0	Commerson, La Réunion	23–27 Jan 1980
7 days	5003.0	Commerson, La Réunion	21–27 Jan 1980
15 days	6433.0	Commerson, La Réunion	14–28 Jan 1980

highest rainfall for different durations of $R = 76.9D^{0.475}$ with that for Cherrapunji, India of $R = 49D^{0.485}$ (R in inches, D is the duration in days).

The totals recorded at La Réunion (Table 4.4) resulted from a combination of intense tropical cyclones together with relief influence. The previous record 24 hr fall of 1168 mm in July 1911 at Baguio in the Philippines was also associated with the passage of a typhoon whose effect was greatly increased by relief (Lockwood 1974).

Discussion of heavy falls raises the question of probable maximum precipitation (PMP). While there is some debate as to the meaning and validity of the term, it can be defined as 'that depth of precipitation which, for a given area and duration, can be reached, but not exceeded under known meteorological conditions' (Weisner 1970). Various methods have been used to estimate PMP including empirical relationships involving meteorological characteristics, storm models and storm transposition (Weisner 1970). Lumb (1971) produced graphs of PMP for areas of 10 km² for coastal and inland stations at various altitudes in East Africa, an example being shown in Fig. 4.9. On the graph, values at Sigona (altitude 2135 m) for a storm in April 1967 have been inserted and it can be seen the figures are very close to the calculated PMP. Values decrease with altitude as the volume of effective precipitable water decreases. Temple and Rapp (1972), in a detailed examination of landslide damage, flooding and erosion caused by a storm of more than 100 mm in less than 3 hr in a highland area of Tanzania, found that the minimum economic losses over a small area were well over US $90 000. However, using the same approach as Lumb (1971) they found that this storm at most was less than 50 per cent of the PMP for a 3 hr period. Clearly, the area might well receive considerably higher falls in future. A number of analyses of PMP for various parts of India, especially for 1-day falls, have been undertaken (e.g. Dhar, Kulkarni and Rakhecha 1981; Dhar, Rakhecha and Mandal 1981b; Dhar, Rakhecha and Kulkarni 1982b; Rakhecha and Kennedy 1985).

Schwarz (1963) suggests that the PMP at a point for a hurricane without orographic influence will be between 1000 and 1100 mm in 24 hr. The extreme fall in Table 4.4 for 24 hr indicates that an orographic component needs to be added to an assessment of PMP.

Chia and Chang (1971) provide a detailed discussion of the effects of a heavy rainfall in December 1969 at Singapore. Two-day totals ranged from 483 mm at Changi in the east of the island to 216 mm in the north, and the total damage due to flooding was estimated at US $4.5 million. The return periods of the maximum 24 hr falls varied from about 20 years in the north to 5000 years in the east. However, because of problems associated with the analysis and doubts of extrapolating for such long return periods these values merely indicate the exceptional

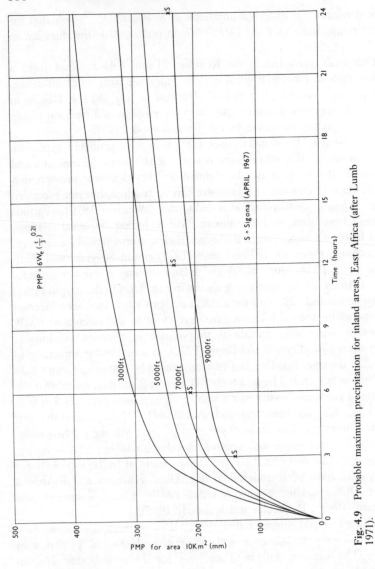

Fig. 4.9 Probable maximum precipitation for inland areas, East Africa (after Lumb 1971).

nature of the storm. A major feature was the prolonged period (10 hr) of very intense rain with a maximum rate exceeding 80 mm/hr. Over three catchments, the average rainfall depths, equivalent surface runoff and percentage this formed of the average rainfall depths were as follows:

1. 337 mm, 227 mm, 67 per cent.
2. 348 mm, 223 mm, 64 per cent.
3. 278 mm, 207 mm, 74 per cent.

The different percentages lost as surface runoff were attributable to the land-use differences, a feature to be discussed in Chapter 9.

4.2 SPATIAL VARIATIONS

Large-scale variations of rainfall in space exert a great influence upon agricultural systems and water supply. Not only do amounts depart from average conditions in individual years but the way in which they vary differs from area to area. For a particular period, rainfall can be well above average in one region but below average in another. The differing patterns of fluctuations in different areas are illustrated by correlations in July rainfall between various south-east Asian stations (Table 4.5, Fig. 4.10). In Table 4.5 a correlation of 0.42 or more indicates a significant relationship between stations. Where a correlation is negative, the stations vary in the opposite sense. Interestingly, Calcutta has negative correlations with the other seven stations. Parthasarathy (1984) for Indian summer monsoon rainfall found high spatial coherency between neighbouring sub-divisions. However, the north-east region and northern west and peninsular Indian sub-divisions had negative correlations. Yoshimura (1971) divided monsoon Asia into seventeen regions in January and twenty in July on the basis of inter-station correlations while Jackson (1972), for annual rainfall, divided Tanzania into nine regions. Other examples where regions have been defined using a variety of techniques include Morris and Rumbaoa (1980) for part of the Philippines, Gregory (1982) for the Sahel region, Suppiah and Yoshino (1984a, b) for Sri Lanka and Jackson (1985) for Tanzania.

Morth (1970) found that for East Africa large departures from average in the same sense covered wide areas, sometimes exceeding the whole of East Africa. Nicholson (1980), using areal averages for semi-arid areas of West Africa, also found strong coherence of variations in the region from 20°–10°N although occasionally there were differences. Despite such evidence for departures to be in the same direction over

Table 4.5 Simultaneous correlation coefficients of monthly precipitation amounts among eight stations for July (after Wada 1971).

	1	2	3	4	5	6	7	8
1. Bombay	1.00	0.52	0.17	0.18	-0.12	0.09	0.22	-0.74
2. Madras		1.00	0.10	0.27	0.10	0.01	0.26	-0.49
3. Bangkok			1.00	0.17	-0.10	-0.07	0.41	-0.34
4. Saigon				1.00	-0.09	-0.34	0.15	-0.20
5. Hong Kong					1.00	0.27	-0.09	-0.19
6. Taipei						1.00	-0.35	-0.11
7. Kagoshima							1.00	-0.39
8. Calcutta								-1.00

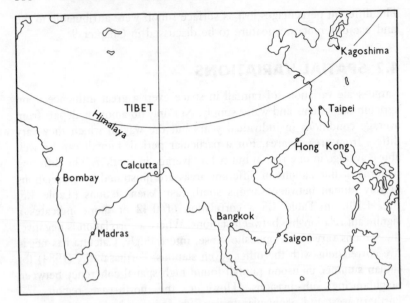

Fig. 4.10   Location of south-east Asian rainfall stations (after Wada 1971).

large areas, it will be shown below that the nature of tropical rainfall is such as to lead to marked contrasts in fluctuation patterns over extremely small distances.

In § 2.10 the often marked variations in rainfall over short distances for individual rainstorms and longer-period means in response to factors such as relief, water bodies and the effects of meteorological conditions such as the trade-wind inversion were indicated. Within a major climatic region, these striking rainfall gradients will result in marked contrasts in agricultural systems. The whole seasonal regime of rainfall can change in a short distance. This is illustrated by Mount Kilimanjaro where on the south and south-east slopes, the south-east monsoon is the main rain-bringer compared with the north-east monsoon on the north-east slopes.

In addition to large-scale variations and effects of 'permanent' influences such as relief, an important aspect is the nature of individual tropical rainstorms. These are often of very limited spatial extent with considerable gradients in intensity and amount. Even when a large area can be described as 'rainy', in reality the rain will comprise cells of often intense rain separated by areas of dry conditions or only light falls.

An example of the complexity of tropical situations is shown in Fig. 4.11. Orchard and Sumner (1970) suggest that for the day, there were four separate storm cells, each independent of the others. Daily rainfall at two stations 3.2 km apart in a level area near Dar es Salaam, Tanzania

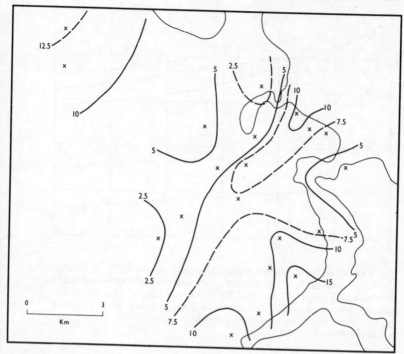

Fig. 4.11 Dar es Salaam: rainfall (mm) for 11 March 1969 (after Orchard and Sumner 1970).

(Jackson 1969b) further illustrate the marked local variations on a day-to-day basis (Table 4.6). Watts (1955) describes equatorial weather as 'local rather than regional' and Dale (1959) states that much of the rainfall in Malaysia originates from single cumulonimbus clouds producing heavy falls over areas varying from < 2.6 km² to 62.2 km². Oliver (1969) refers to the small diameter of rainstorms in the Sudan (< 2.5 km) and how this is relevant to actual evaporation rates (Ch. 5). Sansom (1953) describes a mature thunderstorm as having a number of cells or centres of activity, and while the size of individual cells varies considerably they seldom have a diameter greater than 8 km. The average duration of the mature stage of a cell, when heavy rain is produced, is usually from 15 to 30 min, the whole storm generally lasting 1–3 hr.

Because of the 'patchiness' of rainfall, both Johnson (1962) for East

Table 4.6 Daily rainfall at two stations, 3.2 km apart near Dar es Salaam, April 1967.

	6th	7–8th	10th	12th	13th	24th	25th
Station A	54.9	33.3	2.3	13.5	13.5	14.2	21.1
Station B	100.3	3.3	31.8	64.8	5.1	8.4	0.8

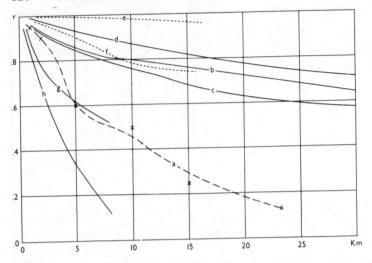

Fig. 4.12 Correlation–distance relationships (after Sharon 1972). (a) Southern Arava, ungrouped data; (b) Lower Jordan Valley, ungrouped data; (c) Lower Jordan Valley, days with <25 mm; (d) East central Illinois summer rain; (e) East central Illinois, summer, low centre rainfall; (f) East central Illinois, summer, air mass storms; (g) Coshocton, Ohio, largest summer storms; (h) Walnut Gulch, Arizona, largest summer storms.

Africa and Hammer (1972) for the Sudan, analyse daily data based upon the percentage of stations in an area reporting rainfall rather than looking at individual station values. Both authors state the tendency for rainfall to develop *in situ* with a lack of movement of rain areas. Riehl (1954) also points out the value of analysing areal integrals of tropical rainfall because of its 'showery' nature. Henry (1974) in analysing rainstorms (> 10 mm) over South-East Asia, Colombia, Panama and Costa Rica found that a typical storm was about 30 km in diameter with a spacing between two storm centres of about 60 km. However, relief and season affected these values. Areas between the storms had little or no rain. Analyses in Africa (Sharon 1974, 1981; Jackson 1974) suggest that rainstorms tend to occur at preferred distances apart, leading to an increase in correlation with distance beyond a certain minimum value.

Johnson (1962) found that there was no single scale on which rain areas occurred in East Africa. Often areas were localised with a diameter of < 30 km, but sometimes there was a broad general area of perhaps 500 km within which rainfall was patchy. Large falls were most likely in areas of widespread rain. At a distance of the order of 100 km, Johnson (1962) states that spatial correlation fell to zero and at greater distances up to 220 km there were negative correlations. Figure 4.12 shows how the change of correlation with distance between rain gauges varies with rainfall type. Large summer storms in Arizona show the most rapid

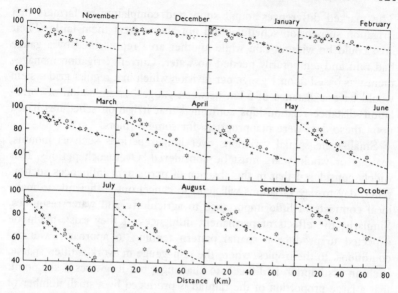

Fig. 4.13 Netherlands: correlation–distance relationships (after Ph. Stol 1972).

decrease in correlation with distance, these localised falls being more akin to tropical rain. In contrast, data for the Netherlands, Fig. 4.13, show a much less rapid decrease of correlation with distance, although the steeper curves in summer indicate the occurrence of more localised rainfall than in winter.

The patchy nature of tropical rainfall is likely to be particularly marked in semi-arid areas and often not detected by the sparse rain-gauge network in such regions. The considerable spatial variation in daily rainfall in the tropics suggests the need for a dense rain-gauge network. Unfortunately, for historical reasons and lack of resources, networks in developing tropical countries tend to be far less satisfactory than those of higher latitudes.

The marked spatial variations in tropical daily rainfall have considerable implications. Availability of water at certain, often short, critical periods in a crop growth season, can greatly influence yields. Since much of the rainfall does occur in a comparatively small number of heavier storms, whether or not an individual one 'hits' a certain area can be vital. Because of the often limited size and marked gradients in tropical storms, considerable contrasts in yield over short distances will result. This is of particular importance to the peasant farmer operating on a small scale of a few hectares. An individual or a whole village may find that their yields are very different to those a few kilometres away. In the Philippines, spatial variations in rainfall over an irrigated area were examined using gauges with separation distances of 550–3300 m (IRRI 1984).

The marked differences found supported complaints by farmers that irrigation suspension schedules did not always treat them fairly. Their area could be without rain while another area represented by a gauge had rain and temporarily needed no water. Current irrigation management was based upon 1 gauge per division which in a regular grid system would have a distance of 5 km between gauges. Examination of correlation–distance relationships confirmed that irrigation schedules based upon these data were inappropriate for some farmers.

Small-scale spatial variations over longer periods such as months, seasons or whole years must be considered. Over such periods, it is widely considered that in the absence of permanent influences such as relief, day-to-day variations will tend to cancel out sufficiently to make local contrasts of little importance to agriculture and water resources. Similarly, the effects of permanent influences, if they exist, might be expected to produce a spatial pattern tending to approach 'average' conditions. In the tropics, where the rain-gauge network is often sparse, such an assumption is difficult to challenge, but it ignores a key point: that a large proportion of the rainfall is produced by a small number of storms. Thus, the essential spatial pattern for considerable periods will be determined by a few storms. While this feature is universal (§ 4.1), in temperate areas where rainfall is more widespread, the spatial contrasts will not be too marked. In areas of very localised tropical rainfall this is not so and considerable variations can persist over small distances for long periods.

The magnitude of spatial variations over small distances, not due to the influence of 'permanent' factors such as relief, can be illustrated with a variety of data from Tanzania. Monthly totals for the two gauges, 3.2 km apart used in Table 4.6, are shown in Table 4.7. Large differences exist for individual months, e.g. April. For the period April–June

Table 4.7 Monthly rainfall (mm) at two stations 3.2 km apart near Dar es Salaam April 1967–March 1968.

Total for April 1967–March 1968
Station A – 1437.3
Station B – 1666.5

Total for April 1967–March 1969
Station A – 2593.3
Station B – 2793.9

	Apr	*May*	*June*	*July*	*Aug*	*Sept*
Station A	315.2	193.4	31.8	131.2	47.5	89.1
Station B	437.6	236.0	44.3	97.3	59.7	83.1

	Oct	*Nov*	*Dec*	*Jan*	*Feb*	*Mar*
Station A	49.7	206.0	134.5	2.6	90.5	145.8
Station B	48.3	180.7	172.4	2.1	82.8	222.2

Table 4.8 Annual rainfall (mm) 1946–54 for two stations 8 km apart (altitude difference 15 m).

Fatemi Sisal Estate – A
Mgude Sisal Estate – B

	1946	1947	1948	1949	1950	1951
A	754.6	973.1	966.5	790.1	1069.6	1188.7
B	859.8	1248.4	1420.9	619.8	1223.3	1127.0
(A–B)	−105.2	−275.3	−454.4	+170.3	−153.7	+61.7

	1952	1953	1954	Average (1942–66)
A	755.1	1108.7	930.7	943.9
B	565.4	859.8	869.8	924.8
(A–B)	+189.7	+248.9	+61.0	

1967, the difference is 177.5 mm (33 per cent of the gauge 1 value). Over the year April 1967 to March 1968, the difference is 229.2 mm (15.9 per cent of the gauge 1 value). Over the 2-year period to March 1969, the difference is 200.6 mm. However, individual months show a reversal of the cumulative difference, e.g. July and November 1967, or very little difference and statistically there is no significant difference between the two gauges. These figures suggest the possible order of magnitude of differences over small distances which, while persisting, cannot be regarded as permanent.

Further data for two stations 8 km apart are shown in Table 4.8. These stations are in a flat area with a difference in altitude of only 15 m some 130 km inland from the previous two stations. Annual averages are fairly close but for individual years the differences are often considerable. The very large differences which are opposite in sign for 1948 and 1953 are especially noteworthy. Data for three pairs of stations are

Table 4.9 Monthly and annual rainfall – Tanzania comparison between pairs of stations (averages 1936–70 (mm))

(A) Dar es Salaam and Pugu (13 km apart)		(B) Lugoba and Fatemi (35 km apart)		(C) Ruvu and Alavi (18 km apart)	
Annual averages		Annual averages		Annual averages	
	DSM = 1078.5		Fatemi = 958.9		Alavi = 970.8
	Pugu = 1049.0		Lugoba = 1026.7		Ruvu = 1006.9
1938	DSM = 1062.2	1957	Fatemi = 1090.9	1939	Alavi = 517.4
	Pugu = 1561.1		Lugoba = 1544.8		Ruvu = 886.2
	(wettest on record)		(wettest on record)	1966	Alavi = 1117.9
1946	DSM = 921.5	1939	Fatemi = 758.2		Ruvu = 1530.4
	Pugu = 501.9		Lugoba = 558.0		
	(driest on record)		(driest on record)		
		1958	Fatemi = 665.2		
			(driest on record)		
			Lugoba = 891.0		

Table 4.9 continued

(A) Dar es Salaam and Pugu (13 km apart)	(B) Lugoba and Fatemi (35 km apart)	(C) Ruvu and Alavi (18 km apart)

(A) Dar es Salaam and Pugu (13 km apart)

January averages
DSM = 59.7
Pugu = 70.1
1969 DSM = 20.6
Pugu = 90.9

March averages
DSM = 137.7
Pugu = 134.1
1964 DSM = 442.2 (wettest on record)
Pugu = 143.8
1968 DSM = 146.6
Pugu = 268.7 (2nd wet. on record)

May averages
DSM = 180.6
Pugu = 164.1
1961 DSM = 61.5
Pugu = 125.5

December averages
DSM = 94.5
Pugu = 90.7
1964 DSM = 309.9 (wettest on record)
Pugu = 87.9
1946 DSM = 114.0
Pugu = 16.3
1969 DSM = 10.2 (driest on record)
Pugu = 56.1

December averages
Fatemi = 91.7
Lugoba = 87.4
1943 Fatemi = 69.6
Lugoba = 8.9
1954 Fatemi = 0.0
Lugoba = 61.7
1966 Fatemi = 98.6
Lugoba = 0.0
1968 Fatemi = 49.5
Lugoba = 118.1

(B) Lugoba and Fatemi (35 km apart)

January averages
Fatemi = 94.0
Lugoba = 97.3
1956 Fatemi = 264.9
Lugoba = 98.3
1965 Fatemi = 37.6
Lugoba = 153.7

March averages
Fatemi = 135.4
Lugoba = 149.6
1949 Fatemi = 30.2
Lugoba = 121.7
1958 Fatemi = 151.9
Lugoba = 362.0

April averages
Fatemi = 208.8
Lugoba = 196.6
1957 Fatemi = 206.0
Lugoba = 432.3 (wettest on record)
1966 Fatemi = 125.0
Lugoba = 290.3

November averages
Fatemi = 94.0
Lugoba = 110.7
1941 Fatemi = 57.4
Lugoba = 208.5
1970 Fatemi = 67.1
Lugoba = 5.1

(C) Ruvu and Alavi (18 km apart)

January averages
Alavi = 62.5
Ruvu = 65.8
1941 Alavi = 11.2
Ruvu = 66.3
1962 Alavi = 28.4
Ruvu = 76.5
1967 Alavi = 36.3
Ruvu = 0.0

March averages
Alavi = 150.9
Ruvu = 150.9
1946 Alavi = 82.0
Ruvu = 7.1
1965 Alavi = 23.1
Ruvu = 58.2
1970 Alavi = 149.1
Ruvu = 64.0

April averages
Alavi = 258.6
Ruvu = 249.7
1960 Alavi = 347.5
Ruvu = 149.6
1963 Alavi = 361.7
Ruvu = 546.1
1968 Alavi = 397.0
Ruvu = 217.4

November averages
Alavi = 107.4
Ruvu = 106.4
1950 Alavi = 186.9
Ruvu = 31.2
1954 Alavi = 13.0
Ruvu = 86.4

SPATIAL VARIATIONS 125

shown in Table 4.9. Differences for individual months or years are often considerable despite the similarity in average values and proximity of the stations. For Burma, in an analysis of seven stations over a 21-year period, Huke (1966) found that variations from average were not consistent year by year. The two wettest stations, Mergui and Akyab for example, showed variation in the same direction only ten times – just as the law of chance would dictate. Only in 1 year did all stations deviate in the same direction from average. In only 8 of 18 years did two dry stations show deviations of the same sign.

Hulme (1984), for the generally dry year of 1983 in central Sudan, also indicates localisation in totals. For example, one station had its second wettest year and another had a total in the second wettest decile. Therefore even in a drought year, some locations may have above average rainfall. This may either ameliorate or intensify the impact of drought depending on the human factors in the area (Hulme 1984). For Upper Volta (Birkina Fasso) Ofori-Sarpong (1983) also provides examples of small-scale variations.

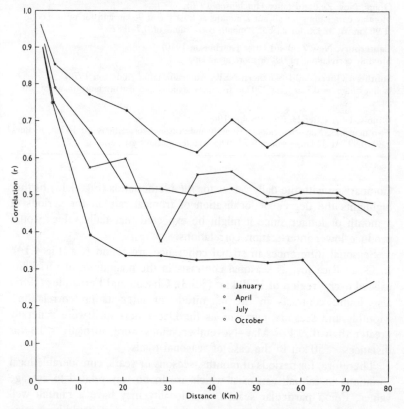

Fig. 4.14 Ruvu Basin, Tanzania: correlation–distance relationships (after Jackson 1974).

Table 4.10 Ruvu Basin, Tanzania. Average correlation coefficients for individual months

| January | 0.458 | October | 0.680 |
| April | 0.286 | July | 0.496 |

Figure 4.14 shows how correlation between rainfall stations varies with distance apart of the stations for monthly rainfall over the Ruvu River Basin, Tanzania. The low average correlation for each month is shown in Table 4.10, and data for Sierra Leone, an area of similar size, on an annual basis produced an average correlation of only 0.182 (data from Gregory 1965). These tropical values may be compared with data for New Zealand, Australia and the USA presented in Table 4.11. The comparison shows that the correlation for any distance category appears to be much less for tropical conditions than elsewhere (Jackson 1974).

Table 4.11

Otago, New Zealand (after Hutchinson 1970)
Monthly correlations for all but 2 months at least 0.5 at 96 km and for most months 0.6 at 96 km. At 32 km for all but 1 month correlations of 0.7–0.82.

Canterbury, New Zealand (after Hutchinson 1970)
Monthly correlations of 0.8 or more at 32 km.

Southern Queensland/Northern NSW, Australia (after Anderson 1970)
Isolines for $r = 0.5$ at least 320 km from key stations and usually more than 480 km away.

Illinois, USA (after Huff and Shipp 1969)
Two networks, May–September. Monthly and seasonal correlations. At 16 km, r values 0.80–0.89. At 32 km, r values 0.72–0.85. Winter r values greater.

Comparison with the daily values for the Netherlands (Fig. 4.13) further highlights the degree of localisation of tropical rainfall for periods of a month or longer since it might be expected that daily values would produce lower inter-station correlations.

Seasonal differences in spatial coherence are found (e.g. Fig 4.14). In Costa Rica just as seasonal contrasts in the magnitude of variability existed over a region of 1770 km^2 (§ 3.3), Chacon and Fernandez (1985) also found contrasts in the magnitude of inter-station correlations. Monthly and seasonal correlations for December–April were generally greater than 0.7. For May–November values were normally <0.3 for distances < 50 km in the case of seasonal totals.

Therefore, for periods of months, seasons or years, considerable local variations in rainfall can exist which are in no way related to average values. For a particular season, one locality may have a rainfall well above average while another area a short distance away could have rainfall below average. These contrasts, together with further examples and their implications are discussed by Jackson (1978, 1985). As has already been pointed out, the small scale of operations of peasant farmers and their limited resources in terms of irrigation, makes such marked local variations particularly significant. They can be of sufficient magnitude to create considerable differences in crop yield over small distances, one farmer or village having a 'good' year while another only a short distance away experiences an 'average' or 'poor' year. The degree of localisation of rainfall may in fact be regarded as a point in favour of fragmentation of peasant holdings. Because of local rainfall differences in western Queensland, sheep properties in a dry year may show large differences in carrying capacity between adjacent properties and sometimes this may be so in different parts of one property (Oliver pers.comm.).

While it is widely accepted that over large areas such as the Indian sub-continent, in a particular year, some regions can have well above-average rainfalls and others below-average or average amounts, the degree of 'local' variation suggested above is not perhaps appreciated. Lack of communication means that localised conditions are not brought to the attention of the authorities and even if they are, satisfactory conditions near by may promote disbelief. This may perhaps be indicated by a quotation from an official report in Tanganyika in 1945 discussing famines and food shortages:

> ... it is only the better class of chiefs who fail to do so [i.e. report a famine]. He only appeals to the administration when affairs go beyond his control. There is, of course, the whining class of chief who immediately upon a partial shortage of food in his area reports a famine.

Whatever the truth of such a statement, it is possible that this indicates a lack of appreciation of the magnitude of local variations in rainfall. While overall a region may have experienced a good season as far as rainfall is concerned, there could be local areas suffering considerable hardship.

The value of utilising areal integrals of rainfall rather than single-station data for daily values, has already been referred to. This method could be extended to monthly or seasonal data, but would mask very considerable local variations. The often low correlations between near-by stations raises the problem of deciding on suitable areas for integration. In East Africa for example, river basins have been used but within an individual basin, some areas can have above average falls and some below average amounts in a particular period. A simple approach to delimiting areas is discussed by Jackson (1985). The magnitude of variations over small distances for considerable periods highlights the need to improve existing rain-gauge networks in tropical countries and that great care must be taken in utilising data for planning purposes.

Although average values may be similar over an area, considerable differences can exist for individual periods.

4.3 DIURNAL VARIATIONS

Time of occurrence of rainfall has already been referred to in § 4.1 when discussing light showers. For larger falls, occurrence at night, when evaporation is low, will allow water more time to penetrate into the soil and become available to plants. In tropical countries, heavily dependent upon agriculture, a concentration of rainfall in large day-time storms can severely inhibit work and transportation.

There is still misconception about the general existence of two main patterns of diurnal variation. The first is a 'maritime' type with a maximum during the night and early morning, the second is a 'continental' type with an afternoon maximum. Many coastal and ocean locations exhibit the maritime type although, as will be seen below, there are exceptions. However, mechanisms responsible are still subject to debate. On the other hand, much evidence indicates the fallacy of general acceptance of the continental regime. The significance of disturbances in creating rainfall (Ch. 2) suggests that the concept of two diurnal variation types is far from the truth. It may be that the intensity of disturbances will show a diurnal variation and hence might produce rainfall or increased rainfall at particular times of day in some cases. However, their varied nature and varied occurrence in time and space (Ch. 2) suggests considerable differences in diurnal rainfall regimes. Seasonal variations in regime at a particular location due to changes in meteorological conditions producing rain may occur. Local factors are often important as was illustrated in § 2.10 using examples such as Lake Victoria, the 'Sumatras' of Malaysia and Mount Kilimanjaro.

Gray and Jacobson (1977) present a survey of literature relating to diurnal variations of tropical rain and cloud. General discussions of the complexity and uncertainty are also presented by Riehl (1979) and Sharma (1983). Three key mechanisms are frequently discussed in the literature: solar radiation, air–sea interactions and atmospheric tides (Sharma 1983). The latter is the subject of much debate, its significance being dismissed by many. During day-time, solar heating of cloud tops may exceed radiational cooling. Solar heating may also reduce liquid water content and suppress cloud development (Sharma 1983). Absence of solar radiation at night and radiational cooling of cloud tops will tend to increase lapse rates and promote convective overturning of the cloud layer, enhancing nocturnal rainfall. Land–sea breeze effects have already been referred to in Chapter 2. Day-time heating over land with resultant sea breeze, ascent over land and subsidence over the sea in theory should favour maximum rainfall during early morning and after-

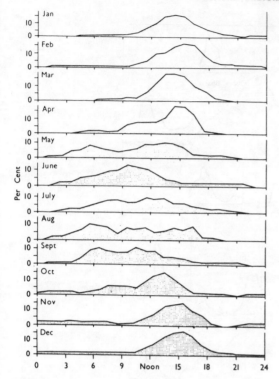

Fig. 4.15 Singapore: diurnal variation of rainfall (after Nieuwolt 1968).

noon when land/sea temperature contrasts are greatest (Sharma 1983). However, a range of other factors such as relief, creating local airflow, interaction between sea breeze and general winds complicate the picture. At night, this circulation is reversed.

The wide variety of patterns of diurnal variation can be illustrated by a number of examples. In many cases 'explanations' of diurnal regimes are uncertain. At Khartoum, a 'continental' location, 77 per cent of the rain occurs between 18.00 hr and 06.00 hr (Oliver 1965). Nieuwolt (1968) for Singapore, indicates how a maritime regime occurs in the south-west monsoon period (May–September) when air comes directly off the sea. During the rest of the year, a continental type is experienced because of the passage of the north-east airstream over the Malaysian Peninsula creating day-time convection (Fig. 4.15). Dale (1959) distinguished three patterns of diurnal variation over Malaysia, an inland type and west and east coast types. The latter two types showed a seasonal change but the former did not. In a more detailed analysis of summer rainfall over Malaysia, Ramage (1964), using August as a typical month, distinguished five types. These he explained in terms of the

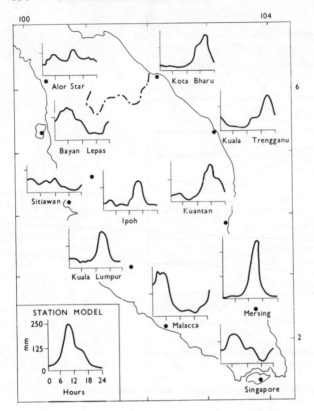

Fig. 4.16   Malaya: diurnal variation of rainfall, August (after Ramage 1964).

interaction between the south-west monsoon, local land and sea breezes, local topographic winds and general effects of relief. Illustrations of the different regimes are shown in Fig. 4.16.

Thompson (1957b) indicates an extremely complex situation in East Africa. While some inland stations were found to have an afternoon maximum (e.g. Nakuru, Table 4.12) many others did not exhibit this simple continental type. Dodoma and Tabora (Table 4.12) are two examples. Both stations are on broad plains distant from mountains or higher plateaux, and during the main rainy periods precipitation can occur at any hour of the day though there is a minimum during the mornings and early afternoons.

An illustration of the significance of disturbances in creating varying diurnal regimes is provided by Griffiths (1972). Disturbances, often called 'line squalls', travel in a general westerly direction over West Africa (Ch. 2). For Nigeria, Griffiths describes varied regimes at six stations:

Table 4.12 Percentage of annual rainfall occurring in each hour of the day

Station	Hour (East African Standard Time)								
	0	1	2	3	4	5	6	7	8
1	1.5	1.0	0.9	0.8	0.7	0.5	0.3	0.2	
2	5.0	5.6	6.1	6.1	5.3	4.3	4.0	3.6	
3	4.8	4.3	4.3	4.6	4.4	3.9	2.8	2.2	

Station	8	9	10	11	12	13	14	15	16
1	0.2	0.1	0.7	0.9	3.5	7.1	13.3	16.3	
2	2.9	2.2	1.8	2.1	2.3	2.7	2.9	3.6	
3	1.7	1.5	1.6	1.7	2.3	4.0	4.9	6.7	

Station	16	17	18	19	20	21	22	23	24
1	16.5	12.6	8.5	5.3	3.0	2.2	2.3	1.7	
2	5.3	6.7	6.4	4.5	3.6	3.8	4.4	4.8	
3	5.9	6.5	5.3	5.2	4.9	5.3	5.7	5.5	

Note: 1 Nakuru, 2 Dodoma, 3 Tabora (after Thompson 1957b).

1. Tiko (4° 06′N, 9° 21′E, altitude 50 m) – on the coastal belt with a diffuse early morning maximum.
2. Port Harcourt (4° 51′N, 7° 1′E, altitude 15 m) – outside the coastal belt with a marked afternoon maximum caused by local thunderstorms and young line squalls.
3. Enugu (6° 28′N, 7° 33′E, altitude 140 m) – diffuse night and early morning maximum in the rainy season, perhaps due to the southern Cameroons line squall source area.
4. Minna (9° 37′N, 6° 32′E, altitude 260 m) – evening and early morning maxima, the former may be due to line squalls, the latter from monsoon rain or distant source area line squalls to the east.
5. Jos (9° 52′N, 8° 54′E, altitude 1260 m) – marked afternoon maximum characteristic of line squall source area, a minor early morning maximum may be of the monsoon type.
6. Sokoto (13° 1′N, 5° 15′E, altitude 345 m) – marked night and early morning maxima associated with line squall source areas to the east.

Oguntoyinbo and Akintola (1983) for Ibadan (07° 20′N, 3° 50′E) ascribe a late afternoon maximum throughout the year to the convectional nature of thunderstorms. They also found a tendency for falls in early morning and at night in mid-season (June–August) which they relate to oceanic influence when the inter-tropical discontinuity has moved north of Ibadan.

For north-east Brazil, Kousky (1980) found that most coastal areas had a nocturnal maximum, thought to be due to convergence between onshore airflow and the land breeze. A day-time maximum some

100–300 km inland was considered to result from the inland advance of the sea breeze. Diurnal variations in the interior were thought to result from mountain–valley breezes. On the north coast, seasonal variations occurred, linked to seasonal variations in low-level mean airflow. Near the mouth of the Amazon, a nocturnal maximum was found between January and May and an afternoon maximum between June and September.

Gray and Jacobson (1977) found an early morning maximum at small island stations in the west Pacific while for large islands there was only a weak morning peak but a pronounced afternoon maximum. They attributed the latter to a 'heat island' effect not found with the smaller islands. However, Schroeder, Kilonsky and Ramage (1978) in an analysis of the Hawaiian islands, found that in most cases, except for the largest island of Hawaii, there was a nocturnal maximum dominance, even although the islands were larger than the Gray and Jacobson 'large' ones. Schroeder *et al.* (1978) suggest that the significance of the trade winds over the Hawaiian Islands accounts for the difference. They explain diurnal variations over the different islands in terms of diurnal variations in the trade-wind layer and local circulations associated with island heating and cooling and general land configuration. They infer that the nocturnal maximum for low islands applies to adjacent ocean areas. Of the mechanisms suggested in the literature, Schroeder *et al.* (1978) believe that radiative destabilisation at cloud top (see above) is responsible for the early morning peak.

Jordan (1980), using ship observations off the west coast of Africa, also found a definite early morning maximum. In contrast, Gray and Jacobson (1977) indicate a late morning/early afternoon maximum and late evening/early morning minimum over the eastern tropical Atlantic. They attribute this to several factors, including disturbances, some of which originate over Africa. Sharma (1983) relates diurnal variations at Nandi, Fiji, to the three factors already indicated. In both wet and dry seasons, an afternoon maximum occurs, but in the dry season there is also an early morning peak.

The examples above and those referred to in Chapter 2 are sufficient to indicate the variety and complexity of diurnal rainfall patterns.

4.4 RAINFALL AND RUNOFF

The rainfall characteristics discussed in this chapter and in Chapter 3 influence the amount and character of runoff. However, the exact nature of the relationship, introduced in § 1.3, is determined by a wide range of factors including evaporation, soil, vegetation, slope, geology, stream catchment size, shape and drainage density.

Runoff components and mechanisms

Runoff can be divided into various components although divisions are to some extent arbitrary. Conflicting views on runoff mechanisms to some extent reflect the different environmental/experimental experiences of authors. Much of the work on mechanisms has been outside the tropics, leading to adoption of models and approaches not always appropriate in the latter. Detailed discussion is outside the scope of this book, aspects being included in most hydrological texts (e.g. Linsley, Kohler and Paulhus 1982; Viesman *et al.* 1977; Branson, *et al.* 1981). Studies of components, processes and influencing factors based on tropical experiments include Bruijnzeel (1983), Bonell, Gilmour and Cassells (1983), Balek (1983), de Oliveira Leite (1985) and a major review of French tropical studies by Dubriel (1985).

A simple division of total runoff is into the components surface, sub-surface and ground water. Surface runoff is that travelling over the ground and in stream channels. Direct precipitation on to water surfaces will only be a small percentage of the total volume of streamflow unless the catchment has a large number of lakes or swamps. However, it may be important in terms of saturation overland flow (see below). Overland flow reaches main stream channels rapidly during and immediately after rain.

In the upper soil layers water which has infiltrated may move laterally. Such sub-surface flow is often referred to as through-flow or inter-flow although some authorities distinguish between the two depending on the layer through which the water moves (Gregory and Walling 1973). Although considerable volumes of water may be involved, in general movement is less rapid than that of overland flow.

Water percolating to deeper layers contributes ground water runoff or base flow. Lateral rate of movement of groundwater is slow and hence there is considerable lag between rainfall and the ground water runoff components. Such runoff is regular and is important during dry spells or seasons when other contributions are absent.

Clearly the distinctions between the different components are to some extent arbitrary. For example, sub-surface flow may again somewhat arbitrarily be divided into quick return flow and deeper delayed return flow. Streamflow is sometimes classified as either direct or base flow, or alternatively quick flow and delayed flow, timing being the essential distinction. Direct or quick flow comprises surface, 'prompt' sub-surface flows and direct precipitation. The key point is that response of runoff to rainfall depends on the path rain follows on reaching the surface.

Runoff can be generated in various ways. 'Hortonian' overland flow occurs when rainfall intensity exceeds infiltration capacity. It is a significant mechanism in tropical and semi-arid climates but where rainfall intensities are low and where soil is deep and well protected by vegetation (§§ 6.3, 7.1) it may never occur or only rarely. However, even in the tropics and semi-arid areas other runoff mechanisms are important. The idea that only part of a catchment may contribute to runoff (of Hortonian origin or otherwise) is widely known and this part will vary depending on antecedent conditions and storm characteristics. Other mechanisms contributing to runoff on a partial area basis include the idea of saturation of a thin topsoil layer overlying an almost impermeable subsoil or rock (Betson and Marius 1969). Even with low-intensity rain, this layer frequently becomes saturated leading to overland flow of rainwater on this layer. Partial area runoff can also be associated with subsurface mechanisms. For example, where a perched water table intersects a hillslope, seepage may occur (sometimes termed return flow) and this, together with rain falling on the resulting saturated area, forms runoff (Dunne and Black 1970). During storms, downslope moving sub-surface flow may emerge at some point and also be termed return flow (Hewlett and Hibbert 1965). Rainfall occurring where the main water table intersects the surface will also produce saturation overland flow.

There is debate as to whether sub-surface flow is fast enough to create a rapid response of streamflow to a rainfall event. Whipkey (1965) found that rapid movement was possible mainly because of passage through cracks, animal holes and decaying tree roots. An alternative is a type of piston effect whereby water entering soil at one location results in emergence of water downslope. Bruijnzeel (1983), for a basin in central Java, reports that sub-surface flow contributes to total quick flow via the mechanism of displacement. In a particular catchment, several mechanisms will operate at the same time and separation of the components is difficult. Bruijnzeel (1983) assessed storm runoff as comprising a mixture of channel precipitation, Hortonian overland flow, saturation overland flow and sub-surface flow in the central Java catchment. He felt that the variable source area concept was applicable to the basin.

Apart from experimental approaches on a plot or catchment basis (§ 9.3), mathematical modelling of rainfall–runoff relationships has received much attention. Discussions of modelling principles are contained in a number of references (e.g. Viesman et al. 1977; Raudkivi 1979; Australian Water Resources Commission 1976; Singh 1982; Morel-Seytoux et al. 1979). A number of models concern tropical conditions (e.g. Balek 1983). Brown et al. (1981) produced a mathematical model of the Upper Nile Basin made up of three sub-models: catchment, lake and channel.

Sellars (1981) produced a flood plain storage model of the Upper Yobe River, Nigeria to estimate evaporation losses from the large flooded open water and swamp area. Hari Krishna (1982) developed a runoff model for small agricultural watersheds in the semi-arid tropics which also allows other components such as soil moisture and evaporation to be estimated.

Factors influencing runoff

The balance between rainfall and evaporation (§ 5.3) gives a rough indication of water available for streamflow. However, the time period considered, rainfall characteristics and factors influencing the disposition of rain at the earth's surface complicate the picture. Aspects of the water balance are considered later, particularly in §§ 5.3 and 8.1. The significance of vegetation, slope, soil type, condition and moisture content in determining the proportion of rain becoming surface flow, sub-surface flow or ground water is discussed in §§ 6.3, 7.1, 8.2 and 9.3. The more these factors encourage infiltration, the less will be the proportion of rain lost immediately as surface flow. The greater the proportion of rain infiltrating, the lower the peak discharges and the more evenly distributed in time is streamflow. Dubriel (1985) presents a review of processes and factors influencing runoff generation in tropical arid, humid and rainforest zones, based upon a range of experiments over a 30-year period. He summarises the influence of factors such as soil, geology and vegetation in tabular form. Experimental rainfall-runoff relationships and shapes of hydrographs (plots of stream discharge against time) are among aspects covered. Shape of the catchment affects the concentration time (§ 4.1). When rain occurs over the whole of a generally circular catchment, streamflow from the various parts is likely to reach the lower main channel at approximately the same time, creating a marked discharge peak. This will not be so for a long, narrow catchment since discharge peaks from the lower tributaries will have left before those of the upper tributaries arrive downstream. Direction of movement of a storm will also influence hydrograph shape. With a storm moving up a catchment, water from the lower reaches may have left the catchment before that from upstream arrives at the outlet. For a storm moving downstream, water from the upper reaches may well reach the outlet at about the same time as that from lower reaches, creating a more marked peak. Catchment size has a number of effects apart from determining the total volume of rain falling on it. The larger the catchment, the greater the variation in factors such as slope, vegetation and rainfall conditions. Rainfall–runoff relations for large catchments will therefore tend to be complex, particularly for rivers covering more than one major climatic division. For large catchments, there will be a considerable lag between storm occurrence and peak flows downstream because of the great distances involved. Individual storms may cover the whole of a small catchment and hence discharge peaks will reflect their occurrence. For large catchments, flood peaks are due to the superimposition of runoff from a number of storms occurring at different times over different parts. Flooding over large catchments is therefore most likely in years or seasons with above-average rainfall when storms are most frequent and base flow between runoff peaks is high. Flood peaks over small catchments, reflecting the incidence of a single large storm, can occur even in dry years.

Drainage density, itself a reflection of factors such as climate and geology, influences rainfall–runoff relations. A large number of closely spaced channels means that overland flow is short and surface runoff quickly reaches main streams, making for higher discharge peaks. A sparse drainage network results in more overland flow, allowing more time for infiltration and hence a greater contribution from sub-surface and base flow. Surface flow from different parts of the catchment is less likely to arrive in the lower main channel at approximately the same time. Peak discharge will therefore tend to be less for a sparse drainage network. Geology has various direct and indirect effects. Indirectly, it influences soil type, slope and drainage density. Rock type and structure also influence the·rate and direction of movement of ground water flow. Hence they affect rate of response of base flow to rainfall. Structure can sometimes result in rain occurring over one catchment being directed as ground water into another catchment.

Tropical characteristics

Hydrological characteristics in tropical areas are reviewed in a number of works such as Balek (1983), Dubriel (1985) and in papers in Keller (1983) for the humid tropics and IAHS (1979) for low-rainfall areas. Balek (1983) presents a classification of tropical rivers and then considers great tropical rivers and basins. McMahon (1979) points out that arid zone streams are much more variable than those in humid areas and extrapolation from the latter to the former is dangerous.

The seasonal nature of rainfall over much of the tropics (§ 3.2) leads to marked seasonality in streamflow. Thus Ledger (1964), in discussing the great seasonal variation in West African river flow states: 'The pattern clearly reflects the seasonal distribution of rainfall to the virtual exclusion of all other factors'. He distinguishes between the north of the region, where the short rainy season means that runoff is concentrated in a few floods in August and September, and the south where runoff is spread more evenly through the year under a longer rainy season. One effect of catchment size is illustrated by the southern coastal rivers of West Africa where a 'little dry season' occurs in July and August separating two rainfall peaks. This is reflected in a double-peaked hydrograph in the smaller coastal rivers. Larger rivers have a single peak because their headwaters extend inland to areas having an August–September rainfall maximum. The upper reaches are in flood at this time when runoff downstream is affected by reduced rainfall (Ledger 1964).

Rivers of the Indian sub-continent reflect the very seasonal nature of rainfall, being characterised by a period of heavy discharge associated with the monsoon and an often lengthy period of low flow. The dry weather flow of the Indian peninsular rivers, derived from ground water, gradually decreases, often becoming a trickle just before the monsoon. Variations of the order of 1–300 in mean monthly flows of these rivers are common (Rao 1968). In the Damodar River, 80 per cent of the total discharge occurs from June to August, with an August maximum and considerable flooding (Sen 1968). The seasonal regime of rivers deriving much of their flow from the Himalayas differs since glaciers and snow contribute significantly to dry weather flow. The period of minimum flow is in winter but does not reach the low levels of peninsular rivers. Even so, the seasonal variation is considerable as is illustrated by the snow-fed Indus River system (Table 4.13). Differences between individual rivers in the system are also apparent in Table 4.13.

Table 4.13 Discharge in the Indus River system (from Gulhati 1968)

River	Mean annual flow (mill. m^3)	Percentage of mean annual flow during			
		Apr–June	July–Sept	Oct–Dec	Jan–Mar
Indus	110 396.46	31	54	8	7
Jehlum	27 876.65	44	36	8	12
Chenab	28 986.78	28	56	7	9
Ravi	7 894.26	30	51	8	11
Beas	15 665.20	15	67	10	8
Sutlej	16 775.33	23	62	9	6
All rivers	207 594.68	30	54	8	8

The tendency for tropical river regimes to reflect marked seasonality in rainfall is modified by the factors discussed earlier. If these factors encourage infiltration, wet season discharge is reduced. Correspond-ingly, the build-up of ground water increases dry season flow, producing a general evening-out of the seasonal regime. As will be seen later, a vegetation cover is extremely important in encouraging infiltration. The vegetation is influenced by rainfall, especially the length of the wet season. In West Africa, the effectiveness of the vegetation cover in retarding runoff decreases from south to north as the length of the wet season declines. Hence, in the north, not only is rainfall concentrated in a short season but the resulting runoff is also concentrated in the absence of the beneficial effects of a continuous vegetation cover. In the south the dry season is shorter and the effectiveness of the vegetation cover also promotes dry period base flow. The larger rivers in this region therefore have an appreciable dry season discharge. However, Cordery and Pilgrim (1983) question whether soil and vegetation characteristics are major influences on flood runoff, except for minor floods from small homogeneous areas and deep forest soils in temperate areas.

The proportion of rain becoming runoff (the runoff coefficient) is of great interest. Ledger (1964) indicates the wide range of coefficients for West African streams in response to the various controlling factors. North of the 750 mm annual rainfall isohyet, in the absence of an effective vegetation cover, runoff coefficients can be 80 per cent for individual storms and 30 per cent for annual rainfall over small catchments. Over larger catchments with lower overall slopes, where water spreads over a wide area and evaporation loss is considerable, coefficients are much lower. In a modelling study in Nigeria already referred to for example (Sellars 1981), evaporation losses from the large flooded area were estimated at about 45% of total rainfall. Basins of more than 1000 km^2 usually have values of less than 6 per cent (Ledger 1964). Ogunkoya, Adejuwon and Jeje (1984) examined runoff coefficients for 15 third-order basins in south-west Nigeria. Wide variations between basins and over the hydrological year were found. Annual coefficients ranged from 1 to 40 per cent of rainfall. High-percentage runoff from a large storm over three Singapore catchments was indicated in § 4.1.

Variability of rainfall over different time periods (§§ 3.3, 3.6) influences streamflow variations on a daily, monthly, seasonal and annual basis. The factors discussed at the start of this section will, however, to a greater or lesser extent, reduce the magnitude compared with rainfall variations. For example, steep slopes on small upper catchments or lack of vegetation promote a rapid response to daily rainfall variations. Such streams will have considerable day-to-day fluctuations in streamflow compared to those in areas of dense vegetation. For the main Indus River, the ratio of total flow in any one year to the mean lies within a relatively narrow range (1.3–0.77), but for the month of May the range is much greater (1.57–0.43) (Gulhati 1968). When an even shorter period is considered, the variation is greater. The Beas discharge (Table 4.13) for example, in the last 10 days of September has fluctuated between 2000 and 300 m^3/sec. In West Africa, the Mayo Kebbi flows from an area of marked relief and therefore experiences considerable day-to-day discharge variations. The Black Volta catchment resembles the Mayo Kebbi in geology, vegetation and climate, but flows for most of its course through an extensive flood plain. The storage effect of the flood plain regulates variations in daily discharge (Ledger 1964). The tendency for sequences of dry or wet years to occur (§ 3.3) and the

possible existence of fluctuations and trends (§ 3.8) have obvious implications for streamflow.

Characteristics of tropical rainfall discussed in § 4.1, particularly the
occurrence of much of the total in a few large storms of high intensity,
have a great influence on rainfall–runoff relations. Much of the rain may
fall at rates in excess of the infiltration capacity of the soil and will tend
to be lost as immediate surface flow rather than building up ground water
producing dry period base flow. Therefore, many tropical streams have
very erratic discharges reflecting the occurrence of intermittent large
storms. Again, factors such as vegetation have a considerable modifying
influence. A rainstorm on a catchment of scanty vegetation will tend to
produce a greater total volume of streamflow with a more marked peak
than if it occurred over an area of dense vegetation. Bonell, Gilmour
and Cassells (1983) report widespread overland flow in undisturbed
tropical rainforest in north-east Queensland. Prevailing rainfall intensities frequently exceed the saturated hydraulic conductivity of the soil
profile below 0.2 m, creating rapid saturation of the top layer and hence
generation of overland flow (see above). Ledger (1964) contrasts the
influence on runoff of two major types of rainfall in West Africa.
Disturbance-line rainfall (§ 2.9), most of which is of high intensity, gives
much more violent runoff than the monsoon rain of lower intensity to
the south. Duration of rainfall is relevant in terms of the concentration
time of a catchment (§ 4.1). In addition, the longer the duration, the
greater the soil moisture and hence the lower the infiltration rate.
Therefore, the longer rainfall continues the greater will be the surface
runoff. Antecedent rainfall, which influences initial soil moisture and
hence infiltration rate at the start of a rainstorm, will also affect the
proportion of rain becoming surface flow.

The often limited spatial extent of tropical rainstorms and marked
gradients in intensity and amount (§ 4.2) mean that conditions over even
small catchments can vary considerably. This complicates rainfall–runoff
relations.

The main point emerging from the above discussion is the tendency
for streamflow in the tropics to be highly variable in time in response
to rainfall characteristics. This is true over various time scales. Rivers
are commonly very seasonal. Concentration of much rainfall in a few
large, intense storms produces considerable short-term streamflow fluctuations. Variability in rainfall on a monthly, seasonal or annual basis
creates streamflow variability. Periods of high discharge, with resultant
danger of flooding and soil erosion, alternate with spells of low flow and
water shortage. This has implications for hydro-electric power generation, irrigation, river and estuarine transport and water supply for
domestic and industrial use.

Flooding can be of an irregular short-term nature associated with individual storms. Examples from Singapore and Tanzania, resulting in substantial damage, were discussed in § 4.1. In this section it was pointed out that heavy falls resulting in flooding can occur in semi-arid areas, an example from North Africa being cited. Flooding can also be on a regular seasonal basis and in many areas is an essential part of the agricultural system, an aspect discussed in §§ 7.3 and 8.2. Even in such areas, periodic excessive flooding causes loss of life and serious damage to crops, communications and settlement. The *aman* (autumn) rice crop of Bangladesh depends on seasonal flooding. However, excessive flooding in 1974 covered 52 000 km^2 (36 per cent of the area of the country) for the first time in a quarter of a century, restricting the sowing of *aman*. Since the floods were 2 weeks early a quarter of the *aus* (summer) rice was also damaged. An estimated 947 580 t of cleaned rice, 255 000 t of fertiliser and 69 660 ha of jute were destroyed.

Irregular flows impose restrictions on hydro-electric power generation without provision of storage facilities. Rao (1968) indicates that dependable power generation along the peninsular rivers of India is not possible without substantial storage facilities for periods of high inflow. Unregulated discharge from Himalayan rivers is appreciable even during the critical winter months, providing some scope for hydro-electric power on a run-of-the river basis, but even small storages would produce a tangibly higher degree of firm power generation (Rao 1968).

Regulation of river flows to combat problems of flooding and create dependable water supply for a wide variety of purposes is obviously important in the tropics. Dams represent one answer but there are problems. The extremely variable nature of flows implies the need for large constructions and hence cost. The latter is a particularly serious problem because of lack of resources in developing countries. For this reason land-use and management practices are also particularly important as streamflow regulators, an aspect considered in § 9.3. However, there are many examples of dam construction, whether for a single purpose such as power generation or multi-purpose. They cover a wide range of scales, two notable examples being the Aswan High Dam and Volta Dam. Aspects associated with such developments are examined in § 9.6. This analysis of rainfall–runoff relationships forms a framework for later material, particularly that in §§ 9.3 to 9.6.

CHAPTER 5
EVAPORATION

5.1 INTRODUCTION

Evaporation, in terms of the hydrological cycle, is the conversion of water from the liquid or solid state into vapour, and its diffusion into the atmosphere. A vapour-pressure gradient between the evaporating surface and the atmosphere and a source of energy are necessary for evaporation. Solar radiation is the main source of energy, although as will be seen, under certain conditions, sensible heat from the air is important. Webb (1975) points out that although the general physical principles have been well understood for many years, in detail there are complexities which in some cases can cause serious errors in evaluating evaporation. Morton (1983) states that difficulties in assessment have led to the significance of evapotranspiration being underplayed in hydrology and meteorology. This chapter has several main concerns. In § 5.2, the aim is to indicate the complexity of factors influencing evaporation from various surfaces. As a result of this complexity, difficulties in estimating evaporative loss are considerable. To indicate this, a range of techniques is introduced in § 5.5. Within the scope of this chapter, detail is not possible, but a fairly wide variety of literature is indicated to allow more information to be obtained.

A key point about the tropics is that evaporative demand of the atmosphere (see below) is high and this, together with the rainfall characteristics discussed in Chapters 3 and 4, makes water a particularly critical resource. In § 5.3, general relationships between evaporation and rainfall in different areas are introduced using simple water balances to show the wide variety of water regimes which exert a great influence on human activity. The question of climatic classification is introduced in § 5.4. This chapter serves as an introduction to later material, particularly §§ 6.3, 7.2, 8.2, 9.3, 9.4.

5.2 FACTORS INFLUENCING EVAPORATION

The wide range of factors is discussed in many references such as Eagleson (1970), Monteith (1973), McIntosh and Thom (1969), Webb

(1975), De Bruin (1983), Brutsaert (1982), Morton (1983), Doorenbos and Pruitt (1977). Meteorological factors include solar radiation, the temperature of the evaporating surface, vapour-pressure gradient, wind and air turbulence. Solar radiation is the dominant source of energy and sets the broad limits of evaporation. Values of solar radiation tend to be high in the tropics (§ 5.3), modified by cloud cover, making the evaporative demand of the atmosphere considerable. The marked diurnal variation in solar radiation is important. It is, however, the net radiation (incoming − outgoing) which matters and hence the albedo of the surface and its temperature are significant. Furthermore, not all the net radiation is used for evaporation, some heating the surface and the air.

Air and surface temperatures, since they are related to solar radiation, would be expected to show a correlation with evaporation, although not so good as that of radiation. The surface temperature of water governs the vapour pressure at the surface, hence influencing the vapour-pressure gradient. The humidity of the air, the other factor influencing the vapour-pressure gradient, is also important.

With no air movement, air in contact with the evaporating surface would soon become saturated. Wind results in the movement of fresh air into contact with the surface. Within about 1 mm of the surface, upward movement of individual water molecules (molecular diffusion) is important, but above this, turbulent air movement (eddy diffusion) takes over. Turbulence varies with the vertical windspeed gradient. Spatial variations in these meteorological factors, particularly in areas of rugged relief and including cloud cover influence on solar radiation, may create considerable differences in evaporation.

In addition to the meteorological factors which create an evaporative demand of the atmosphere, the nature of the evaporating surface and availability of water are important. The surface itself will modify the meteorological factors by creating differences in net radiation due to albedo variations and turbulence because of surface roughness.

The basic surfaces can be considered as open water, bare soil and vegetation, evaporation from the latter being in the form of intercepted water and transpiration. Given an adequate water supply, contrasts in water loss from these surfaces are less than might be expected. Water quality, the depth, size and shape of a water surface affect the rate of evaporation. For example, in moving over a small water body, moisture content of the air and hence vapour-pressure gradient are little changed. The same is not so over a large water body and hence the 'average' evaporation will be less from a large water body than from a small one. This has implications for methods of assessment using evaporation pans and lysimeters (§ 5.5) and is also related to advection effects considered below. The influence of the shape of the water body can be seen in Fig. 5.1.

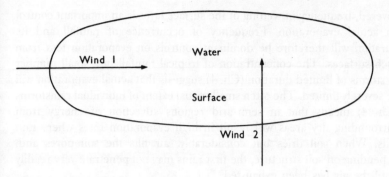

Fig. 5.1 Influence of the shape of a water surface and wind direction on evaporation: higher average evaporation for wind 2.

Advection of energy by air from outside an area may greatly increase evaporative demand. In arid and semi-arid regions, considerable energy may be advected from surrounding dry areas over irrigated zones or evaporation pans. Evaporation from the latter will then be much greater than if they were surrounded by moist conditions in a humid area. Advected energy may equal net radiation and methods of assessment of evaporation which ignore the former can be much in error for semi-arid areas. Transference of advected energy downwards to the evaporating surface is often termed 'the oasis effect' and transference to a crop by air blowing through it is termed 'the clothes-line effect'.

Oliver (1969) points out that marked oasis effects over areas as large as the cotton-growing areas of Gezira, Sudan, can create water losses up to twice those calculated using standard meteorological formulae (§ 5.5) which ignore advection. In the case of small irrigation schemes in tropical semi-arid areas, in particular, this is extremely important. Oliver (1969) indicates how moderately high windspeeds in the Sudan accentuate advection effects and also the significance of diurnal variations in windspeed, the latter tending to be lowest when rain most commonly falls.

Evaporation from bare soil is often less than from open water because water is not always freely available. Therefore, soil-moisture content is critical. In the case of saturated soils, evaporation can be greater than from an open-water surface due to differences in aerodynamic roughness. Soil colour, influencing albedo and hence net radiation, will affect evaporative demand.

Evaporation from bare soil decreases markedly once a shallow surface layer dries out. As a rough indication, Oliver (1969) cites surface layers of about 10 cm for clays and 20 cm for sands. Thus, in arid/semi-arid areas with much bare soil and sparse vegetation or in the early stages of growth of a crop when only a small proportion of the ground is covered, frequency of wetting of the surface layer is an important control on actual evaporation. Frequency of occurrence of rainfall and its duration will therefore be dominant controls on evaporation loss from such surfaces. The concentration of tropical rainfall in a small number of storms of limited duration (Ch. 4) suggests that actual evaporation will be severely limited. The often small spatial extent of individual rainstorms (Ch. 4) means that in semi-arid regions advection of energy from surrounding dry areas will result in high evaporation rates where rain falls. When soil dries out considerably, air fills the soil pores and, depending on soil structure, the first rains may not penetrate very readily until the air has been exhausted.

Storage of water at or near the surface will vary with soil type, Oliver (1965, 1969), for example, contrasting coarse-textured sands and gravels with clay soils in the Sudan. Because of the high infiltration capacity and permeability of the coarser soils not only is there little runoff, except under very intense storms, but rapid penetration of water to lower depths protects it from evaporation. The more impermeable clays not only create more runoff, but also since the water is held near the surface it evaporates at a high rate. In the Sudan, a given amount of rainfall on clay is considered to have an effectiveness only two-thirds of that on sands. Deeper-rooted trees and bushes can tap the water below the surface on sands and silts which therefore support more of this growth than clays. Shallow-rooted grasses are able to flourish in the brief rainy season on clays, but water is inadequate for deeper-rooted vegetation. Gentle slopes and depressions can lead to the concentration of surface water which increases the value of the limited rainfall in semi-arid areas (§ 8.3). However, on clay soils, the rate of evaporation of this water will be very high.

Transpiration of water taken in by plants occurs through leaf stomata. There is a resistance to movement of water through the soil and also through plant tissues. Kramer (1983) discusses the transpiration process, the factors controlling it, as well as its measurement. The plant factors influencing it include leaf area, orientation, size, shape, leaf surface characteristics, root–leaf ratio and stomata. The internal resistance (stomatal resistance) is the primary plant control, the degree of opening of stomata being influenced by various factors such as light, atmospheric humidity, temperature and water deficit. There is also a separate external resistance to molecular diffusion because of frictional drag of air over the leaf. It is argued by Penman and Schofield (1951) that transpiration will be less than evaporation because of the higher albedo of vegetation than water, stomatal resistance and stomatal closure at night. The albedo varies between vegetation types and seasonally, Bultot and Griffiths (1972) for the Congo for example, quoting values of 0.13 for equatorial forest, 0.23 for *Paspalum notatum*, and for savanna, 0.18 in the rainy and

0.23 in the dry season. Webb (1975) in reviewing a range of experiments cites results suggesting transpiration is roughly the same for green grass and a range of other types of vegetation, being about 0.6–0.8 of open-water surface evaporation. Other experiments however indicate that transpiration from forests may exceed that from grass by up to 20 per cent. While this difference is often attributed to the lower albedo of forest (i.e. a higher net radiation) than grass, Webb (1975) suggests this is not so important as the much greater evaporation from forest than grass when the foliage is wet. Webb (1975) indicates studies showing that forest evaporation can be several times greater when foliage is wet than when it is dry, a point also referred to in § 6.3 when discussing interception. For example, one study in Kenya indicated a 1 month evaporation total of 242 mm when the canopy was mostly wet, implying that on some days, evaporation must have exceeded 20 mm.

Other factors, however, have a marked impact upon the relation between open-water surface evaporation and water loss from a vegetation surface. The height and 'roughness' of the surface influence airflow turbulence which is a significant factor in evaporation. Williams and Joseph (1970) indicate the importance of total surface area of leaves, citing Hutchinson, Manning and Farbrother (1958) who show how estimated crop-water requirements of cotton follow the development of leaf area. Factors such as these mean that water use by a crop can at times exceed that of open-water surface evaporation. Hutchinson *et al.* (1958), for example, estimated peak values of transpiration by cotton of 1.5 times open-water evaporation with maximum leaf area development. Conversely, Van Bavel, Newman and Hilgeman (1967) show that stomatal resistance of orange trees greatly decreases rates of water use even when water is plentiful. Water requirements of crops will be discussed further in Chapters 6 and 7.

A further indication of the complex influence of the evaporating surface is presented by Cooley and Idso (1980) in assessing the effect of lily pads on evaporation from a tank in Arizona. They review previous studies indicating uncertainty as to whether vegetation on a water body increases, decreases or leaves unaffected the evaporation loss. This is relevant to the question of the effects of removing vegetation (e.g. reeds) from swamps (§ 9.4). Their experiment gave a reduction in evaporation due to the presence of pads which was matched by a similar reduction in measured net radiation (compared with an identical tank with no pads). Calculations also indicated that the reduction was due to changing radiative properties of the surfaces and was in agreement with results for a wooded Ontario swamp. Cooley and Idso (1980) concluded that the introduction of any kind of vegetation cover over an extensive water body would similarly reduce evaporation. However, they stressed that this applies to extensive areas since introduction to a limited area could result in an increase due to an 'oasis' effect (really a clothes-line effect – see above). Use of lily pads which lie flat on the surface means that they receive no more advected energy than the water. Hence, although the experiment was for a small area tank (2.1 m diameter) in a highly advective environment, it was felt that results were representative of an extensive area.

Evaporation from all surfaces, including plant transpiration, is commonly termed 'evapotranspiration'. If water is freely available then the term 'potential evapotranspiration' is used, defined by Penman *et al.* (1956) as 'evaporation from an extended surface of short green crop, actively growing, completely shading the ground of uniform height and not short of water'. Given these strict conditions, Penman argues that evapotranspiration will be primarily determined by climatic factors with some differences due to different albedos and stomatal resistance. In practice, these conditions are rarely met, the nearest equivalent being short, dense, grass surface. Vegetation will often not completely cover the ground, nor be short and of uniform height, these factors influencing available energy and turbulence. If the surface is not 'extended', particularly in drier areas, advection effects may be important. Under these different conditions, given that water is freely available, it is still possible to think of a 'potential' water use by a particular vegetation surface, but its value would differ from that under conditions specified by Penman.

Often, water is not freely available and this introduces the term 'actual evapotranspiration' or 'water use'. Water available to plants is generally considered to be that between field capacity and permanent wilting point, but between these two limits, water may not be equally available. Actual evaporation can then be much less than the potential value. This complex issue will be discussed in Chapter 6. Clearly, in arid and semi-arid areas, actual evaporation will often be considerably less than the potential value because water is not available.

Climatic factors therefore induce an 'evaporative demand' of the atmosphere, but the actual evaporation resulting will be influenced by the nature of the evaporating surface(s) as well as the availability of water. The surfaces will themselves influence evaporative demand by their albedo and surface roughness, the latter affecting turbulence. Much confusion arises in the literature because the conditions are not always defined, this aspect being considered for example by Ward (1971) and Venkataramana and Krishnamurthy (1973). Increasingly, there is a tendency simply to use the term 'evaporation' to denote loss from any surface. Differences in assessment of evaporation using various methods (§ 5.5) result not only from inherent weaknesses in the techniques but also because they are concerned with water loss under varying conditions.

One aspect, the significance of which will be considered in Chapter

7, is that while rates of water loss from different surfaces will vary, since they are determined by climatic conditions, the ratios between them at different places and times are often assumed to be constant. For example, evergreen forest in Kenya has been found to have a water-use value of 0.9 of open-water surface evaporation as estimated by the Penman method (§ 5.5) (Blackie 1965). This can be denoted as $E_t/E_o = 0.9$, where E_t = water use or requirement by the vegetation, E_o = open-water surface evaporation (Penman). Maize in the early stages of growth when it covers only a small portion of the ground has a ratio $E_t/E_o = 0.45$, but in the later stages of growth, the ratio reaches unity (Dagg 1965a). These ratios assume plenty of water, but if this is not freely available then actual values would be less.

This brief introduction indicates something of the complexity of factors influencing evaporation. More detailed treatments are contained in the references indicated at the start of the section.

5.3 EVAPORATION AND ITS RELATION TO RAINFALL

The large amount of energy available for evaporation in low latitudes is a dominant factor. Figure 5.2 shows how net radiation for the earth's surface (R) is much greater in low latitudes than in high. Potential evaporation is much more constant from year to year than is rainfall because of the small variation in key factors such as solar radiation. For example, Norman, Pearson and Searle (1984) cite a coefficient of variation for

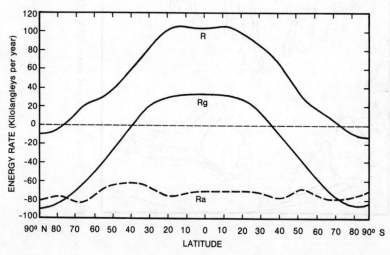

Fig. 5.2 Average annual latitudinal distribution of the radiation balances of the earth's surface (R), the atmosphere (Ra), and the earth–atmosphere system (Rg), in kilolangleys per year (after Sellers 1965).

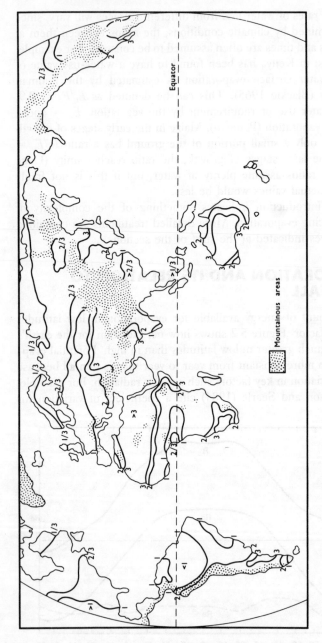

Fig. 5.3 Radiational index of dryness (*RN/PL*) (after Budyko 1956).

evaporation of only 5 per cent on an annual basis. The seasonal variation reflects variations in altitude of the sun, cloud cover, windspeed and humidity. Diurnal variations in evaporative demand in relation to diurnal variations in rainfall occurrence (Ch. 4) are also important. At night, not only is energy for evaporation much less, but Oliver (1965) points out that in the Sudan, for example, lower windspeeds and stable atmospheric conditions reduce turbulent diffusion of water vapour. Because of stomatal closure transpiration is minimal at night, being for some plants perhaps only 3–5 per cent of that during the day (Oliver 1969).

The high values, seasonal and diurnal variations only become meaningful when related to available water. Except for irrigation, the latter is related to rainfall although – as will be seen in Chapter 6 – this relationship, operating through the atmosphere–soil–plant water system, is very complex. Figure 5.3 shows the ratio RN/PL over the world, where RN is the net radiation available to evaporate water from a wet surface, PL is the latent heat of water vapour. Values > 1, where 'evaporative demand' exceeds rainfall, can be classified as 'arid' and values < 1 as 'humid' (Budyko 1956). Figure 5.3 shows that only a small part of the tropics has rainfall in excess of evaporative demand. This annual picture is misleading; seasonal variations in both evaporative demand and even more so in rainfall will in most cases produce periods when the latter is in excess of the former and vice versa. This is illustrated in Fig. 5.4 over Africa for the months of April and July where the difference between precipitation and potential evaporation is mapped (Davies and Robinson 1969). These maps also show the marked spatial contrasts as well as the small proportion of the continent having positive values in any month. On an annual basis, only 10 per cent of Africa has positive values.

Seasonal variation in rainfall and evaporation can be illustrated by means of simple water-balance studies of the kind used by Thornthwaite and Mather (1955). More complex forms and their applications will be considered in Chapter 8. When rainfall exceeds potential evaporation, soil-moisture reserves are recharged. When soil-moisture capacity is reached (a value varying with soil type as well as with rooting characteristics), any further rainfall is classed as surplus, either surface runoff or drainage beyond root range. This is an oversimplification, since surface runoff can occur if rainfall intensity exceeds infiltration rate or with surface saturation (§ 4.4).

When rainfall is less than potential evaporation, soil-moisture reserves are utilised. Once soil moisture drops below field capacity, however, water may not be freely available (Ch. 6). Thornthwaite and Mather (1955) assume that evapotranspiration is proportional to available water.

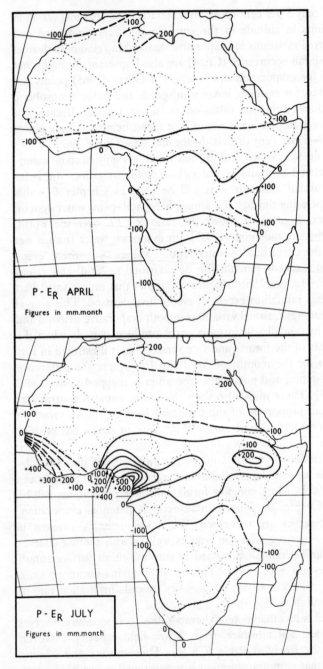

Fig. 5.4 Africa: $P-E_R$ April and July (after Davies and Robinson 1969).

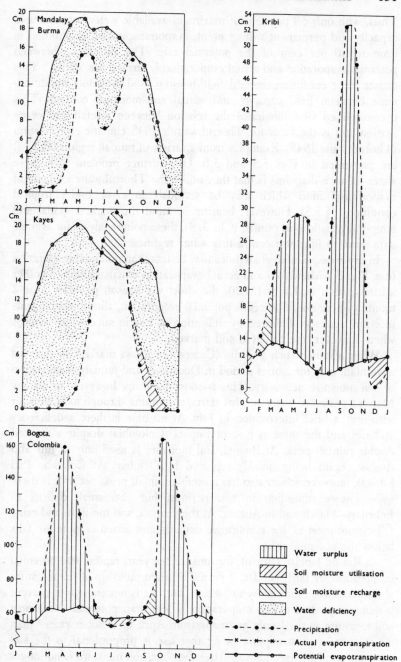

Fig. 5.5 Water balance: Mandalay, Kayes, Kribi, Bogotá (from Thornthwaite Associates, Publications in Climatology 1962, 1963, 1965).

Thus, with only 50 per cent of maximum available water between field capacity and permanent wilting point, evapotranspiration is assumed to drop to 50 per cent of the potential rate. The difference between potential evaporation and actual evaporation is termed the 'deficit'. This causes some confusion since 'deficit' is often used to denote the difference between field capacity and actual soil-moisture content. This oversimplified view illustrates the relation between rainfall and evaporation and is the basis for Thornthwaite's 1948 climatic classification (Thornthwaite 1948). Examples from a variety of rainfall regimes (Ch. 3) are presented in Figs 5.5 and 5.6. One further problem with these water-balance diagrams is that they utilise the Thornthwaite estimate of evapotranspiration which may be seriously in error under tropical conditions (§ 5.5). However, bearing in mind the marked spatial and temporal variations in rainfall (Chs 3, 4), these point analyses for average data serve to illustrate contrasting water regimes.

In dry areas, rainfall at all times may be less than potential evaporation (e.g. Fig. 5.5 Mandalay) and actual evaporation equals rainfall. At other stations, such as Kayes (Mali), the short wet season provides 1 or 2 months where rainfall exceeds potential evaporation, allowing some soil-moisture recharge followed by utilisation of this in succeeding months when, however, the deficit is still marked.

While a station such as Kribi (Cameroon) shows marked fluctuations in rainfall, only for a brief period in December and January is utilisation of soil moisture necessary. This is soon made up, however, so that by March a period of water surplus starts, continuing through to November with only a brief interruption in July. At no time is there a deficiency at Kribi and the same is true of Bogotá (Colombia) despite a marked double rainfall peak. At Bogotá, soil moisture is used only in July and August, again being quickly replaced by October. At Colombo (Sri Lanka), however, which also has a double rainfall peak, not only is there soil-moisture utilisation in two periods, but also small deficits in February–March and in August. At these times, soil moisture and rainfall cannot meet all the evaporative demand and actual evaporation falls below the potential value.

At Rio de Janeiro (Brazil), for most of the year, rainfall and potential evaporation are fairly similar. A period of slight build-up of soil moisture about March is followed by utilisation, but this is not enough to prevent a deficiency when actual evaporation is less than potential. Note how soil moisture is used only gradually from April to October rather than all at once, the assumption being that use is proportional to the soil moisture available (see above). After October, soil moisture is no longer available and the actual evaporation equals rainfall. At a station such as Rio, year-to-year variations in rainfall will considerably affect the periods of soil-moisture recharge, utilisation and deficit.

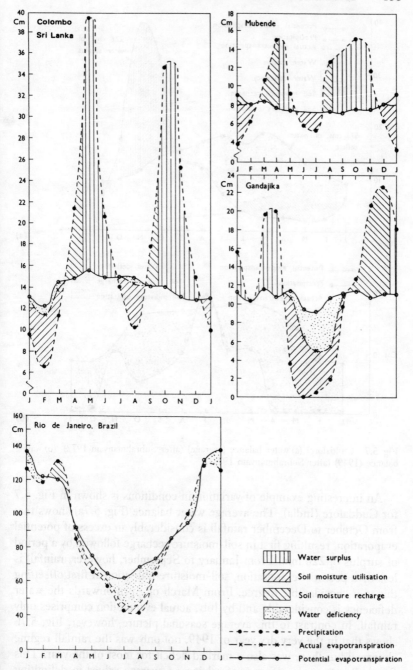

Fig. 5.6   Water balance: Colombo, Rio de Janeiro, Mubende, Gandajika (from Thornthwaite Associates, Publications in Climatology 1962, 1963, 1965).

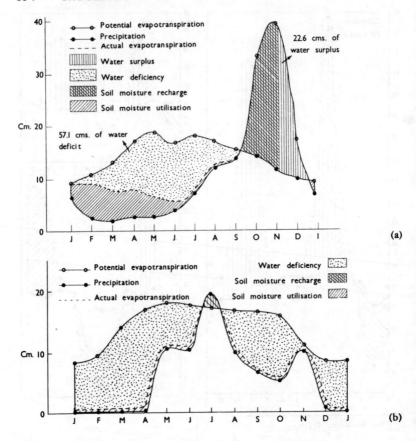

Fig. 5.7 Cuddalore: (a) water balance (average) (after Subrahmanyam 1972). (b) water balance (1949) (after Subrahmanyam 1972).

An interesting example of variation in conditions is shown in Fig. 5.7 for Cuddalore (India). The average water balance (Fig. 5.7a) shows that from October to December rainfall is considerably in excess of potential evaporation, resulting first in soil-moisture recharge followed by a period of surplus of 226 mm. From January to September, however, rainfall is less than potential evaporation, soil-moisture utilisation at first offsetting this to a considerable degree. From March or April onwards the water deficiency is considerable and by July, actual evaporation comprises only rainfall. In contrast to this average seasonal picture, however, Fig. 5.7b shows that in the very dry year of 1949, not only was the rainfall regime abnormal but there was also a marked deficit for most of the year. This figure illustrates the dangers of relying on average values in delimiting seasonal conditions.

Mubende (Uganda) and Gandajika (Zaire) illustrate the possible

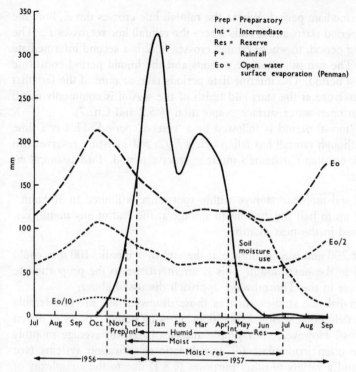

Fig. 5.8 Songea: water-balance diagram.

complexity of the seasonal water regime. At times, following soil-moisture recharge, there is a considerable surplus. This is followed by periods of soil-moisture utilisation, and particularly in the case of Gandajika by a water deficiency from May to August. Such conditions, where at certain times of the year there is a surplus, perhaps resulting in flooding and soil erosion, and a deficit at other times, are typical of considerable parts of the tropics. The graphs in Figs 5.5, 5.6 and 5.7 relate rainfall and evaporation over periods of a month. For monthly periods, a statement that potential evaporation exceeds rainfall is correct. However, for the short time that rain is actually falling, rain will generally exceed evaporation, hence being available for surface runoff or penetrating the surface and perhaps becoming ground water which is unavailable for evaporation (§ 4.4).

An alternative method of depicting rainfall–evaporation relations has been used by Cochemé and Franquin (1967) in agrometeorological investigations of parts of West Africa. An illustration of their approach is shown in Fig. 5.8 for an East African station. The period when rainfall lies between $E_o/10$ and $E_o/2$ is the preparatory period when soils may be prepared for planting. When rainfall is between $E_o/2$ and E_o is the intermediate period. Where the rainfall line crosses the E_o line, the humid period starts and it ends where the rainfall line recrosses E_o. The following period, to where rainfall crosses $E_o/2$ is a second intermediate period. The two intermediate periods and the humid period constitute the moist period. The intermediate periods take account of the fact that crop-water use at the start and finish of the season is commonly much less than open-water surface evaporation (§ 5.2 and Ch. 7).

The 'moist' period is followed by a 'reserve' period. This is a time when although rainfall has fallen below $E_o/2$, soil-moisture reserves can be used, creating Cochemé's moist + reserve period. Two assumptions are made:

1. That soil-moisture storage within root range is limited to 300 mm.
2. That up to half the amount in storage at the end of any month can be used in the next month.

Thus, if 200 mm was in storage at the end of 1 month, 100 mm could be used in the next month. This is an alternative to the proportionate rate of use in the Thornthwaite approach discussed above.

Water-balance studies such as those discussed are of considerable value in delimiting seasonal conditions, growing periods, times of deficit and excess. However, such simple analyses based on average monthly data are often inadequate for agroclimatic classification systems (see below) and a variety of other purposes (§ 8.1) due to the complexity of the system (§ 6.3).

5.4 CLIMATIC CLASSIFICATION

Discussion of rainfall–evaporation relationships raises the question of climatic classification. The idea of using rainfall alone was introduced in § 3.2 and rainfall regimes are related to broad agricultural systems in § 8.3. However, the water supply–demand situation is important (§§ 3.2, 3.4, 5.3), complexities being indicated in § 6.3. Two useful overviews of climatic classification are provided by Oldeman and Frere (1982) and ICRISAT (1980).

Purposes of classification

The significance of agroclimatic classification is discussed by Oldeman and Frere (1982). Frere (1980), on the basis of papers presented in ICRISAT (1980) suggests three lines of classification depending upon objectives:

1. To differentiate broad climatic types using mean monthly and annual data.

2. To evaluate agricultural potential. Clearly climate will only be one aspect of evaluation. Data used will again often be monthly averages but include derivation of extra parameters such as length of rainy season, crop-growth days, biomass production, etc.
3. Classifications for technology transfer. In order to modify and transfer land and water management and cropping systems principles, an understanding of the climatic limitations of different regions is necessary. Data need to suit particular purposes and are likely to be more specific than with the previous two categories. Analysis for periods less than a month (e.g. 5–10-day periods), taking variability of factors such as rainfall into account and how this relates to variations in agricultural production, is likely. Such approaches are important in extrapolation of scientific experiments between areas. For example, Cochrane and Jones (1981) discuss the location of representative sites for testing new grass and legumes and the definition of ecologically comparable regions for the development and transfer of pasture technology in tropical South America.

Problems in classification

Problems with climatic classification for a particular purpose and environment are considerable. Spatial and temporal variations in variables present difficulties. Numbers and distribution of stations will often be inadequate to represent spatial variations. The problem of scale needs to be addressed and inevitably maps will be generalised. However, since boundaries represent transition zones and there are variations from year to year (§ 3.2) such problems are not perhaps so great as they first appear. Variability in, say, rainfall means that it may be more appropriate to define a probable year rather than an average one (Meher-Homji 1980). Reddy (1983a) suggests that classifications based on average monthly, seasonal and annual data have limited practical application in agricultural production studies. It may also be necessary to represent the sequence of climatic events during the year, for example using a cumulative water balance.

Approaches to classification

Traditional approaches assign ranges of values to particular variables. Ranges are decided upon in various ways, many earlier systems being based on a relation with vegetation (Gadgil and Joshi 1983). Stations are then assigned to a particular class. Such an approach is termed divisive (Grove 1980). An alternative, receiving attention because of availability of computing power, is the agglomerative approach (Grove 1980, 1985). Variations in climatic factors (e.g. monthly values to represent seasonal patterns) are used to obtain groupings of stations showing similar characteristics by means of techniques such as Principal Components Analysis (Gadgil and Joshi 1983; Grove 1985). Since the approach does not begin with predetermined ranges in variables it may be of limited use in, for example, assessing suitability of an area for a particular crop. However, since it groups 'similar' stations, it is useful in, say, the transfer of technology. Gadgil and Joshi (1983) found that climatic groupings so obtained for India did correspond closely to the vegetation distribution.

Although concerned with the semi-arid tropics, various papers in ICRISAT (1980) considered a wide range of approaches. A number of the papers stressed the need to take rainfall probability into account and the value of water-balance approaches (e.g. Virmani, Sivakumar and Reddy 1980; Krishnan 1980; Sarker and Biswas 1980). Grove (1985) discusses a range of indices used to define arid environments, including Budyko's radiation index of dryness, RN/PL (§ 5.3). Gadgil and Joshi (1983) used an agglomerative approach on Indian data for monthly rain, a moisture index (the ratio of precipitation to potential evaporation) and minimum temperature. Gadgil and Joshi (1980) also used an agglomerative approach for pentade rainfall in India. Oldeman and Frere (1982) refer to a number of systems based on either monthly rainfall, rain and evaporation, or rain, evaporation and soil moisture. They review a system developed by FAO to define agro-ecological zones which is global in nature and has been applied to Africa, West Asia, South-East Asia, South and Central America. The system makes use of length of growing season when water availability allows crop growth and temperature is non-limiting. The climatic maps are superimposed on soil maps to delimit agro-ecological zones. Oldeman and Frere (1982) also discuss an agro-climatic classification for rice-based cropping patterns. They use a water-balance approach to demonstrate the validity of a wet month defined as having more than 200 mm and a dry month with less than 100 mm as criteria to delimit climatic classes.

In a series of four papers, Reddy (1983a,b, 1984a,b) discusses agro-climatic classification for the semi-arid tropics. In the first paper, computation of classificatory variables is discussed (Reddy 1983a). The method is a modification of that of Cochemé and Franquin (§ 5.3) and Hargreaves. Weekly data on rainfall and potential evapotranspiration are used to define an 'available effective rainy period', wet and dry spells and likely percentage of crop failure years. In the second paper (Reddy 1983b), a range of agroclimatic variables is identified and their agronomic significance investigated in the third paper (Reddy 1984a). A soil-water balance simulation and agronomic data are used in the analysis. In the final paper (Reddy 1984b), analysis of variables already defined for India, Senegal and Upper Volta uses divisive and agglomerative techniques.

The above discussion indicates the varied purposes and problems associated with classification. This is reflected in the range of approaches, none being universally applicable. Therefore, within the broad survey of the tropics presented in this book it was felt that a simple classification based on rainfall regimes (§ 3.2), despite its limitations, was appropriate. References cited above will serve as a guide to approaches suiting a range of needs and problems.

5.5 METHODS OF ASSESSMENT

Problems of assessment of evaporation and crop-water use are not confined to the tropics. However, the high evaporative demand in the tropics and importance of water resources make accurate assessment essential to an evaluation of the potential of an area, lengths of growing seasons, periods of excess and deficit and irrigation needs. Unfortunately, the complex set of factors influencing evaporation rates means that assessment faces great problems. Here the aim is to indicate something of the very wide range of approaches. This is more useful than discussing a small number in greater detail. References are given to allow particular techniques to be followed up. Where methods involve direct measurements of water loss, replication of real-world conditions is difficult. Where evaporation is indirectly assessed, measurement problems lead to certain factors being ignored or inaccuracies in their determination or the use of questionable assumptions. The fact that there is a large number of methods is some indication that they all have limitations. In the tropics, sparse meteorological networks, human resources and financial limitations increase the problem. Because of this, de Bruin (1983) stresses the need to pay special attention to simplification of existing methods so that a minimum of data is required.

Most methods of assessment can be grouped under the following headings.

1. *Direct measurement* using evaporation pans, atmometers and lysimeters.
2. *Meteorological formulae*
    (a) aerodynamic methods;
    (b) energy budget methods;
    (c) combination methods (a combination of (a) and (b));
    (d) empirical formulae.
3. *Moisture/water budget methods.*

However, a number do not fall so readily into these groups, including the use of porometers, tritiated water as a tracer, complementary approaches and the possible value of remote sensing data from satellites. A brief mention of these approaches is made after the above groups are considered.

Overviews of methods, with bibliographies, should be referred to for more details. For example, Webb (1975) briefly indicates a wide range of methods before discussing physically based ones (aerodynamic, energy budget, combination – see below) in more detail. Refinements and complications due to vegetation canopies, spatial variations in the nature of the surface are discussed. The measurement of the variables needed is discussed, together with a review of experimental investigations with some emphasis on forest cover. A particularly important review of the shortcomings of a range of methods is presented by Morton (1983) who then goes on to discuss the value of complementary relationship approaches (see below) to estimate areal evapotranspiration. Since it focuses on the humid tropics, a discussion by de Bruin (1983) is valuable. The focus is on the Penman (combination) approaches, developments/simplifications but reference is also made to water balance and complementary approaches to estimation when water supply is limited. Tomar and O'Toole (1980a) concern themselves with assessment for wetland rice conditions. They review a range of approaches before presenting details of a microlysimeter (see below). Other major reviews include Doorenbos and Pruitt (1977), Brutsaert (1982) and Kotoda (1986).

1. Direct measurement

Atmometers measure water evaporated from surfaces such as porous filter papers (Piché type), porous porcelain spheres (Livingstone type) or plates (Bellani type). All are oversensitive to windspeed. They do not indicate evaporation from an open-water surface and are generally only of value in relative comparisons between identical instruments exposed similarly. Chang (1968) however, indicates how they can be used to give results of more practical value. Webb (1975) indicates the possible value of pairs of atmometers (unshaded/shaded or mounted at different heights) in assessing net radiation and evaporation. Kyaw Tha Paw and Gueye (1983) demonstrate a relationship between a Piché instrument and potential evapotranspiration from an individual leaf.

Evaporation pans are a very useful field method, being relatively inexpensive and easy to handle. The amount of evaporation from the pan must be adjusted to take any rainfall into account. Simple as the concept is, the evaporation varies considerably with pan size, depth, material, colour, exposure and water level. Even the brand of paint used has been found to affect evaporation (Dagg 1968b).

Size of the pan is important, especially in drier areas, and experiments suggest that while a pan of 3.7 m in diameter will produce almost the same rate of evaporation as a 730 ha lake, a pan of only 0.46 m would have a value 50 per cent higher. Exposure, material and colour influence energy available for evaporation. The general effect of the various factors is to make pan evaporation usually greater than that from a large open-water surface. A correction factor (pan coefficient) must be applied to estimate open-water surface evaporation, but unfortunately these coefficients are not constant for a particular pan type, tending to show a seasonal variation and to be affected by exposure. Furthermore, ratios between evaporation from different pan types are not constant and hence comparisons using dissimilar types are dangerous. Many types of pan are in use but the US Weather Bureau class A pan is perhaps the most widely adopted. Webb (1975) raises the possibility of using pan records to estimate incoming radiation.

Advection effects (§ 5.2) are not serious in a humid area, but in an arid area they can result in the pan seriously overestimating evaporation from a large water surface or large irrigated area. In central Iraq, Wartena (1959) found that a class A pan located outside a field of irrigated rice had a daily evaporation rate of about 16 mm, while a pan inside the field had a rate of only 5 mm. Readings from the pans were realistic since the potential maximum evaporation rate would decrease as air accumulated moisture over the irrigated area. The difference does indicate the problems involved in location of pans and that in arid regions a 'guard' area, composed of, say, well-watered grass around the pan is necessary. The fact that pans to some extent integrate effects of advected energy in dry locations means that they are advocated by some for such situations (Tomar and O'Toole 1980a).

Other problems also exist but enough has been said to indicate that pan results must be treated with considerable caution. Pan data can be used not only to estimate open-water surface evaporation but also evapotranspiration and crop-water use by the application of other correction factors. For a type of raised pan in Kenya, the correction factor for mature pine, cypress and bamboo forest was 0.86. Using a different type of pan in South Africa, factors for sugar cane were as follows:

Young, 0.35; Mature, 1.40; Average, 1.09 (Chang 1968).

Such factors, which will vary with pan type, have practical value in estimating crop-water use and irrigation needs. For wetland rice, a ratio of 1.2 for the season is suggested by Tomar and O'Toole (1980a). A comparison was made between elevated class A pan evaporation in a wetland rice location and estimates of crop use based on regression relationships using pan evaporation data from a standard grassed dryland surface (IRRI 1983). Values were virtually identical.

Lysimeters measure evaporation from a soil or vegetation surface. They consist of a tank(s) containing soil and vegetation. The approach is a water-balance one where

rainfall + irrigation = evaporation + drainage + soil-moisture storage change

With the drainage type, soil moisture is assumed constant and drainage from the tanks is collected and measured. Here:

evaporation = rainfall + irrigation − drainage

Assumption of constant soil moisture necessitates the soil water being kept at field capacity by irrigation if necessary and here there are problems over short time periods. The drainage lysimeter is only really useful to determine potential evaporation over long time periods.

To assess actual evaporation and for short-period measurements, the weighing lysimeter is used, changes in soil-moisture storage being detected by weighing the tank. Weighing presents problems with large instruments but recent hydraulic methods have helped. The tank rests on a flexible water-filled cushion connected to a manometer. Changes in the water level of the manometer monitor weight changes.

Despite the simple principle, there are many practical problems, including those associated with evaporation pans such as advection effects and size. The tank must be large enough to avoid restricting root development, and physical conditions such as soil texture and structure inside the lysimeter must be comparable with those outside. The lysimeter crop must be the same height and density as the surrounding areas to avoid different surface roughness and area exposed to radiation and wind. In arid/semi-arid areas, the size of the 'guard area' needs to be considerable. The high cost and immobility generally preclude the use of lysimeters as routine field instruments. They are primarily a research tool to check other methods, establish E_t/E_o ratios and crop-water use. Lysimeter design principles are discussed in a variety of texts (e.g. Chang, 1968) and references cited earlier in this section. Tomar and O'Toole (1980a, b) present a microlysimeter for use with wetland rice. Instead of weighing to determine soil-moisture change, the change in submerged water level is measured. Other examples of particular designs are presented by Dagg (1970), Heatherly, McMichael and Ginn (1980) and Sammis (1981a). Related in principle to the lysimeter approach is the use, given suitable conditions, of plot experiments to assess water use. On uniform, level plots from which no surface flow occurs, water use can be assessed using measurements of rainfall and soil-moisture change within root range when there is no drainage loss.

2. Meteorological formulae

(a) Aerodynamic methods

These consider the factors influencing removal of water vapour from the evaporating surface; the vertical gradient of humidity and the turbulence of the airflow (§ 5.2). Turbulence can be assessed by measuring the vertical gradient in windspeed. Equations derived, such as that of Thornthwaite and Holzman (1939), make assumptions not always valid and all such approaches need accurate, frequent readings at two heights. Since such data are not usually available this method is not for routine use.

Webb (1975) presents a useful discussion of the bulk aerodynamic approach using windspeed at only one height to estimate evaporation from lakes and including refinements of the basic methods. Measurement problems preclude use over vegetation surfaces. This approach, together with the energy budget approach (below) have been used by Omar and El-Bakry (1981) to estimate evaporation from Lake Nasser.

Another aerodynamic method is the eddy correlation technique. This needs very detailed simultaneous measurements of fluctuations in vertical wind velocity, air density and specific humidity. Theoretically the technique is very sound but there are considerable instrumentation problems. As well as providing estimation over short time periods, other information relevant to plant studies is produced (Tomar and O'Toole 1980a). Lloyd *et al.* (1984) discuss a microprocessor-controlled data acquisition and processing system which calculates fluxes of heat, water vapour and momentum using the eddy correlation approach.

(b) The energy budget approach

If the energy available for evaporation can be determined, then the value of evaporation can be calculated. For an area, the following equation can be written:

$$Q_n + H = S + A + E + C_h + Ph$$

where Q_n = net radiation = incoming − reflected − terrestrial radiation; H = horizontal advection of sensible and latent heat; S = heat flux (flow) into the soil; A = heat flux into the air; E = evaporation energy; C_h = heat storage in the crop; and Ph = photosynthesis; C_h and Ph are small compared to the other terms and are ignored.

S is commonly measured for land surfaces or for a waterbody evaluated from changes in water heat content (Webb 1975). Alternatively, since again S is often small, it is ignored. Advected energy can be considerable in arid/semi-arid areas (§ 5.2) but since it is difficult to evaluate, most methods ignore this factor. The original equation reduces to

$$Q_n = A + E$$

Despite this simplification, measurement problems exist and certain assumptions must be made in order to evaluate the proportion of Q_n used for evaporation.

(c) Combination methods

The widely used Penman method includes aerodynamic and energy budget terms in the equation. It attempts to eliminate the measurement problems of both (*a*) and (*b*). Useful discussion of the basic concept, together with developments, simplifications and limitations is presented by Webb (1975), Stewart (1983) and de Bruin (1983) among others. Confusion exists because there are different forms of the original equation, which has been modified by various workers. Indeed de Bruin (1983) cites the confusion as a drawback of the approach. Some forms are concerned with open-water surface evaporation, but others, by the inclusion of various factors, are concerned with assessing evaporation from a short crop. The general form of the equation is

$$E = \frac{\triangle \, Q_n + \gamma \, E_a}{\triangle + \gamma}$$

where Q_n = net radiation, expressed in evaporation units; E_a is a function of windspeed at a height of 2 m, the saturated vapour pressure *at mean air temperature*, and the actual vapour pressure at mean air temperature; $\triangle$ = the slope of the saturated vapour pressure temperature curve (Fig. 5.9); and γ = the psychrometric constant.

The constants used to determine Q_n and E_a are empirically assessed and have been modified by various workers. Alternatives to the original formula include the Penman–Monteith form whereby a canopy or surface

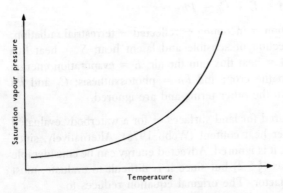

Fig. 5.9 Saturated vapour pressure–temperature curve.

METHODS OF ASSESSMENT 165

resistance is introduced to characterise transfer of water vapour between stomata and the atmosphere (de Bruin 1983), a special form of the Penman equation to estimate potential evapotranspiration from a reference grass cover and the use of crop factors (Doorenbos and Pruitt 1977). The Priestley–Taylor simplification is based on net radiation (Webb 1975; de Bruin 1983), a further simplification for the humid tropics being proposed by de Bruin (1983). Stewart (1983) discusses relationships between different forms of the combination equation, including assumptions and input needs. He concludes that choice of a form for a particular use is a compromise between data needs and physical reality.

Thompson (1982) compared accurate estimates using a range of meteorological and crop physiological values·in the Penman–Monteith equation with those from much simpler concepts based on the same formula but suitable for routine use. Bachelor and Roberts (1983) used this equation to estimate transpiration from wetland rice and found good agreement with estimates using the Penman equation and locally derived crop factors. After using change in soil moisture to assess evaporative loss from various Indian crops, Wallace, Batchelor and Hodnett (1981) then used the values in the Penman–Monteith equation to assess surface conductances.

De Oliveira, Da Mota and Da Silva (1980) derive linear regression equations relating Penman evaporation to altitude, latitude and minimum distance to the Atlantic Ocean in Brazil. The equations could then be used to estimate potential evaporation at any point. Monteny et al. (1981) derived a regression relationship between net radiation and solar radiation in two regions of the Ivory Coast. They used this to estimate daily evapotranspiration as a function of daily solar radiation and also suggest that the relation could be used in the Penman formula. Wind functions in Penman equations are discussed by Stigter (1980).

(d) Empirical formulae methods

There are many of these, using various meteorological parameters. They are not based on physical principles but on observed relationships between evaporation (e.g. from an evaporation pan) and one or more meteorological variables. Since these relationships have often been derived in one location, they may be unsatisfactory elsewhere. Doorenbos and Pruitt (1977) discuss several methods and Tomar and O'Toole (1980a) refer to attempts to calibrate methods against actual field measurements for rice. Two examples are the Thornthwaite method and the Blaney–Criddle formula.

The Thornthwaite method utilises only mean monthly temperature together with an adjustment for the day lengths of the months. It is therefore popular. While it appears to work well in temperate continental

North America where it was derived, it seems to be less satisfactory in some other areas.

The Blaney–Criddle formula is

$$U = KT\,p$$

where U = monthly consumptive water use by the crop; K = a crop coefficient; T = mean monthly temperature; and p = monthly percentage of day-time hours in the year.

By including a crop coefficient, the significance of crop type is recognised. K increases with height of crop and completeness of ground cover. Duru (1984) points out that dependence of the Blaney-Criddle equation on temperature is a disadvantage in countries such as Nigeria where this factor is relatively constant. He adopts an earlier equation, the Blaney–Morin, which incorporates a humidity factor which varies widely in Nigeria. Constants for the equation are derived using class A pan values with a conversion factor of 0.7. Values were compared with those derived from the Penman approach and Duru (1984) suggests better accuracy and consistency than the latter under Nigerian conditions.

3. Moisture/water budget methods

The water balance of a stream catchment or lake can be examined:

$$E_t = R - Q + \triangle S + \triangle G + L$$

where E_t = evapotranspiration from the catchment; R = rainfall; Q = streamflow; $\triangle S$ = water storage change within root range; $\triangle G$ = water storage change beyond root range; and L = outflow/inflow other than past streamflow measurement points.

The approach is also used to examine the impact of land-use changes on the hydrological cycle. It will be examined in more detail in Chapter 9. Accurate assessments of R and Q are necessary and because of problems in assessing $\triangle S$ and $\triangle G$ it is desirable to adopt a yearly period such that they are negligible, the end of the dry season being a suitable start and finish. In its simplest form, where $\triangle S$, $\triangle G$ and L are assumed to be zero, then, $E_t = R - Q$. Reddy (1983) refers to a range of meteorological water-budget approaches to estimate evapotranspiration before deriving a model to estimate water loss in fallow and cropped situations. Estimated evaporation and soil moisture storage compared well with observed values.

Sill, Fowler and Lagarenne (1984) present a water-vapour-budget approach to lake evaporation estimation. Upwind and downwind horizontal water-vapour flux in the atmosphere for air moving over a water body is measured, the difference equalling surface evaporation.

Other approaches

The above discussion indicates something of the wide range of approaches. Some techniques do not readily fall into any classification. The complementary relationship for estimation of areal evapotranspiration is discussed in detail by Morton (1983) with some reference to the approach by de Bruin (1983). The basis is a complementary relationship between areal and potential evaporation based upon the interaction between evaporating surfaces and air moving over them.

The use of tritiated water as a tracer in the measurement of transpiration and biomass in a tropical pine plantation is discussed by Sansigolo and Ferraz (1982). Azam-Ali (1983) discusses the use of a porometer to measure stomatal resistance in millet together with measurement of other crop and environmental variables to assess water use. One advantage is that evaporation can be estimated over periods of a day or less. Stomatal resistance in two tropical forest species was measured using a porometer by Whitehead, Okali and Fasehun (1981). They then used the Penman–Monteith equation to calculate transpiration using the stomatal conductance values and environmental variables. Kramer (1983) presents a useful discussion on porometers. Webb (1975) and de Bruin (1983) refer to the possible future value of remote sensing as an aid to evaporation assessment.

Some comparisons

The problems associated with assessment can be illustrated using a variety of case studies. Figures 5.10 and 5.11 compare estimates using different methods at two contrasting sites in East Africa (Blackie 1965). While the Thornthwaite and Blaney–Criddle methods define potential evaporation from an extensive, evergreen, well-watered surface and the rest from an open-water surface, it is generally accepted that rates from the two surfaces are of similar magnitude and directly related (Blackie 1965). However, Figs 5.10 and 5.11 show that estimates differ widely and furthermore, differences are not consistent at the two sites. The Penman and pan methods agree reasonably through the year at Kericho but only in the wet season (April–July) at Atumatak. Blackie (1965) shows that pan and Penman values differ more with decreasing altitude. He attributes this to advection effects at hotter, drier, low altitudes which affect pan rates but are not taken into account by the Penman method. For Katumani, Kenya, Porter (1984) found good correlation

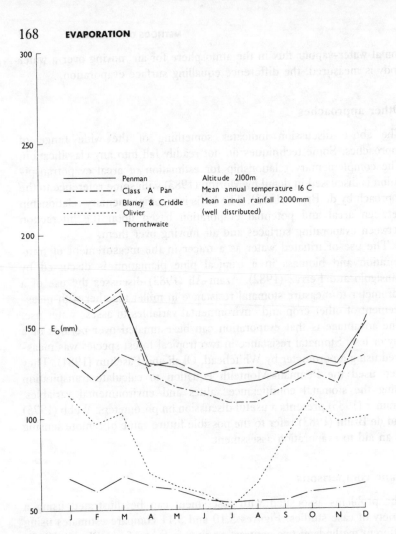

Fig. 5.10   Kericho: mean monthly evaporative demand (after Blackie 1965).

between open-water surface Penman values and class A pan measurements ($R^2 = 0.95$). However, high pan values were underestimated by high Penman estimates. On a seasonal basis, Penman underestimated pan values during the warm dry high solar radiation periods by about 5 per cent and overestimated pan values during low sun and rainy periods by about 7 per cent. On an annual basis the pan value was about 0.98 of the Penman value.

A comparison between potential evaporation (E_t) from mature rainforest at Kericho, assessed using the water-balance approach and rates using other methods (E_o) is shown in Table 5.1. The Penman method gives the most consistent results in terms of E_t/E_o ratios. On energy-

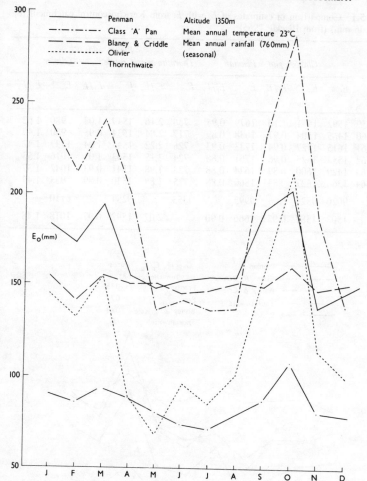

Fig. 5.11 Atumatak: mean monthly evaporative demand (after Blackie 1965).

balance considerations, ratios for the Thornthwaite and Olivier methods (2.1 and 1.5, respectively) would seem to be seriously in error. Figure 5.12 compares potential evaporation (E_t) from mature Kikuyu grass, determined by lysimeter with E_o values from other methods. Again, the Penman method appears to be most directly related to E_t values with a fairly consistent E_t/E_o ratio, averaging 0.75.

For a number of Egyptian stations, Omar and Mehanna (1984) compared potential evaporation values calculated from Blaney–Criddle, radiation, pan evaporation, Penman (adjusted constants for Egypt), and Penman (adjusted by Doorenbos and Pruitt (1977)). The comparisons demonstrated that relationships between the different methods are not

Table 5.1 Comparison of estimates of E_o with E_t from Kericho control catchment (E_t and E_o in mm) (from Blackie 1965)

Water year	Class A pan			Penman		Thornthwaite		Blaney and Criddle		Olivier	
	E_t	E_o	E_t/E_o	E_o	E_t/E_o	E_o	E_t/E_o	E_o	E_t/E_o	E_o	E_t/E_o
1958–59	1592	1441	1.10	1679	0.95	738	2.16	1549	1.03	980	1.62
1959–60	1465	1504	0.97	1638	0.89	717	2.04	1534	0.95	980	1.49
1960–61	1615	1727	0.94	1733	0.93	726	2.22	1547	1.04	1137	1.42
1961–62	1553	1625	0.96	1761	0.88	724	2.15	1543	1.01	1016	1.53
1962–63	1429	1600	0.89	1624	0.88	723	1.98	1541	0.93	1042	1.37
1963–64	1366	1627	0.84	1560	0.88	725	1.88	1540	0.89	955	1.43
Total	9020	9524		9995		4353		9254		6110	
Mean	1503	1587	0.95	1666	0.90		2.07	1542	0.97	1018	1.48

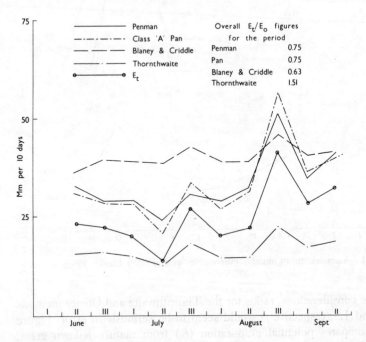

Fig. 5.12 Muguga: comparison of E_o estimates and measured E_t from Kikuyu grass (after Blackie 1965).

constant. Riou (1984) used a variety of methods in central Africa including pans, lysimeters, Penman, water balance and a number of empirical formulae, including Thornthwaite. The latter gave very erratic monthly distributions and Riou (1984) stresses the danger of using empirical formulae developed under different climates. Howell, Phene and Meek (1983) in a semi-arid environment compared evaporation from a standard class A pan and one 'screened' with a 50 mm mesh with potential evaporation estimates from two combination equations. The 'screened' pan had values about 10 per cent less than the standard pan, both being highly correlated with the combination estimates. Pan coefficients were 0.81 and 0.91 for the standard and 'screened' pans respectively. Benacchio *et al.* (1983) for Venezuela used pan data to adjust Thornthwaite potential evapotranspiration values. However, the relationship varied seasonally, sometimes the Thornthwaite value being greater than pan values while at other times the reverse was true.

CHAPTER 6
WATER AND PLANTS: A GENERAL REVIEW

6.1 INTRODUCTION

While plant growth in the tropics, as elsewhere, is influenced by a wide range of physical factors including various climatic elements such as temperature, radiation, wind and humidity, in many cases, water availability is a major control. Only in the humid tropics does water shortage not have an impact, but even here the characteristics of tropical rainfall and the fact that many crops require dry periods at certain growth stages (Ch. 7) have a marked influence on agriculture.

Before examining the implications for water and agriculture of the nature of tropical rainfall and evaporation demand (Ch. 7) it is necessary to have some understanding of plant–water relations. This includes the significance of water to plants and the way in which plants obtain and lose water in the atmosphere–soil–plant system. These aspects are reviewed in more detail in a number of texts (e.g. Slatyer 1967; Salter and Goode 1967; Chang 1968; Rose 1966; Penman 1963; Marshall 1959; Russell 1963; Sutcliffe 1979; Kramer 1969, 1983; Humphries 1981).

6.2 THE SIGNIFICANCE OF WATER TO PLANTS

The major ways in which water is important to plants are as follows:

1. It is a major constituent of plant protoplasm, often making up more than 90 per cent of the total weight of the plant. Most organic substances in protoplasm, including carbohydrates, proteins and nucleic acids are hydrated in their natural state and removal of water affects their physical and chemical properties. When protoplasm is dehydrated it stops being active and below a certain water content is killed. Among the processes affected by dehydration directly is photosynthesis, the significance of this being discussed later.

2. Water takes part directly in a number of chemical reactions. Among these, it is a source of hydrogen atoms for the reduction of carbon dioxide in photosynthesis and is a product of respiration. It also takes part in hydrolytic processes such as starch digestion.

3. Water acts as a solvent in which many other substances are dissolved and in which they undergo chemical reactions.
4. A considerable part of water in plants occurs in large vacuoles within the protoplasts and it is largely responsible for maintaining turgidity (rigidity) of cells and plants as a whole. Maintenance of turgor is essential for cell enlargement and growth and is important in movements of leaves, flower petals and various specialised plant structures. If leaves lose their turgidity, the stomata guard cells close, thus preventing further uptake of carbon dioxide for photosynthesis.
5. It is the medium whereby dissolved substances move between cells and from organ to organ.

Details of the significance of water to plants can be found in Sutcliffe (1979) and Kramer (1983). Almost every process occurring in plants is affected by water availability but the links are complex. 'The relationship varies with plant characteristics, stage of development, soil and climatic conditions' (Chang 1968). In simple terms, water is absorbed by the plant roots and lost by the evaporative process termed transpiration. About 95 per cent of water taken in by plants is transpired with 5 per cent or less being used in metabolism or growth (Kramer 1983). It is the balance between water intake and loss which is important. If water cannot be absorbed by the roots to compensate for transpiration loss then a water deficit develops in the plants (although certain xerophytic plants store water to overcome this).

For reasons listed previously, a water deficit has a great impact upon many aspects of plant growth, development and functioning (see Humphries 1981; Kramer 1983). Water stress does have some beneficial effects, especially in terms of quality of plant products (§ 7.3). The magnitude of the deficit which has an impact depends on plant type, stage of development and the process being considered. The extent of the impact will vary between processes. Of particular significance is the reduction of photosynthesis for direct and indirect reasons. Different species vary greatly in their ability to withstand a water shortage before their photosynthetic rate is seriously reduced. They also differ in their ability to recover when the water shortage ceases. Kramer (1983) provides a detailed discussion of the impacts of a plant-water deficit on photosynthesis.

Many studies have examined the influence of soil moisture on rate of photosynthesis, some indicating a decline when soil moisture is very high, others that the rate is only slightly affected until almost permanent wilting point. Kramer (1963) however, points out problems in design of experiments to assess the relation between soil moisture and photosynthesis which make results uncertain. Also, it is the plant-water deficit, not soil moisture, which influences photosynthesis and this involves both rates of absorption (i.e. from the soil) and transpiration. While these two processes are interrelated, certain combinations of environmental conditions can favour one process rather than the other. Even with very moist soil, a high transpiration rate (in reponse to evaporative demand) can lead to a plant-water deficit and, for example, stomatal closure, if soil moisture cannot be absorbed rapidly enough. Therefore, analysis of the relation between soil-moisture tension and photosynthesis is an oversimplification. This is perhaps particularly so in view of the uncertainties about transpiration rate in relation to soil-moisture tension which are discussed later. Overall, the relationships between water stress, photosynthesis and respiration are rather complex and some aspects are not fully understood. Even the measurement of water stress in plants presents considerable problems (e.g. Squire, Black and Gregory 1981). These problems, including what to measure as well as how, are the cause of some inconclusive and contradictory results (Kramer 1983).

The variable response pattern of photosynthetic rate to variation in soil moisture has practical implications for irrigation. Crops which show a decline in photosynthetic rate with only a slight fall in soil moisture would need more frequent irrigation (of moderate amounts) to ensure maximum yield than a crop which was little affected until almost permanent wilting point. The stage of growth and development at which water stress occurs can exert great influence on the final yield of some crops, especially in the case of annual cereals which seem very sensitive to moisture shortage during the formation of reproductive organs and flowering (Ch. 7). Many plants have a period in their life cycles when they are very sensitive to water stress.

A number of other factors influence plant–water relations. Stark and Jarrell (1980), for example, point out that for many crops, salinity induces plant adjustments which assist with maintenance of a favourable water balance. Radin and Ackerson (1981) found that nitrogen nutrition had a strong impact on stomatal sensitivity to water stress in cotton. Humphries (1981) discusses experiments indicating the complex effects of soil fertility on plant–water relations. Past history of exposure to water stress (acclimation) both in terms of frequency and degree has been found to influence stomatal response and degree of reduction in photosynthesis (Ackerson 1980; Kramer 1983; Mathews and Boyer 1984).

Transpiration is often described as a necessary evil. Exposure to the atmosphere of a large area of moist cell surfaces is necessary to allow carbon dioxide absorption by leaves and this results in loss of water by evaporation. If plants had a cuticle allowing the former but not the latter then 'presumably they would not transpire' (Sutcliffe 1979). Some plants have adaptations to reduce transpiration, but only at the expense of a reduction in carbon dioxide intake and hence slower growth. Closure of stomata at night, a common feature, serves to conserve water at a time when photosynthesis has ceased because of absence of light.

The use of antitranspirants to reduce loss has received attention in recent years (Sutcliffe 1979; Branson et al. 1981; Kramer 1983) and is discussed in § 9.2. Some aim to allow passage of carbon dioxide and oxygen but not water vapour. Others aim to produce partial stomatal closure. This will have much more effect on water vapour than carbon dioxide but certainly there will normally be some reduction in photosynthesis and yield. Another group attempts to increase reflectivity, thereby reducing heat load on leaves.

Probably the most common cause of plant death is the lack of sufficient water to replace that lost by transpiration. Even a temporary water deficit can sometimes be fatal. Plants have a number of ways of avoiding the build-up of water deficits. They may increase their efficiency of water intake, for example by extending their root systems. In irrigation control, it is often considered advisable to allow the development of some deficit, at non-critical periods, to encourage plants to do this. Effective rooting depth tends to increase with decrease in soil moisture and roots are finer and have more and longer branches under moisture stress. Frequent irrigation leads to the development of shallow, horizontally spread root systems which cause problems during dry conditions. Examples of experiments investigating the effects of plant water stress on rooting systems, including implications for irrigation, are presented by Meyer and Ritchie (1980), Robertson et al. (1980), Stanley, Kasper and Taylor (1980). Plants may also reduce water loss (transpiration) in a number of ways, including shedding of leaves, changing leaf orientation and form, increasing diffusion resistance of leaves to transpiration by stomatal closure (e.g. Hall, Foster and Waines 1979; O'Toole and Cruz 1980; Humphries 1981; Kramer 1983; Norman, Pearson and Searle 1984; Wilson and Witcombe 1985; Prasad et al. 1985). Water storage is normally of small importance in avoidance of water deficit in annual crops. However, it is significant in a plant such as pineapple (Hall, Foster and Waines 1979; Kramer 1983). For various crops in the Lebanon, Fuehring et al. (1966), found that partial closure of stomata began as early as 2 days after irrigation, indicating stress on plants even when most 'available' soil moisture was still present (§ 6.3). They found a linear relation between degree of stomatal opening and length of time after irrigation.

In general, plants have three broad ways in which they cope with drought conditions (Hall, Foster and Waines 1979; Humphries 1981; Kramer 1983; Wilson and Witcombe 1985): (a) drought escape – by completion of their life cycle before a serious plant-water deficit develops; (b) drought avoidance – by increasing water uptake and reducing water loss (transpiration) and by water storage; (c) drought tolerance – mechanisms allowing maintenance of turgor and functioning under water-deficit conditions.

Transpiration has some beneficial effects. Severe leaf heating may inhibit photosynthesis, and in such circumstances the latent heat used in evaporation could be important in cooling the leaves. However, this is probably only significant under extreme conditions. It was once believed that mineral salts were absorbed passively by roots and carried into the shoot via the transpiration stream. Evidence now suggests that absorption of water and absorption of dissolved substances at the root surface are largely independent processes, and once mineral salts enter the root cells adequate supplies are readily available to all parts of the plant even at very low transpiration rates (Sutcliffe 1979). A plant may deplete the soil of inorganic nutrients in the immediate vicinity of its roots. A rapid rate of movement of water into the plant because of transpiration helps to prevent this by bringing dissolved substances to the root surface.

Considerable attention is focused on the observed parallelism between transpiration and net photosynthesis or net assimilation rate (gross photosynthesis – respiration). Net photosynthesis represents the excess of dry matter gain by photosynthesis over loss by respiration. Since photosynthesis increases with light intensity and solar radiation is the major factor influencing transpiration, the correspondence between net photosynthesis and the latter is not unexpected. *In the field* a linear relationship between transpiration and net photosynthesis has been found over a wide range of experiments (Chang 1968; Hanks, Gardner and Florian 1969). However, this considerably understates the complexity of the observed relationship, a key aspect being the extent to which both transpiration and photosynthesis are affected by plant-water deficits.

Stomatal closure resulting from a plant-water deficit will reduce both transpiration and carbon dioxide intake, the latter in turn limiting photosynthesis. However, there is much uncertainty about the significance of the impact of stomatal closure on carbon dioxide intake and in turn on photosynthesis. Water deficits have an impact on the photosynthetic machinery and biochemical processes themselves in addition to any impact on reduced carbon dioxide intake via stomatal closure. Kramer (1983) provides a review of this complex area. While there is evidence to suggest that stomatal closure does lead to a direct reduction in photosynthesis, many studies indicate that non-stomatal effects of water deficit on photosynthesis are more important. Generalisation is difficult because effects vary with a considerable number of factors including plant type and stage of development. Whatever the complexities of the subject, the relationship between transpiration and net photosynthesis has a practical application in irrigation and water management. The relationship applies only when plants are actively growing and may be different, for example, during ripening.

The classification of plants on the basis of their water relationships is discussed by Chang (1968) and Sutcliffe (1979). Hydrophytes usually grow in water or swamps. Hygrophytes, such as mosses and some ferns, are land plants in damp locations with high humidity and often saturated soil. Since they usually grow in shade they are adapted to photosynthesise efficiently at low light intensities. Most field crops belong to the mesophyte group, although these can be sub-divided into those which wilt permanently after losing 25 per cent of their water content and those which wilt after losing 25–50 per cent (sometimes called xerophytic mesophytes). Xerophytes wilt permanently only after losing 50–75 per cent of their total water content. It should be emphasised that this grouping is a very general one and, as has been indicated above, the question of the effect of water deficits on plant growth is very complex.

6.3 THE ATMOSPHERE–SOIL–PLANT WATER SYSTEM

Many factors influence the proportion of rainfall which is eventually available to plants, as well as the period of availability. A simplified system is shown in Fig. 6.1. In both the soil and plant sub-systems, only water inputs and outputs will be considered, details of movements through the systems being ignored. Each of these sub-systems is itself complex.

Detailed discussions of the system are found in Slatyer (1967), Rose (1966), Sutcliffe (1979), Jury (1979), Kramer (1983), Branson et al. (1981). Balek (1983), Oldeman and Frere (1982), Norman, Pearson and Searle (1984) deal specifically with tropical conditions. Jury (1979) discusses the classification of models of the system in terms of varying degrees of complexity. They range from complex deterministic models of all aspects of water movement through the system to simple ones such as empirical correlations between, say, rainfall and crop response. Jury (1979) describes a range of actual models and their application to dryland farming systems. Sutcliffe (1979) and Kramer (1983) describe movement of water as analogous to electricity with a driving force (the water potential) and resistances in the system. The relative importance of soil and plant resistances has been the subject of much discussion. While it has been suggested that soil resistance becomes limiting in relatively moist soil, most experiments indicate that plant resistance exceeds that of soil until soil-water content has decreased to about permanent wilting point (see below) (Kramer 1983). Blizard and Boyer (1980) state that suggestions that soil resistance is the major one have ignored the

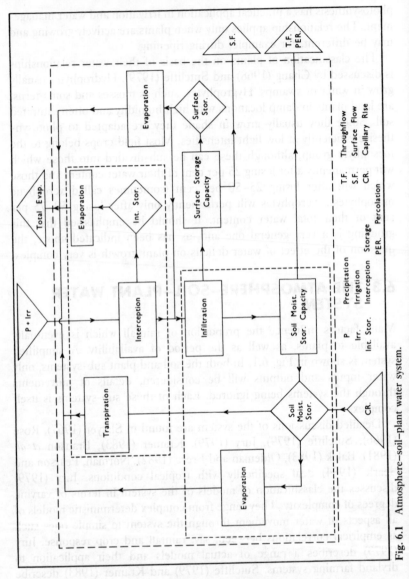

Fig. 6.1 Atmosphere–soil–plant water system.

increase in plant resistance and/or have assumed root densities were too low. Sometimes shrinking soil and roots may produce an increase in resistance at the soil–root interface (Kramer 1983).

Water inputs

These include precipitation in its various forms – solid and liquid. Although most water is taken in by plants through roots, under certain conditions small amounts of water and solutes are taken in through leaves and stems. Kramer (1983) also points out that rain and sprinkler irrigation can leach minerals out of leaves. In some areas, horizontal precipitation, often termed 'fog drip', is significant. This is the deposition of tiny water droplets in cloud or mist upon vegetation or other solid objects which may then drip on to the ground. Measurement and analysis of fog drip is difficult, and therefore there is disagreement about both amounts and significance. Fog drip may contribute directly to water supply, but its effect is also indirect by reducing transpiration since it uses energy in evaporating. Norfold Island pines in Hawaii, with an annual rainfall of 2600 mm, produced an additional 760 mm of moisture from heavy cloud (Ekern 1964). For the leeward Koolau Range, Oahu, Hawaii over a 12-month period, Ekern (1982) found direct interception of cloud water to be < 305 mm compared with rainfall of 2835 mm. Juvik and Perreira (1974) demonstrated the importance of fog drip in the mountain forest of Mauna Loa, Hawaii, especially during dry summer periods. Absolute amounts were greatest at 1580 m, with 638 mm over a 7-month period. The relative contribution increased with height up to 2500 m where fog drip was 65 per cent of direct rainfall. Nagel (1956) demonstrated the considerable amounts of water which may be involved under special circumstances such as those of Table Mountain, South Africa. For a period of 1 year, measured fog precipitation amounted to 3294 mm compared with a rainfall of 1940 mm. It is not possible to extrapolate these instrument measurements to a natural surface such as vegetation, but they indicate that amounts may be large. Kramer (1983) refers to the fact that the effects of fog drip on the coastal fog belt of the Pacific states of the USA are reflected in changes in vegetation species. In general, fog drip would seem to be an important moisture source in some areas.

Dew, the condensation of water on plants, can be considered an input to the system, but this feature is one which has aroused even more controversy than fog drip. The agrometeorological aspects of dew have been summarised by Wallin (1967). To a considerable extent, much of the controversy arises from the problems of measurement and conflicting results obtained from the large numbers of very different instruments and techniques. Three ways in which dew may be significant to plants are commonly suggested:

1. Dew water may drip off plants on to the soil, but the amounts involved are small and would wet only a few millimetres of soil.
2. Dew water may be absorbed by the leaf surface and directly used by the plant; Slatyer (1967) points out that such water movement would be slow even in the case of wilted plants, although it could assist restoration of turgor.

3. As in the case of fog drip, evaporation of dew may reduce early morning transpiration and hence plant-water deficit.

Wallin (1967) concludes his review by stating that the benefit 'seems greatest in the arid and desert areas of the world' where other moisture sources are absent. Pereira (1973) suggests that in tropical climates, dew can be locally important if harvested rapidly, before evaporation, as by grazing animals at dawn. He reports Tanzanian herdsmen collecting drinking water at dawn from dew on tall grasses. In the Negev Desert stones are placed around plants. These cool by radiation at night and concentrate dew. One disadvantage is that the infection process of many foliar plant-pathogenic fungi requires a water film on the plant (Wallin 1967). Newton and Riley (1964) for example, found that cotton boll surfaces wet with dew promote the development of the cotton boll rot fungi.

Irrigation water must be regarded as an input. Within the tropics, it is rainfall which is the most important form of precipitation, although hail is significant in terms of crop damage in some areas (Ch. 9). The characteristics of tropical rainfall and some implications were discussed in Chapters 3 and 4. Implications are considered in more detail in Chapter 7.

The interception process

The fog drip process discussed above is an interception process, but there is also interception of vertical precipitation, particularly rainfall, by vegetation. This process involves a number of variables such as plant type and spacing, wind conditions, evaporation, rainfall intensity, duration, amount and interval between storms. Its significance is related to three aspects:

1. The effects of drip points and stemflow on the spatial distribution of water reaching the ground and hence the pattern of soil-moisture variation under a vegetation canopy. In § 4.1 the way in which maize plants concentrate intercepted water at their roots was mentioned.
2. The significance of the evaporation of intercepted water. Opinions vary considerably, some authorities regarding it as a loss in water input but others feeling that it will be related to a corresponding decrease in evaporation and transpiration from the canopy because of utilisation of energy. It seems unlikely that there is a simple equivalence and there is evidence to suggest that evaporation of intercepted water proceeds at a rate several times that of free-water surface evaporation (Rutter 1963).
3. A vegetation canopy is important in protecting soil against raindrop impact (see below and § 7.1). However, Wiersum (1985) points out

that if intercepted drops concentrate into larger ones and then fall from a sufficient height, their new erosive power may exceed its initial level. Wiersum cites studies from tropical areas such as Colombia, Malaysia, Australia and Indonesia where this has been observed and also where stemflow washes away litter and soil. Wiersum also found that litter is the most important vegetative factor protecting the soil (§ 7.1).

In view of the considerable number of controlling variables, the proportion of rainfall which is intercepted varies considerably. For tropical rainforest, Jackson (1975) found that as much as 85 per cent of a 1 mm shower may be intercepted, but only about 12 per cent of a fall of 20 mm. Therefore, storm size distribution will have a considerable impact on the proportion of rain intercepted. Jackson (1971b) found that over a 6-month period, about 16 per cent of rainfall was intercepted by tropical forest. Pereira (1952) in Kenya found that over a 6-year period, figures for cypress and bamboo were 26 per cent and 20 per cent respectively. Hopkins (1960) found that about 35 per cent of rain was intercepted by forest in Uganda, this being similar to the results of Vaughan and Wiehe (1947) in Mauritius. Freise (1936) for sub-tropical forest in Brazil calculated that 20 per cent of precipitation evaporates in the crown space and 18 per cent either disappears into the bark or evaporates.

For a forest plantation in Java, Indonesia Wiersum (1985) found that interception averaged 12 per cent but as with other studies showed great variation with rainstorm characteristics. For a number of Himalayan forests, Pathak, Pandey and Singh (1985) found that interception varied between 8 per cent and 25 per cent of gross rainfall, with differences occurring between tree species.

Lockwood and Sellers (1982) used a multi-layer model to examine interception loss from a range of tropical and temperate vegetation ranging from trees to crops to grass canopies. They present evidence that in low-windspeed equatorial environments, changing vegetation type would have little impact on interception loss and runoff. Herwitz (1985) investigated interception storage capacities of different species in a tropical rainforest environment in Queensland. Highly significant differences were found between species. The importance of bark in interception storage capacity indicates that the emphasis on the role of leaf surfaces in past interception studies needs modification for tropical rainforests (Herwitz 1985).

Penman (1963) quotes results for some agricultural crops determined by Wollny (1890) in Germany. Some of these are reproduced in Table 6.1 showing the considerable effect of plant density and type. For wheat, Butler and Huband (1985) found that after stem extension, interception

Table 6.1 Percentage interception by crops (after Wollny 1890)

Crop	Plant density (plants m³)	1–31 July 1880 rainfall 175 mm	1–25 Sept 1880 rainfall 65 mm
Maize	9	17	50
	36	32	77
Soyabeans	9	0	28
	36	17	47
Beans	25	25	35

loss was about 40 per cent. It must be emphasised that the problems of measurement and analysis of interception are considerable (Jackson 1971b, 1975). The above data indicate, however, that a considerable part of the rainfall may not reach the ground.

The infiltration process

Infiltration has already been referred to when discussing rainfall–runoff relationships (§ 4.4). The proportion of water reaching the ground surface which infiltrates into the soil depends to a considerable extent upon the relation between rainfall intensity and the infiltration capacity of the soil. As long as the former is less than the latter, all the rain reaching the surface will infiltrate. The infiltration capacity is a variable depending upon soil type, condition and moisture content. Both condition and moisture content will vary with time. The impact of raindrops, for example, may lead to the compaction (or capping) of the soil surface, thereby decreasing the infiltration capacity (Ch. 7). With the occurrence of rainfall, soil moisture will be increased, leading to a decrease in infiltration capacity.

However, it is not only the soil surface which matters. The characteristics of the lower layers are important since the rate at which water is moved away from the surface layers (permeability) will influence the rate of infiltration at the surface. The upper layers become saturated first before appreciable quantities of water pass to the deeper layers. It will thus be some time before water penetrates to the latter. Only the surface layers may be moist, resulting in a low infiltration capacity and perhaps surface runoff, while the deeper layers remain dry. When the rain is mainly short, heavy showers or prolonged drizzle, not much water will penetrate to any appreciable depth even through a permeable soil profile (Mohr and Van Baren 1959). Once the upper layers of soil are saturated, downward transport begins and from that time it is no longer the infiltration capacity which dominates but the permeability (Mohr and Van Baren 1959). Movement of moisture in the soil is discussed by Slatyer (1967), Rose (1966), Marshall (1959), Jury (1979), Kramer (1983).

A further factor which influences the proportion of the rainfall moving into the soil is the time it has to infiltrate. This will depend upon the nature of the ground surface. On steep slopes, water will move rapidly over the surface and hence have little time to infiltrate, whereas on flat surfaces or depressions water will be held until it infiltrates or evaporates. The nature of the surface cover will also have an effect upon the movement of surface water. A vegetation cover tends to increase the proportion of rainfall infiltrating, not only by retarding surface flow but also by shielding the soil surface from raindrop impact, thereby decreasing compaction effects (Ch. 7). In addition, root systems increase soil permeability. The characteristics of the vegetation and crop type vary greatly in their effect upon the infiltration process. Surface litter can also have a great effect upon infiltration. Human activity has considerable influence upon infiltration through modification of the surface cover, slope and soil characteristics. This is discussed in Chapter 9.

Further discussion of the infiltration process is contained in Slatyer (1967), Rose (1966), Norman, Pearson and Searle (1984), Marshall (1959), Ghosh (1980), Roy and Ghosh (1982), Hardy *et al.* (1983), Kramer (1983).

Soil moisture

The water available to plants depends mainly on the soil water potential and hydraulic conductivity, both of which decrease as soil water content decreases. Discussion of the principles of soil moisture retention and movement lies outside the scope of this book, being discussed in references already referred to under infiltration. In broad terms, water available to plants is that between field capacity (FC) and permanent wilting point (percentage) (PWP) [1]. FC is a soil characteristic. Wilting is a plant characteristic occurring when leaves lose their turgor (§ 6.2) and depends on the plant and factors influencing water loss (transpiration) and water intake (from the soil). It can occur even when soil moisture is near FC if plant water intake cannot match transpiration loss because of high evaporative demand. Most plants have an osmotic potential of -1.5 to -2.0 MPa (Kramer 1983) and therefore a value of -1.5 MPa is commonly taken as the water potential at which soil moisture becomes severely limiting (PWP) since a water potential gradient is necessary for movement.

The relationship between water potential and soil water content for a range of soils is shown in Fig. 6.2. While there is evidence that a number of plants, particularly in arid plant communities, may reduce soil moisture content well below a potential of -1.5 MPa, as can be seen from Fig. 6.2, a large change in potential near this level is related to only a small change in soil water content and perhaps of little practical

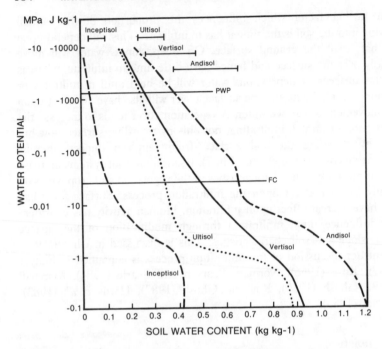

Fig. 6.2 Relationships between soil water potential and water content in the surface soil of a sandy inceptisol, an oxidic ultisol, a clayey vertisol and an andisol (after Norman, Pearson and Searle 1984). PWP – permanent wilting point; FC – field capacity. Available plant water shown by horizontal lines at the top of the figure.

importance. More detailed discussion of PWP is found in Slatyer (1967), Russell (1963) and Kramer (1983).

Figure 6.2 indicates how the amount of water available to plants varies widely between different soils. In Barbados, Hudson (1969) found that cultivation significantly increased the water-holding capacity of the soils.

Even more critical than the water-holding capacity of different soils per unit depth is the extent to which use can be made of this capacity. This involves the rooting habits of the crop, both root density and rooting depth. These are likely to be far more important than the question of soil type. As the soil begins to dry out below field capacity, the rate of water movement in the soil decreases rapidly and the distance in the soil from which the roots can extract moisture decreases. The density of the roots in a particular soil depth will therefore influence the degree to which all the 'available' moisture can be extracted.

Rooting depth varies considerably between plant types. Maize in Kenya was observed to root to a depth of 2 m (Dagg 1965a), Kikuyu grass to a depth of up to 6 m (Dagg 1968a) and evergreen forest to more than 13 m (Dagg 1965b). The amount of water available to the plants will vary greatly in these three cases. However, the situation is not a simple one-way relationship, the most successful plants being those which best adapt their rooting systems to suit moisture conditions. Reference has already been made in § 6.2 to the response of root systems to changing environmental conditions. Robertson *et al.* (1980) found that corn, soyabeans and peanuts in sandy soil varied in their rooting response to plant water stress and irrigation treatment. Stanley, Kasper and Taylor (1980) found that root tolerance to water table level fluctuations changed during the growth cycle of soyabean. The response of crops varies with soil type. Thompson, Gosnell and De Robillard (1967) found that sugar cane in Natal exploited soil moisture to about 2 m in sandy soils but only 1 m in clay, even with maximum irrigation. Under dry land conditions, in sandy soils, the crop exploited soil moisture to the wilting point in the surface 30 cm of soil, but in clay the cane extracted moisture held at tensions of more than 1.5 MPa in the top 1 m.

Gerard and Namken (1966) found that on medium-textured soils cotton develops an extensive root system which can extract water to depths of 120–150 cm. On fine-textured soils, moisture extraction was largely restricted to the surface 60 cm. More frequent irrigation was necessary on the fine-textured than on the medium-textured soils. In low-rainfall areas, permeable soils accentuate aridity for shallow-rooted grasses since water penetrates to lower levels. These soils are more favourable to deeper-rooted bushes and trees such as acacia.

Measurement of soil moisture presents problems (e.g. Squire, Black and Gregory 1981; Kramer 1983) including those of sampling because of marked vertical and horizontal variations. Choice of method depends partly on the objective which is often to assess moisture available to plants. The latter depends primarily on the water potential (see above) but relationships between potential and moisture content (Fig. 6.2) are important. Soil water content can be measured directly using the laborious and destructive gravimetric (weighing) method or indirectly using neutron scattering or gamma ray attenuation for example.

Total soil water potential comprises several components. One component, the matric potential, can be measured in the field using electrical resistance units and tensiometers for example, although they suffer considerable limitations. While some other components can normally be neglected, in heavily fertilised soil or arid areas where salt accumulates, the osmotic potential may be important. In the field, measurement of the latter presents problems. In the laboratory, techniques exist to measure total potential, matric potential and osmotic potential.

This brief outline indicates the difficulties in measuring available water which in turn have a bearing on points made in this section and § 6.2, particularly as far as experimental results are concerned. It is also relevant to the use of water-balance approaches to assess evaporation (§ 5.5) where change in soil moisture is one component. However, problems with soil-moisture measurement mean that a water-balance approach using climatic data to estimate soil moisture is of considerable value (§ 8.1). McGowan and Williams (1980a, b) suggest that errors associated with the neutron probe method (see Haverkamp *et al.* 1984, Vauclin *et al.* 1984) can be made small enough for most water-balance studies.

Evaporation of water from the soil was discussed in Chapter 5. Some of the water entering the soil may drain beyond root range or be lost by lateral sub-surface flow. McGowan and Williams (1980a, b) point out that in water-balance studies, estimating drainage water loss causes problems and presents a simple graphical method to distinguish between it and evaporation. Water can rise from the water table by capillary action and become available to plants, or the plant roots themselves may be able to tap the ground water.

Crop-water requirements

Discussions of water uptake by plants and movement through plants are presented by Jury (1979), Sutcliffe (1979), Kramer (1983). The relation between evaporation, transpiration and crop-water use was discussed in Chapter 5. The term crop-water use includes all the evaporation from the area considered (i.e. evaporation from the soil and vegetation surfaces and plant transpiration). The proportion of the total water loss from the area which is transpiration will vary considerably, particularly in response to the proportion of the ground covered by the crop. Thus, Norman, Pearson and Searle (1984) state that transpiration is 50 per cent of evapotranspiration with a leaf area index [2] (LAI) of 2 and 95 per cent of evapotranspiration with a LAI of 4.

Since crop-water use will vary from place to place in response to changing evaporative demand, no set figures exist. It is possible however to express the requirement (E_t) as a proportion of potential evaporation (E_o) (e.g. open-water surface evaporation) derived by one of the methods discussed in Chapter 5. The assumption is that this proportion (or ratio E_t/E_o) will be constant. Examples of ratios for particular crops and stages of growth as well as absolute water-use data are presented in Chapter 7 and their application and significance discussed in Chapter 8.

Such figures and ratios of crop-water use (requirement) represent 'potential' values, i.e. they assume water to be freely available. Under many circumstances this may not be true. A major complication is that the water between field capacity and PWP, while technically 'available' (qualified by the statements made previously about PWP) may not be equally available. When soil moisture drops below field capacity, plants may experience difficulty in obtaining water. Figure 6.2 shows how soil water potential decreases as water content decreases. Since movement of water from soil to roots depends upon the water potential gradient between the two and the lower limit of potential for crop plants is -1.0 to -2.0 MPa, water will be progressively less available (Kramer 1983). A considerable number of experiments analysing this aspect have been carried out. Stern (1967) found that the ratio E_t/E_o for cotton declined rapidly when available water in the soil fell below 60 per cent. Ekern (1966) found that Bermuda grass maintained a high rate of water use until soil-moisture tension reached 0.1 MPa after which it decreased. Denmead and Shaw (1962) found that when potential transpiration rates were high (6 or 7 mm/day), the transpiration rate fell when moisture was still close to field capacity (tension of about 0.03 MPa). However, with low potential transpiration rates the full rate was maintained almost to wilting point. Fuehring et al. (1966), for a number of crops in the Lebanon, also found that under conditions of high evaporative demand, moisture use decreased when much of the 'available' water was still present in the soil.

Slatyer (1967) points out that the question of equal availability of water to plants for growth must not necessarily be equated with the rate of transpiration. He quotes Richards and Wadleigh (1952) who, in reviewing the experiments suggesting 'equal availability' of water, showed that in every case, while transpiration may have been relatively unaffected by increasing soil-moisture tension, plant growth was progressively and severely inhibited as PWP was approached (an indication that the plants were experiencing difficulty in obtaining water). Kramer (1983) refers to abundant evidence that growth and other plant functions are progressively reduced with decrease in soil water potential. It has already been stressed (§ 6.2) that it is the plant water deficit which is important and that this is influenced by evaporative demand as well as by factors controlling intake. Plant water deficits may be considerable in 'moist' soil under high evaporative demand or conversely, small under low evaporative demand even in a relatively 'dry' soil.

Evidence of 'equal availability' in terms of transpiration may be due to two factors. Slatyer (1967) indicates how the moisture content between various tension levels, combined with errors in determining soil-water content, can suggest equal availability down to PWP. He also points out that following moisture recharge, extraction is most rapid from the soil zone of maximum root density nearest the soil surface. These zones may dry out to almost PWP, while in the deeper layer the soil is close to field capacity. The zone of maximum absorption will gradually move down. Thus, crops with a high root density and considerable rooting depth, may be subject to only small internal water deficits even though much of the soil is near PWP.

While in certain cases transpiration rates may appear to be little

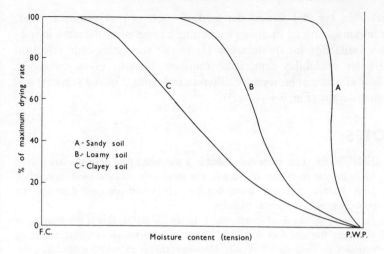

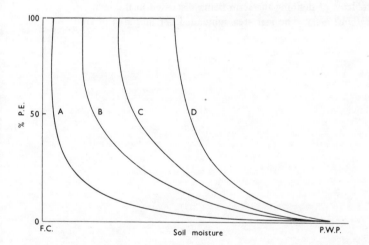

Fig. 6.4 Drying rate of three types of soil (upper); effect of rooting depth on potential evapotranspiration (lower) – curves A to D correspond to increases in rooting depth (after Chang 1968).

affected by decreasing soil water potential, in other cases this is not so. Factors influencing the relation between transpiration and soil-moisture tension include soil texture, the hydraulic conductivity of the soil, rooting depth and density and potential transpiration. Two particularly important features can be mentioned. At high potential transpiration rates, transpiration falls below the potential rate at an earlier stage of moisture tension than in the case of low potential transpiration (Fig. 6.3 (upper)). Similarly, as root density increases, the potential transpiration rate is maintained until much greater soil-moisture tensions are reached (lower). In Fig. 6.3 (lower) the marked effect of root density is clear, the response ranging from one suggesting a progressive decrease in soil-water availability for transpiration (1) to one suggesting little effect on soil-water availability until large moisture tensions (4) are reached. Similar effects can be seen for different soil types (Fig. 6.4 (upper)) and rooting depths (Fig. 6.4 (lower)).

NOTES

[1] *Field capacity*. The water remaining in a soil when free drainage has ceased following the saturation of the soil. For most soils, this state will be reached 2 or 3 days after saturation, but for others with delayed drainage the process may take several months.
Permanent wilting point (percentage). In simple terms, this is the amount of water in the soil below which plants will be unable to obtain further supplies and will therefore wilt. However, this is an oversimplification, the problems of defining the term being discussed in the text.
[2] *Leaf area index*. The leaf area subtended per unit area of land.

CHAPTER 7

WATER AND PLANTS: THE TROPICAL SITUATION

7.1 THE IMPLICATIONS OF THE NATURE OF TROPICAL RAINFALL

The characteristics and implications of the nature of tropical rainfall were discussed in Chapters 3 and 4. Here some aspects are amplified.

Impacts of high intensities and large storms

High rainfall intensities are of considerable agricultural significance and are perhaps the main reason why it is necessary to adopt agronomic and mechanical measures for soil and water conservation in the tropics (Ch. 8). A major review of soil erosion by water in the tropics is presented by El-Swaify, Dangler and Armstrong (1982), including the extent of erosion, its impact, prediction and control. Two distinct processes are involved: detachment and transportation. The former results from either raindrop impact or surface flow. The latter is normally due to overland flow but is also produced by raindrop splash.

Soil aggregates, generally too large to be easily removed by runoff, tend to be broken up by raindrop impact and the fine particles detached can be more easily carried away in suspension. The kinetic energy of high-intensity tropical rain is considerable (Ch. 4), making the significance of the above process very great. However, Wustamidin *et al.* (1983) found large differences in the water drop energy required to cause the breakdown of aggregates taken from 17 soils. The fine particles detached will tend to block the soil pores, creating a compact surface which considerably reduces the rate of infiltration. This 'sealing' effect (compaction) is often deeper than just the surface layers because the water entering the soil through cracks and pores will have its load of silt filtered out at lower levels (Webster and Wilson 1966). Norman, Pearson and Searle (1984) point out that many tropical soils exhibit surface sealing. This effect can also occur with irrigation. The influence of different sprinkler intensities, drop diameters and travel distance on surface sealing has been examined by Ragab (1983).

The resulting decrease in infiltration capacity will lead to an increase in surface runoff and soil erosion and reduce the proportion of rain entering the soil to build up soil-moisture reserves and be available to plants. The breaking up of the soil aggregates and movement of the particles by splash will lead to an increase in soil erosion by water. A vegetation cover will protect the soil surface from the effects of high-intensity rainfall – an indication of this, together with the considerable significance of the erosion process, being provided by Hudson (1957). On land with a 4.5 per cent slope in Zimbabwe, the following treatments were compared:

1. Permanent dense sward of grass, giving protection from raindrop impact as well as impeding runoff.
2. Two layers of mosquito gauze on a frame 15 cm above a bare soil surface, giving complete protection from raindrop impact, but not impeding surface flow.
3. Bare soil.

Over a period of 3 years, the bare soil lost 882.1 t of soil per hectare, the grass plots 8.4 t and the gauze plots 7.5 t, indicating the very great importance of raindrop impact in causing erosion. The small difference between plots 1 and 2 shows that impedance to flow by the grass cover is of minor importance compared to the protection against raindrop impact. However, the protective effects of vegetation may not be so simple as generally thought since, for example, the interception process in forest may result in an increase in erosive power of drops (§ 6.3). Further comments on vegetation influence are made below and in §§ 8.2 and 9.3.

Fournier (1967) cites an experiment in the Upper Volta which indicates that solid particles can be transported at least 1.2 m simply as a result of splash. Spores of micro-organisms may also be transported to new infection sites by raindrop splash (Stedman 1980). Fournier (1967) demonstrates the significance of rainfall intensity, used as a measure of kinetic energy (Ch. 4). He found that in Upper Volta rainstorms with a maximum instantaneous intensity of less than 1.5 mm/min were very rarely erosive. In 2 successive years (1956 and 1957), only 4.7 and 2.5 per cent of the total erosion occurred under such storms. Above a maximum instantaneous intensity of 1.5 mm/min, erosion frequently occurred, and it always occurred above a figure of 2.0 mm/min; 88 per cent of the total erosion occurred under storms with a maximum instantaneous intensity of 2 mm/min or more. The intensities quoted by Fournier will vary with factors such as slope and soil type. Tamiahe et al. (1959) found that rain becomes erosive on sandy loams for slopes of only 3.5 per cent if the intensity exceeds 0.1 mm/min. In Chapter 4, a critical intensity of 25 mm/hr was quoted (Hudson 1971).

Apart from its direct influence on infiltration, surface runoff and soil

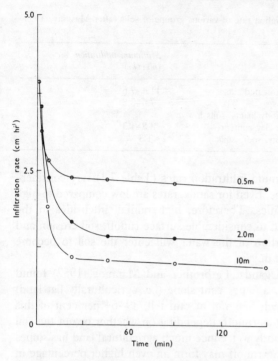

Fig. 7.1 Infiltration characteristics of a red earth soil at points 0.5 m, 2.0 m and 10 m from the trunk of an isolated tree (after Slatyer 1962).

erosion, high rainfall intensities are much more likely to be in excess of infiltration capacities, thereby increasing surface runoff and soil erosion. The variability of infiltration capacity was discussed in § 6.3. The infiltration rate decreases with increase in soil moisture and will fall with time after the onset of rainfall. This is illustrated in Fig. 7.1. The rate of decrease also falls with time and finally a fairly stable minimum rate is usually achieved. However, Roy and Ghosh (1982) for Gangetic alluvium (India) found that even for long durations, the infiltration rate did not attain a constant minimum nor did the plot of the two become asymptotic with the time axis. The initial rate of decrease over, say, the first 15 min is very rapid. The curve shows that it is not only the intensity of rainfall which matters but also its duration. Due to low terminal hydraulic conductivity, many tropical soils have a low long-term infiltration rate (Norman, Pearson and Searle 1984). Also pedologically different soils with similar initial infiltration rates may have very different long-term rates. Initial soil-moisture conditions, depending on previous rainfall, influence infiltration rates. The cracking clays of, for example, the Gezira of the Sudan and western Queensland at the start of the wet season have high infiltration rates. The clays then swell and seal the

Table 7.1   Minimum infiltration rate of various groups of soils (after Musgrave 1955 – taken from Slatyer 1967)

Soils	Minimum infiltration rate (mm/hr)
Deep sands, deep loess, aggregated silts	11.4–7.6
Shallow loess, sandy loams	7.6–3.8
Many clay loams, shallow sandy loams, soils low in organic matter, soils of high clay content	3.8–1.3
Soils of high swelling capacity, sodic soils	<1.3

lower soil layers. Minimum infiltration rates (Table 7.1) vary considerably between soil groups. Even for sands, rates are low compared to high tropical rainfall intensities. Therefore, high rainfall intensities in the tropics are likely to lead to considerable surface runoff, soil erosion and a relatively low proportion of the water will enter the soil to become available for plant growth.

At Namulonge in Uganda, Farbrother and Manning (1952) found that on plots with only a 2 per cent slope (i.e. agriculturally flat land) during a period in which 576 mm of rain fell, 39–64 per cent of this was lost from bare soil as runoff, depending on whether or not the soil surface was kept constantly wet. Since much agricultural land has slopes of more than 2 per cent, runoff may form an even higher percentage in many cases. Temple (1972), in a review of African experiments, cites a number of cases of high-percentage runoff.

The concentration of tropical rainfall in a small number of large, high-intensity storms influences soil-formation processes (see below) and the proportion of rainfall becoming surface runoff. Since soil moisture will increase and the infiltration rate therefore decrease during a storm, if a considerable amount of rain does occur over a fairly prolonged period, then towards the end of the storm a very high proportion of the water may be lost as surface runoff. If, on the other hand, rainfall occurred in smaller amounts but of similar intensity and therefore shorter duration, provided the soil was fairly dry to begin with, much of it would enter the soil because of the high infiltration capacity during the storm.

Antecedent rainfall will influence soil moisture and hence the proportion of a storm which infiltrates. An illustration of this is provided by Fournier (1967) at Lake Alaotra, Madagascar (Table 7.2). The total

Table 7.2   The effects of two successive downpours at lake Alaotra, Madagascar (after Fournier 1967)

Date	Rainfall (mm)	Duration (min)	Erosion (t/ha)	Runoff (%)
23 Dec 1959	26	30	1	5.4
24 Dec 1959	24	90	3.3	39.4

amounts of rainfall for the two storms were similar. Despite the fact that the second storm was only one-third as intense as the first (compare durations), it caused a much higher proportion of runoff and erosion than the first. The reason was that it occurred on soil which had already been wetted by the first storm. Temple (1972) cites a storm at the end of a wet season in Tanzania which produced eight times as much runoff as a storm at the start of the season, despite the fact that it was only 5 per cent heavier. This was due to the soil moisture at the end of the season inhibiting infiltration. Under a heavy, prolonged storm, percolation beyond rooting depth and lateral sub-surface flow are more likely than under moderate-sized storms since storage capacities are more likely to be approached.

Therefore, high intensities and concentration in a few large storms result in a large proportion of rain being lost as surface runoff, considerable erosion and only a limited proportion of water becoming available to plants. Much of the erosion occurs in a short time associated with a few heavy storms. Hudson (1971) reports that at Mazoe, Zimbabwe in nearly all seasons more than half the total erosion occurred during the one or two heaviest storms of the year. In one case, three-quarters of the yearly loss took place in 10 min. Fournier (1967) found that 88 per cent of the annual erosion in Upper Volta occurred in approximately 14 hr and 6 hr during 1956 and 1957, respectively. El-Swaify, Dangler and Armstrong (1982) discuss the wide variety of forms of water erosion, including sheet (inter-rill), rill, and gully as well as more specialised types such as piping, pedestal, puddle and vertical erosion. All types occur in the tropics. Although tropical rainfall characteristics create high potential erosion hazard, the magnitude of the problem will be influenced by other factors such as soil type and condition, slope and length of slope, vegetation, agricultural system and cropping type, cultivation practices, soil and water conservation measures (El-Swaify *et al.* 1982; Goodland, Watson and Ledee 1984; Norman, Pearson and Searle 1984; Quansah 1981; Ruangpanit 1985; Wiersum 1985). For example, the steeper and longer the slope, the greater the erosive power of overland flow. Soil and water conservation measures are discussed in § 8.2.

Two recent studies of erosion and runoff in tropical forests, focusing on the role of vegetation, are of interest. Ruangpanit (1985) related runoff and erosion to rainfall intensity, duration and percentage crown cover in mountain forest in Thailand. Both elements increased with amount, intensity and duration of rain and decreased with crown cover up to 70 per cent. Increasing cover beyond 70 per cent had no real effect. Litter and high organic matter were important in increasing infiltration and decreasing runoff and soil loss.

For plantation forest in Java, Wiersum (1985) found that direct soil cover (litter) was the most important cover protecting the soil. The additional protection of herbal undergrowth was relatively small. Direct canopy effect was even smaller due to two conflicting influences. The erosive power of water drops may be increased by the interception process (§ 6.3), this being opposed by reduced amounts of rain (through-fall). The net effect depends on soil surface conditions. Wiersum (1985) cites other workers whose findings agree with his. Thus, Ruangpanit (1985) and Wiersum (1985) demonstrate the importance of the protective litter layer. If it is not present (e.g. because of cropping between trees) then erosion risk can be considerable if interception increases the erosive power of drops. However, Wiersum (1985) found that with bare soil, a negative relationship between erosion and erosive power of rain occurred. This might be explained by soil character, which quickly produced a crust under rainsplash action. With a protective cover the crust formed less easily. Both studies indicate the complexity of the forest situation (§ 9.4).

Understandably there is considerable interest in the quantitative prediction of rainfall erosion. Developments are discussed by El-Swaify *et al.* While the Universal Soil Loss Equation (USLE) is the most widely used model to predict sheet and rill erosion, it has been misused. The basic equation is $A = RKLSCP$, the six parameters influencing erosion being rainfall erosivity (R), soil erodibility (K), cropping management (C), erosion control practice (P), length of slope (L), steepness of slope (S). Use and misuse of the approach, potential errors and problems associated with application in the tropics, including lack of data, are considered by El-Swaify *et al.* (1982). Lewis (1985) provides one example of tropical application.

A simplification of the USLE model, the Soil Loss Estimation Model for Southern Africa (SLEMSA) has been proposed by Elwell (1977, 1981) for use in developing countries. The basic framework of the SLEMSA has been used to produce a prediction model for Zimbabwe (Elwell and Stocking 1982). The authors discuss the problems and advantages in using this model in developing tropical countries, stressing the need to adapt it to local circumstances, including the available data base. They and El-Swaify *et al.* (1982) stress the need for verification of such models in a range of situations.

Impacts of other characteristics

The impact of the marked seasonality of rainfall over much of the tropics on the water balance and crop types is discussed in Chapter 8. In the seasonally arid tropics, soil moisture may approach PWP at the end of the growing season and only begin to accumulate again at the start of the next rainy season. Perennial vegetation will utilise this as soon as it is available and there will be no time at the start to build up moisture.

However, annual crops with their lower moisture requirements at the start of the growing season (§ 7.2) may allow moisture reserves to build up. Therefore, Bunting (1961) suggests that a region considered arid from the point of view of perennials may be effectively humid for annuals. He also points out that in regions with 3 or more months of dry weather, bush fires are a regular feature and that this has prevented the establishment of evergreen forest and thicket species. Bunting considers that evergreen forest has been held back in both Uganda and the Congo in this way.

Under a seasonal rainfall regime, the rainfall characteristics at different times of the season are very important. At the start of the rains, light showers will do no more than wet the surface layers of the soil and evaporate quickly, contributing little to soil-moisture build-up or plant growth. However, in § 4.1, the possible benefits of light showers were suggested. Following the dry season, vegetation cover may be sparse or absent and therefore not protect the soil surface from the impact of high-intensity rainfall. Heavy, intense storms at the start of the wet season would cause compaction of the soil surface, leading to a high proportion of runoff, erosion and little build-up of soil moisture. However, the development of cracks in the soil during the dry season and general low moisture content may allow considerable downward movement at first. Heavy storms in the middle and end of the wet season, when vegetation protects the soil, will not have such an effect on infiltration characteristics. Jackson (1986a) found a tendency for rainfall to occur in fewer raindays with higher mean daily intensities than monthly totals would suggest for tropical locations with a limited rainy season, particularly at the start of the season as well as dry months sometimes having fairly high mean daily intensities. These are all situations where protective vegetation cover may be limited. Figure 7.2 summarises many of the influences of tropical rainfall characteristics discussed previously, including the tendency for series of wet and dry years to occur (§ 3.3).

Relationships between erosion and rainfall totals can be formulated in a very general way. In arid areas, despite lack of protective vegetation, lack of rain normally limits water erosion although rare heavy storms can cause considerable problems (§ 4.1). Sparse vegetation and dry soil will promote wind erosion. Semi-arid areas and those with a short rainy season having limited protective vegetation may be liable to considerable water erosion, particularly since intensities can still be high in such locations. In addition, less intensively weathered soils in semi-arid areas may be more susceptible to erosion (El-Swaify et al. 1982). Despite the erosive capacity of water in wetter areas, the protective natural vegetation cover limits erosion. However, loss of this cover may result in large erosion. Therefore, sound land use and management practices in such areas are critical (§ 8.2).

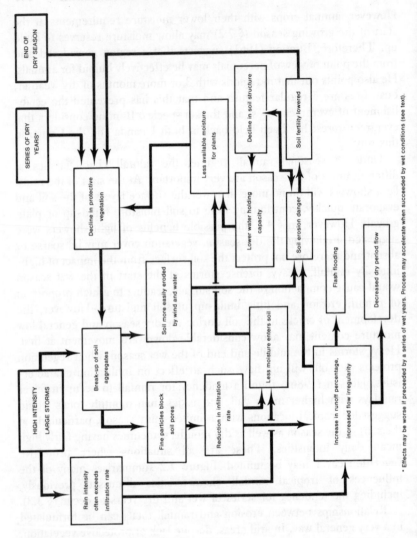

Fig. 7.2 Influences of tropical rainfall characteristics on hydrology and soil (Jackson 1986b).

El-Swaify *et al.* (1982) provide a comprehensive inventory of water erosion in the tropics and Stocking (1984) presents an overview for Africa. Both works stress the significance of vegetation in determining erosion rates, its absence being the key factor in large losses. Stocking (1984) indicates the often localised nature of high rates and argues that long-term control in the African environment must be through vegetation and fertility enhancing farming systems rather than mechanical protection (§ 8.2). El-Swaify *et al.* (1982) examine impacts of population and economic pressures on the extent of erosion. They also examine consequences of erosion, including loss of soil productivity, damage from uncontrolled runoff, siltation of channels and storage reservoirs and environmental impacts of sediment deposition. These aspects are considered in Chapters 8 and 9.

The implications of the marked spatial variations in rainfall for varying time periods were discussed in Chapter 4. They can create considerable differences in yield over small areas even where average rainfall is uniform. In addition to the marked effects on peasant agriculture this has implications for experimental plot yields.

As was shown in § 3.3, variability of rainfall over different time scales must be considered. Daily rainfall characteristics for example, vary from year to year. Bunting (1961) points out that at Tozi in central Sudan, conditions varied greatly in two successive wet seasons, 1954 and 1955, despite the fact that the seasonal totals were similar (749 and 722 mm, respectively). In the 1954 season, 56 per cent of the rainfall (55 rainy days) fell in storms of more than 47 mm, while in 1955, only 39 per cent of the rainfall (78 rainy days) fell in storms of more than 25 mm. In 1955, therefore, the rain was spread over a larger number of rainy days and less fell in heavy storms than in 1954. Since intensity and concentration are important, their variability from year to year is significant.

It is all too easy to attempt an 'explanation' of yield fluctuations and the occurrence of food shortages and famines in terms of rainfall variations. Often, closer examination indicates little relation between such occurrences and rainfall amounts over particular periods (say, monthly figures). This is to some extent a result of the very complex relationship between the nature of rainfall and the supply of moisture to plants (Ch. 6). Rainfall totals over a period of a month may not be very good indicators of the amount of moisture made available to the plant or when it is made available. In addition, however, as well as other physical factors, a wide range of socio-economic factors must be taken into account. A food shortage may result because in a particular year, for a number of reasons, cash cropping rather than subsistence food crops was given more emphasis. If cash crop yields, demands or prices are low then peasant farmers will be unable to purchase sufficient food and at the same time their own production may be inadequate for their needs.

Movement of soil water is important in soil formation and is determined by the permeability of the soil. The link between permeability – the rate at which moisture can be carried away from the surface – and infiltration capacity has already been discussed. If permeability is low then the water cannot be moved away from the surface layers and hence infiltration capacity falls rapidly. Percolation to deeper layers does not really begin until the surface layers are wet (Ch. 6). Hence, under light rainfall conditions little or no moisture penetrates to the deeper layers. Depending on the nature of the ground surface and vegetation, the same can be true for heavy showers of short duration. The moisture in the surface layers will be built up, but because of high intensities much can be lost as surface runoff. Before water has time to penetrate to deeper layers, the rains may cease. Therefore, it is only in prolonged heavy showers that percolation to deeper layers is likely unless showers are so frequent that the surface layers always have a high moisture content. Therefore, it is not only the permeability of the soil which will determine percolation to deeper layers but also the nature of the rainfall.

Rainfall characteristics, particularly the seasonal regime, through their influence on soil moisture and soil moisture movement will have an effect on soil character. Tropical soil formation, including the influence of climatic factors such as rainfall, is discussed in detail by Mohr, Van Baren and Van Schuylenborgh (1972).

7.2 THE IMPLICATIONS OF THE HIGH EVAPORATIVE DEMAND

A rainfall more than adequate for agriculture in temperate latitudes can be insufficient in the tropics because of the high evaporative demand of the atmosphere and therefore of crop-water requirements. Only in the humid tropics where rainfall is considerable in every month does the existence of the high demand not impose restrictions on agriculture. However, as was indicated in Chapter 3, even in humid areas, irregular dry spells occur. Simple comparison of rainfall and evaporation on an annual or an average basis, however, is insufficient. The seasonal nature of rainfall can mean that at certain times of year water supply may be adequate, but not at others. This was illustrated using simple water-balance studies in § 5.3. It is often water availability at certain critical short periods which has considerable influence on agricultural yields (§ 7.3). Also, it is not rainfall as such which supplies the needs of the plants but soil water, and the complexities of this supply system have already been indicated (§ 6.3).

Water requirements vary from crop to crop and also with the stage of growth of the crop, the latter influencing the aerodynamic roughness and proportion of the ground covered (§ 5.2). From the agricultural point of view, therefore, apart from the overall complexity of the atmosphere–soil–plant water system, comparison of rainfall and evaporation is, at best, only a crude indication of the potential of an area.

Data for different crops are presented in Tables 7.3 to 7.5. They show something of the high rates of evaporation from crop surfaces, the range for particular crops, differences between crops and seasonal variation. In the case of the data on rice produced by Kung (1966)

Table 7.3   Water requirements (E_t) – Absolute values

1. Maize
Mugaya, Kenya (mm) (Dagg 1965a)

	Mar	Apr	May	June	July	Aug	Sept	Oct
Local variety	30.0	67.2	61.7	85.6	105.1	78.7	98.8	40.1
Imported variety	30.0	68.7	68.7	92.5	105.1	64.3	46.2	—
Rainfall (70% probability)	47.0	185.8	164.2	38.2	16.5	19.3	18.7	44.3

2. Tea
Kenya (Dagg 1970)
Monthly rates for young tea varied from 40 to 115 mm/month

3. Coffee
Kenya (mean rates, mm/day)   Jan, Feb, Mar   3.8
(Wallis 1963)   Apr, Oct, Nov, Dec 3.3
                             May, June, Sept   2.5
                             July, Aug   2.0

4. Common bean
Brazil (mm) (Norman *et al.* 1984)
Total for 75 day crop 220:

Germination–flowering	*Flowering–pod development*	*Pod development–maturity*
3.2 mm/day	3.2 mm/day	1.7 mm/day

5. Soyabean
(Norman, Pearson   3–7.5 mm/day by closed
and Searle 1984)   canopy

6. Cassava
(Norman *et al.* 1984)   Highest rate 4–5 mm/day

7. Cacao
West Africa (mm) (Opeke 1982) Minimum rate 100–125 per month

8. Taro (flooded)
Florida (mm) (Shih and Snyder 1985) 1.5–7.2 per day, average 4

9. Barnyard grass
Philippines (mm)   (O'Toole and Tomar 1982)   2.8 per day

Table 7.4   Seasonal requirements (mm)

(a) (after Doorenbos and Pruitt 1977)

Bananas	700–1700	Sisal	550– 800
Beans	250– 500	Sorghum	300– 650
Cocoa	800–1200	Soyabeans	450– 825
Coffee	800–1200	Sugar cane	1000–1500
Cotton	550– 950	Maize	400– 750

(b) Taro (flooded)
Florida (after Shih and Snyder 1985) 1200

(c) Cacao
West Africa (Opeke 1982) minimum of 1350

(Table 7.5), the loss by percolation varies widely because of high loss in sandy soils compared with paddies on clay. The data of Robertson (1975) on seepage and percolation also indicate the influence of soil

Table 7.5 Water requirements of rice

(a) (Kung 1966 from a review of work in Asia)	*Average (mm)*	*Extremes (mm)*
Water requirement from transplanting to harvest	800–1200	520–2549
Transpiration	200– 500	132–1180
Evaporation	180– 380	107– 797
Percolation	200– 700	32–1944

(b) (O'Toole and Tomar 1982) Philippines
   7.2 mm/day average

(c) (Robertson 1975) Philippines (mm)
   Evapotranspiration: rainy season (frequent cloud):   4 per day
                         dry season (clearer skies):   5.5 per day
   Seepage and percolation: higher in dry season (lower water table) range 0–6.5 per day depending on soil and other factors.
   Wet season range 0–2.5 per day.
   Surface drainage in wet season may be high.
   Total requirements:
      1000–3000 for wet season crops;
      700–2500 for dry season crop requiring 120 days
      from planting to maturity.

and other factors as well as differences between wet and dry seasons. Of the 1240 mm Kung quotes as the requirement for the complete growth cycle in parts of South-East Asia, 40 mm are for the nursery, 200 mm for land preparation and 1000 mm for irrigation of the paddy. Useful as such figures are as an indication of magnitude, since they will vary from place to place depending on climatic characteristics the water requirements of crops and vegetation are perhaps better studied using E_t/E_o ratios (Chs 5, 6). Once such ratios have been established (using a lysimeter to determine E_t for example) it is then possible, given an E_o estimate for any location to assess the water requirement of a crop at different growth stages. The application of these figures in water-balance studies will be considered in Chapter 8.

A range of ratios for different crops and vegetation types is presented in Table 7.6. For a number of reasons, care must be taken in examining these ratios.

1. Here, E_o is determined by different methods, being most commonly an open-water surface estimate derived from the Penman formula or the US Weather Bureau class A pan. As indicated in the table, however, other methods are also used.
2. In some cases ratios are peak figures occurring over a few days, in others they are monthly averages.
3. The experimental soil-moisture condition varies. The effect of soil moisture on E_t values was discussed in § 6.3. The variation in ratios with different moisture regimes is indicated by data for cotton in Alabama (Doss, Ashley and Bennett 1964) and bananas (Norman, Pearson and Searle 1984) in Table 7.6.

Table 7.6 E_u/E_o Ratios – for different crops, stages of growth, different E_o estimates

1. Cotton

(a) Mubuku, Month
Uganda Ratio (Penman E_o)
(Rijks and Harrop 1969)

1	2	3	4	5
0.33	0.50	0.80	1.20	1.20

(b) Alabama, USA (Doss, Ashley and Bennett 1964)   Ratio (class A pan)

First 2 weeks: 0.5
Max. (1.00 no irrigation
from (1.18 irrigation when 80% available water
top (removed
61 (1.30 irrigation when 30% available water
cm (removed

(c) Gilat, Israel (Stanhill 1962)   Ratio (class A pan)

Young	Mature	Ripening
0.20	0.85	0.10–0.40

2. Sugar cane

(a) Tanzania (Blackie and Bjorking 1968)   Ratio (Penman E_o)

Planting	After 4 months	Until dried out after harvesting
0.3	1.1–1.2	1.05

(b) Natal (Thompson and Boyce 1967)   Ratio (class A pan)   Ratio (Penman E_o)

Full canopy 1.0
Full canopy >'1 under advective conditions
12 month ratoon and 24 month virgin cane data

(c) (Doorenbos and Pruitt 1977)*

				Ratios Relative humidity			
Crop age		Growth stages	Wind:	RH min > 70%		RH min < 20%	
12 months	24 months			Light-mod.	Strong	Light-mod.	Strong
0–1	0–2.5	Planting → 0.25		0.55	0.6	0.4	0.45
1–2	2.5–3.5	0.25–0.5		0.8	0.85	0.75	0.8
2–2.5	3.5–4.5	0.5–0.75	full canopy	0.9	0.95	0.95	1.0
2.5–4	4.5–6	0.75–	full canopy	1.1	1.1	1.1	1.2
4–10	6–17	peak use	full canopy	1.05	1.15	1.25	1.3
10–11	17–22	early senescence	full canopy	0.8	0.85	0.95	1.05
11–12	22–24	ripening		0.6	0.65	0.7	0.75

Table 7.6 (continued)

3. Maize

(a) Muguga, Kenya (Dagg 1965a) — Ratio (Penman E_o)

	Mar	Apr	May	June	July	Aug	Sept	Oct
Local variety	0.45	0.48	0.63	0.87	1.00	0.92	0.67	0.45
Imported variety	0.45	0.49	0.70	0.94	1.00	0.75	0.47	—

(b) Griffith, NSW, Australia (Downey 1971) — Ratio (class A pan)

Planting				Anthesis			Maturity	
6 Nov	23 Nov	23 Dec	3 Jan	20 Jan	1 Feb	19 Mar	7 May
	0.17	0.25	0.40	0.77	0.76	0.14	0.15

4. Tea

(a) Kenya (Dagg 1970)

Ratios:

$$E_t = E_o [0.9a + (1 - a) 0.9n]$$

E_o (Penman), a = fraction of soil covered by crop at noon, n = fractional number of rain-days per month.

0.22 in dry month with little tea cover
0.92 in very wet July 1968
Young tea planted May 1966. 1968 onwards range 0.70–0.83 with 1968 average 0.76.
At end of May 1969 still only 75% cover – not therefore a final value.

(b) Kenya (Dagg 1969b) — Ratio (Penman E_o) Mature tea 0.85 (annual value)

(c) Doorenbos and Pruitt 1977* — Ratio for bushes in full production: non-shaded with > 70% ground cover 0.95–1.0; with shade trees 1.05–1.1 (humid periods); 1.1–1.15 (dry periods)

5. Coffee

(a) Kenya (Blore 1966) — Ratio (sunken pan) 0.86 during rains or irrigation but varying with soil moisture and canopy cover.

(b) Kenya (Dagg 1969b) — Ratio (Penman E_o) Mature coffee—0.80 (annual value)

(c) Doorenbos and Pruitt 1977* — Ratio: mature coffee; no shade; clean cultivation; heavy grass mulching 0.9, if significant weed growth 1.05–1.1

6. Evergreen forest
Kenya (Dagg 1969b) — Ratio (Penman E_o) Annual value – mature plants – 0.90

7. Bamboo forest
Kenya (Dagg 1969b) — Ratio (Penman E_o) Annual value – mature plants – 0.75

8. Kikuyu grass
Kenya (Dagg 1969b) — Ratio (Penman E_o) Annual value – mature plants – 0.75

9. Bermuda and other grasses

(a) Bermuda, Hawaii (Ekern 1966)   Ratio (class A pan) Mature 1.0 (for moisture tensions up to 1 bar then decreased)

(b) Bermuda, Alabama (Doss et al. 1964)   Ratio (class A pan) Young 0.45   Mature 0.75

(c) Arizona (Kneebone and Pepper 1982)   Ratio (class A pan) Mean annual:

3 species Bermuda grass	0.46
Zoysiagrass	0.46
St Augustine grass	0.58
Tall fescue	0.64

Ratios ranged from 0.42 to 0.80 depending on management (irrigation) and grass

10. Pineapple

Hawaii (Ekern 1965)   Ratio (class A pan) Mature 0.35

11. Bananas

(a) (Doorenbos and Pruitt 1977)*   Months following planting:

1	2	3	4	5	6	7	8	9	10	11	12	13	14	15
Ratio:														
0.4	0.4	0.45	0.5	0.6	0.7	0.85	1.0	1.1	1.1	0.9	0.8	0.8	0.95	1.05

suckering — shooting — harvesting

(b) (Norman, Pearson and Searle 1984)   Lower values after 10 months reflect rapid decline in active leaf area of mother plants. Low values in early months apply for heavy mulching with bare soil and frequent rain, values are 0.8–1.0

$E_c = 0.9 E$ pan x Wa where E pan = pan evaporation, Wa = available soil water as a fraction of FC

12. Pearl millet

(Norman et al. 1984)   Ratio (pan) average 0.82, 1 from floral initiation to anthesis

13. Common bean

Brazil (Norman et al. 1984)   Ratios (pan)

Germination–flowering	Flowering–pod development	Pod dev.–maturity
0.62	0.77	0.38

14. Cacao

(Doorenbos and Pruitt 1977)*   Close tree spacing, no cover crop or shade trees: 0.9–1.0   With shade trees and undergrowth: 1.1–1.15

* Data cited by Doorenbos and Pruitt (1977) are said to apply to four methods of determining a reference crop evaporation value: Blaney–Criddle, Radiation, Penman, Pan.

4. In a number of cases, the influence of advected energy on ratios was indicated. Stern (1967) found that evaporation from a cotton crop was using about 12.5 per cent more energy than was locally available through net radiation, and therefore transport into the area must have been substantial to maintain a mean ratio (E_o, Penman) of 1.2. The impact of advected energy has also been referred to by Thompson and Boyce (1967), Doss *et al.* (1964), and Fuehring *et al.* (1966).

5. In the early growth stages, when the crop canopy covers only a small part of the ground, a high proportion of the water loss is from bare soil. Evaporation from bare soil varies with frequency of wetting and soil type (§ 5.2) and hence the latter factors influence ratios. For soyabean, Norman *et al.* (1984) state that soil evaporation is half total evaporation when the soil is wet but 0.25–0.5 of the total during the dry season.

6. Sometimes experimental components are separated, for example, Robertson (1975) indicates that lysimeter and pan were 1.5 km apart for an experiment in the Philippines and 15 km apart for an Australian experiment.

7. Differences between different varieties of a particular crop exist, both in terms of length of growing season and requirements during it. For example, Doorenbos and Pruitt (1977) refer to differences in varieties of rice and sugar cane.

8. Although, as indicated in Chapter 5, the basic concept of E_t/E_o ratios is that since both are influenced by climatic factors, the actual ratios are often assumed to be constant, Doorenbos and Pruitt (1977) indicate that they vary in some cases between geographic locations and climatic conditions such as windspeed and humidity. This is shown for example in Table 7.6 for sugar cane and Table 7.7 for rice.

Table 7.7 E_t/E_o Ratios – paddy rice

(a) (after Doorenbos and Pruitt 1977)					
Humid South America					
	Planting	Harvest	1–2nd month	Mid-season	Last 4 weeks
Wet season	Nov–Dec	Apr–May			
Wind: light–moderate			1.1	1.05	0.95
strong			1.15	1.1	1.0

Doorenbos and Pruitt also present values for humid Asia and North Australia under varying seasonal and wind conditions. However, most values lie within the range 0.95–1.15. The two exceptions (1.25, 1.35) are for the dry season in humid Asia. For upland rice, the same values apply since topsoil layers are maintained near saturation. Only during the initial crop stage will values be reduced by 15–20%.

(b) (Robertson 1975) Philippines (growing period average):	
Ratio (class A pan)	0.91
Open water (no crop)	0.61
Sunken pan	1.2

Australia: Ratio (class A pan) 25-day average with crop at full height 0.98, daily range 0.77–1.47 depending on roughness length of crop.

It must however be remembered that limitations in assessing both E_t and E_o values exist (Ch. 5) and furthermore a range of other factors influence the ratios (see point 9 below). As was suggested in Chapter 5, the relative performance of the various methods of assessment of E_t and E_o varies geographically and with meteorological factors. Given these limitations, whether or not minor differences in ratios have any practical significance is questionable.

9. In the tables, ratios are quoted for a variety of conditions which have an impact in addition to those mentioned above, e.g. presence or absence of mulch, shade, undergrowth, and plant spacing.

Despite these and other problems, some general characteristics emerge. The considerable variation in ratios depending on the stage of growth is apparent. In the early stages of growth, plant-water use is only a fraction (often 0.2–0.5) of free-water surface evaporation and this is important in assessing the agricultural potential of an area or irrigation requirement. Use of an open-water surface evaporation figure for these purposes can be very misleading. Laycock and Wood (1963) refer to the tendency to over-irrigate tea nurseries because of a misunderstanding of water requirements in the early stages and water less heavily later on when in fact the water requirement is greater.

As the crop reaches maturity with full canopy development, the water use (E_t) increases to a maximum. The effect of proportion of the ground covered by the crop, its height and roughness have already been referred to and are mentioned by a large number of workers (e.g. Dagg 1965a, 1970; Stern 1967; Blore 1966; Doss *et al.* 1964). The ratios at maturity will vary somewhat depending on the precise stage, roughness and the completeness of canopy cover. Peak water-use periods are indicated by Salter and Goode (1967) and information for a few crops is provided in Table 7.8. At the stage of maximum water use, the requirement often approximates that of open-water surface evaporation. Ratios vary from about 0.75 to 1.30 (Table 7.6). Pineapple is an exception, with a ratio of 0.35.

Table 7.8 Period of maximum water use (after Salter and Goode 1967)

Crop	Period
Wheat	Various results quoted: but heading, flowering and grain formation all quoted.
Maize	Various results: (*a*) flowering and milk-ripe stage: (*b*) tasselling to hard-dough stages.
Sorghum	Boot to heading stages
Rice	Heading and flowering
Groundnuts	Flowering and pod development, when most dry matter being accumulated.
Cotton	Start of flowering and during boll development.

During the fruiting stage, requirements are usually reduced, the reduction being relatively small during the wet fruiting stage but more pronounced during the dry fruiting stage (Chang 1968). The fall-off in water requirement in the later stages of growth can be seen in Table 7.6 (e.g. maize, sugar cane).

In view of the high values of open-water surface evaporation in the tropics, the requirements of crops at certain growth stages will also be high. This, together with the seasonal nature of rainfall in many tropical areas, means that water exercises a very marked control over agriculture.

7.3 WATER–YIELD RELATIONSHIPS

In Chapter 6, the complexities of the effects of water deficits on the growth and development of plants and the supply of water to them were discussed. Before reviewing examples of yield responses to water, some basic points need clarification. Yields are affected by a wide range of factors and separation of the influence of water from the others can be extremely difficult or impossible. Salter and Goode state that:

> An accurate interpretation of crop responses to different moisture
> conditions can only be made in the absence of other limiting factors and
> when the nature of the interactions between water and other factors are
> properly understood.

Undoubtedly, this has meant that in some cases, conclusions may be of doubtful validity. Furthermore, 'yield' has various meanings, e.g. the economic yield of grain, of forage (grasses) or dry weight of the entire plant including the roots.

In § 6.2, it was stressed that the important factor is the plant-water deficit not the soil-moisture deficit. The latter is important only inasmuch as the plant-water deficit results from the relationship between water intake from the soil and transpiration. The interpretation of experimental data is influenced by two groups of factors:

1. Those influencing the response of plants to soil moisture because they influence the transfer of water from the soil, via the plant to the atmosphere, i.e. soil, weather and plant factors such as rooting habit.
2. Those involved in measuring plant response to experimental conditions.

Salter and Goode (1967) stress the need to make a clear distinction between crops which have 'true' moisture-sensitive stages of growth and other crops which may at times show responses to moisture conditions not related to any particular stage of growth. In the case of the latter category, high evaporative demand or low soil moisture during the dry season can, for example, influence plant growth and yield whatever the growth stage. 'True' moisture-sensitive periods are those when the plant is, or appears to be, more sensitive to moisture conditions than at other stages of growth or development. Often it is not possible to distinguish the two, and this may account for experimental contradictions between different workers in some cases.

A basic distinction is often made between annual and perennial crops, the former usually having a moisture-sensitive stage. Perennials are in most cases sensitive over a wide range of stages.

Annuals seem more likely to respond to variations in available soil moisture. Perennials can adapt to environmental conditions very considerably, particularly by root development which is a key factor in sensitivity to moisture supply. Mature trees are generally less sensitive to moisture conditions than young trees. It is necessary, in the case of perennials, to distinguish between immediate and long-term effects. Although the impact of a drought can be immediate in terms of vegetative growth, it may be three years or more before this is reflected in fruit crop yields. For practical reasons, comparatively few experiments consider really long-term effects. The response to water conditions at one stage of development will depend on conditions in previous stages and years. In the case of perennials, poor conditions at one stage may not be overcome by later good ones.

The various parts of plants and different physiological processes react differently to moisture stress, and the use of only one factor as an indicator of plant responses can be misleading. It is generally reported for example, that the growth rate of vegetative organs is more sensitive to water shortage than that of reproductive organs. Similarly, fresh weight is more affected by water shortage than dry weight (Salter and Goode 1967). Moisture stress may have adverse effects on vegetative growth but favour the yield of marketable product. Burr *et al.* (1957) report, for example, an increase in the sugar content of cane under dry conditions. Soil moisture can also affect quality, as for example in the burning characteristics of tobacco leaf (Van Bavel 1953), which may be as important economically as gross yield. Care is therefore needed in distinguishing between vegetative growth, total yield and marketable yield, since for different crops different parts are marketed. Kramer (1983) discusses possible beneficial effects of water stress but this is a complex area and results are sometimes conflicting.

Salter and Goode (1967) point out the invalidity of equating transpiration rate and yield as being comparable indications of response to soil-moisture stress. These authors also point out that the time of observation is important since the rate of many physiological processes fluctuates throughout the day as a result of diurnal changes in plant-water content and internal water deficits. It is necessary to distinguish between plant growth and plant development in assessing the effect of

Table 7.9   Cereal crops. Stages of development (after Slater and Goode 1967).

Stage	Details
Germination	The appearance of the radicle
Tillering	The formation of tillers, i.e. branches produced from the base of the stem
Jointing	The stage when two nodes can be seen, i.e. the beginning of shooting
Shooting	The stage of elongation of internodes
Booting	The end of the shooting stage and just prior to the emergence of the ears
Heading ⎫ Earing ⎭	The emergence of the ear from the tube formed by the leaf sheaths
Flowering	The opening of the flowers. In the case of maize this stage is often divided into 'tasselling' and 'silking', being the time of appearance of the male and female flowers, respectively
Grain formation	The period of grain development from fertilization until maturity. The period can be sub-divided as follows: 'milk-ripe' – grain contents have a milky consistency 'soft-dough' – grain contents have a doughy consistency 'waxy-ripe' – grain contents have a waxy appearance 'full-ripe' – grain contents hard 'dead-ripe' – grain ripe for cutting

water. *Growth* is an increase in size without any profound qualitative changes in the growing parts, but *development* is the progression through a series of internal qualitative changes with or without external changes. Development stages are shown in Table 7.9.

Kramer (1983) discusses water deficit effects at three development stages: seed germination and seedling establishment, vegetative growth, reproductive growth. He reports, for example, that vegetative growth, especially leaf expansion, seems to be severely inhibited by even moderate water stress and the reproductive growth stage is particularly water sensitive.

It is clear from the above points and others not dealt with here, that evidence relating yields to water conditions must be considered very carefully. Nevertheless, this is a matter of great importance. In § 3.1, the point was made that new high-yielding crop varieties may be more sensitive in terms of yield to fluctuations in environmental variables such as water. The use of irrigation is increasing, not only as a means of ensuring the survival of a crop or a reasonable yield but also as a means of producing yields once regarded as impossible. In most cases, however, limited available water and economic considerations demand high efficiency in water use, application being mainly at times when it is of most benefit. This necessitates an understanding of the effect of irrigation at various stages of growth. Similarly, studies of the effect of droughts at different stages are important in matching planting and harvesting to climatic conditions or in determining when irrigation is most necessary.

Work relating water to plant growth and yield is reviewed below for various tropical crops which are treated under the two broad headings of annuals and perennials. This approach is necessary to illustrate the great differences between individual crops. Detailed discussions are provided by Salter and Goode (1967), Slatyer (1967), Kramer (1983), Oldeman and Frere (1982), Berger (1969) – fibre crops; Williams and Chew (1980) – tree and field crops of humid tropics; Opeke (1982) – tropical tree crops; Angus *et al.* (1983) – post-monsoon dryland crops; Norman, Pearson and Searle (1984) – tropical food crops; Humphries (1978, 1981) – tropical pastures and fodder crops.

Annual crops

(a) Maize

Claasen and Shaw (1970) state that water deficit at different stages of growth has different effects on yield. Soil moisture during flowering and early grain formation (see Table 7.9) seems particularly critical in determining yield (Salter and Goode 1967; Fuehring *et al.* 1966; Wrigley 1969). Lomas and Herrera (1984) in Costa Rica found that rain amount and distribution during the reproductive stage was the main meteorological factor influencing yield. East African work suggests that there are three periods when water is most essential – germination, fertilisation and grain filling (Semb and Garberg 1969). These authors state that after germination, maize can survive with very little water for some time. Wrigley (1969) indicates that water stress in early growth delays flowering. Stress-induced delay in silking (Norman *et al.* 1984; Kramer 1983) leads to loss of synchrony in development of silks and tassels (Table 7.9) with particularly adverse effects. Williams and Chew (1980) refer to planting ideally requiring soil moisture near FC and a particular sensitivity to stress at tasselling and fertilisation. Lack of water at fertilisation usually lowers yields (Semb and Garberg, 1969) and Wrigley (1969) suggests that the water level in the soil should be maintained to the hard dough stage. Hearn and Wood (1964), in Malawi, found evidence that over-irrigation reduced yield. Benacchio *et al.* (1983) in Venezuela also found in addition to lack of water in early growth affecting yields, excess soil water later on when water demand is reduced had negative effects on the filling out period.

The benefits of early planting of maize in East Africa have been indicated by a number of workers (e.g. Semb and Garberg 1969; Gray, 1970; Gwynne 1964; Turner 1965). Planting at the start of the rains or even before the rains was found to give the highest yields. With early planting, maize developed a better root system and was therefore able to use soil water more effectively (Gwynne 1964). Turner (1965) at Ilonga, Tanzania, found that later planted maize received much less rainfall from the tasselling stage to harvest than early planted maize, and therefore suffered from lack of water during grain formation with resultant low yields. Gray (1970) found that in south-west Kenya, the optimum planting date varied with altitude, being earlier at higher elevations where maize grew more slowly. While such results to some extent reflect the particular climatic conditions of East Africa, evidence of the benefits of early planting can be found elsewhere. This aspect is also examined in Chapter 8. However, the benefits are probably not only in terms of actual water supply.

There is evidence that the low yields of late planted maize result both from limited water and from inadequate supplies of nutrients, nitrogen in particular. At the start of the rains the available nitrogen in the soil is greater than later on (Birch 1960) and hence yields may be greater. Semb and Garberg (1969) found that the benefits of planting early were related to increased water availability rather than nitrogen, but Gray (1970) found that in the low-yield season of 1967, there was an interaction between nitrogenous top dressing and planting date, the application of nitrogen preventing a fall in yield because of late planting.

The usual explanation of a decrease in optimum planting density for maize and other crops as the climate becomes drier is that the wider spacing decreases competition for water, but Allison and Eddowes (1968) have suggested that such a simple interpretation may be misleading. Ochse et al. (1961) suggest that in very general terms, optimum rainfall conditions for maize are a little rain at the start of the growth period, soaking rains every 4 to 5 days from the end of the first month up to about 3 weeks after flowering and a gradual tapering off of rain until harvesting.

(b) Wheat

Salter and Goode (1967) conclude from examination of a large number of studies that the plant seems especially sensitive to moisture conditions during the shooting and earing stages (Table 7.9) when the development of the reproductive organs is taking place. A shortage of moisture then has the greatest effect upon yield and in many experiments this seemed irreversible. Optimum moisture conditions later do not overcome the loss. Irrigation and rain have maximum beneficial effects at these stages.

Other workers, however, have arrived at slightly different conclusions. Day and Intalap (1970) examined three growth stages in Arizona and found that the jointing stage was critical. Water stress at that time resulted in fewer days from planting to flowering and reduced yields. They concluded, however, that moisture stress at any stage of growth decreased grain yield. On the other hand, Hearn and Wood (1964), in Malawi, found a consistent, significant, negative response to high irrigation rates and the plants appeared to remain healthier when considerable moisture stress was allowed to develop. For wheat in the Sudan, El Nadi (1969a) found that the flowering phase and stage of grain filling and maturation were more sensitive to drought than the vegetative period. El Nadi found that yields were not reduced when the crop suffered from cycles of water stress induced during the vegetative period if the crop received favourable water regimes afterwards. Chang (1968), quotes Lehane and Staple (1962) who found that the critical moisture-sensitive stages were heading (earing) and filling. For dwarf wheat in India, Singh (1981) found that with no water deficit during the vegetative stage, yields were sensitive to a deficit during the booting/heading period. However, if there was some conditioning stress in the vegetative stage, yields were relatively insensitive during the later period. Kramer (1983) reports that water stress inhibits spike elongation and spikelet formation and if it occurs at early anthesis, flowers are injured and both size and number of seeds are reduced. The considerable importance of water in influencing yields has been demonstrated in India by Gangopadhyaya and Sarker (1965). They analysed the impact of rainfall occurring at different stages of growth and found that 75 per cent of the total variation in yield could be accounted for.

The different conclusions reached indicate something of the complexity of the subject. Responses of different varieties of wheat would seem to differ considerably. Derera, Marshall and Balaam (1969) examined the drought tolerance of fifteen varieties and found that the main factor seemed to be differences in the earliness of plant maturity. However, other factors such as root development and water-use efficiency also appeared to be significant.

(c) Rice

There are two main systems, dry (upland) rice and the wet (paddy) system where the land is inundated. Deepwater rice (grown in 1–6 m of water) is sometimes considered as separate from paddy (grown in < 1 m of water) (Norman *et al.* 1984). Most work has been on the wet (paddy) system. Varieties have been adapted to a wide range of conditions. In South-East Asia, only about 20 per cent of the area under rice grows under dryland conditions, 20 per cent is irrigated often using diversion–dam systems and the remaining 60 per cent is under rainfed conditions (Robertson 1975). The rainfed rice in particular is subject to the full impact of seasonal drought and floods. Chang (1968) pointed out that it has been found that when either upland or paddy rice is adequately watered the water use is the same, and when either completely covers the ground, potential yields are comparable.

Previously, it had been thought that both water need and yield potential of upland rice were substantially lower than paddy.

Paddy rice is physiologically adapted for growth in standing water. Nutrient uptake is reduced if soil moisture falls below field capacity. Williams and Joseph (1970) state that water use is similar to that of other cereals and the most favourable condition is a saturated soil. One of the main functions of extra water in rice paddies seems to be weed control (Ch. 8). Where dryland production is practised continuously, grasses tend to take over the land (Robertson 1975). Additional benefits may be more even soil temperatures and insurance against drying out. The control of water in cultivation, water movement and unevenness of the soil surface generally mean that considerably more water must be used than is required by the crop for transpiration and functioning (Williams and Joseph 1970). Irrigation water often supplies much of the crop nutrient requirement but may contain toxic substances causing damage.

Rice is highly sensitive to a moisture deficiency at heading and flowering (Salter and Goode 1967). This is the stage of maximum water use (Table 7.8) and a deficiency then can have an adverse effect on yield. Matsushima (1966) in Malaysia found that the critical period for water deficit or excess was 20 days before to 10 days after heading. Williams and Chew (1980) state that the most sensitive period is from the start of flowering to the beginning of grain ripening and Robertson (1975) that it is from the latter part of vegetative growth to about 5 days after heading. Jaw-Kai Wang and Hagan (1981), based on work in the Philippines, suggest that for improved varieties, length of time the plant is subjected to stress is more important in determining yield than the particular stage stress occurs. Jones (1981) for dryland rice in Hawaii found that stress causing stomatal closure for up to 8 days during a critical period from 20 days before flowering to 10 days after flowering had little effect on percentage of filled grain. However, longer periods of stomatal closure could seriously affect yields.

Transplanted crops are generally considered to outyield broadcast ones, although there is evidence from Trinidad that yields from broadcast rice are equal to transplanted where snail and weed control is efficient during early growth (Wrigley 1969). Uncertainty about the start of the rains, especially in monsoon areas, favours the transplanting method because of greater flexibility, plants being kept in the nursery until water is sufficient (Williams and Chew 1980). Cultural techniques are discussed in Chapter 8. The systems used are often very complex and experimental evidence on the effects of inundation is often conflicting.

Where growers do not have ready access to irrigation water and rains are uncertain, Wrigley (1969) points out that farmers store the rain in their paddies in case it is needed. The water height cannot therefore be varied to suit the different growing periods. While results from the Philippines suggest that this is no disadvantage, in Japan a 25 mm depth of water outyielded a 125 mm cover and, provided that the soil is completely covered, trials in Malaysia show that yields are inversely related to water depth (Matsushima 1966). Robertson (1975) states that optimum depth from a practical viewpoint is about 50 mm, one danger of shallower conditions being the problem of maintaining a flooded condition, particularly with high evaporative demand. Since rooting depths tend to be shallow (about 150 mm) water stress can develop quickly (Robertson 1975). Wrigley (1969) states that for a maximum yield the soil should be dried out at least 1 month before transplanting to increase nitrogen content.

In Nigeria, Enyi (1963) examined the effects of waterlogging at different stages of growth and continuous waterlogging. He found that the critical period for waterlogging was for 4 weeks beginning 4 weeks after transplanting, when higher grain yields were obtained than if waterlogging was earlier or later. He found no difference between waterlogging at this time and continuous waterlogging, and suggested that where water supply is limited this was the most beneficial time with the soil being kept moist for the remaining period. If there was enough water for 8 weeks then waterlogging immediately after transplanting was suggested.

In Ghana, Halm (1967) found a marked difference between irrigation keeping the soil only at field capacity on the one hand, and two other treatments – saturated and submerged – on the other. He did not find a difference between the latter two treatments and concluded that provided the water level was above the surface, there would be no difference in yield. Oelke and Mueller (1969) in California found that a shallow (4 cm) water depth all season gave higher yields than intermediate (8 cm), deep (18 cm) or fluctuating (4–18 cm) water depths. They found that relative fertility and varietal responses were similar for all water-management systems. In India, however, Pande and Panjab Singh (1970) found that two varieties of rice, while both showing lower yields under rainfall than irrigation, behaved differently under various irrigation treatments. One variety showed no yield variation under three treatments of submergence, cyclic submergence and cyclic wetting and drying, but the other variety did. They explain this by saying that the first variety completed its life cycle during the period of low evaporative demand. Many higher yielding tropical varieties require more controlled water use and do not perform well with deep flooding (Williams and Chew 1980). The problem of water control has been a limiting factor in using high-yielding varieties and better control will not only improve the situation during the wet season but also perhaps allow up to three crops a year. Patrick (1967) found that the yield response to different flooding regimes depended upon the nitrogen status of the soil. Alternate flooding and drying on nitrogen-deficient soils reduced growth and yield. However, this irrigation regime increased yield on a soil with plentiful nitrogen.

Robertson (1975) states that the greater diurnal temperature variation in shallow water seems to stimulate tiller formation and ultimately yield. Shallow water also favours organic matter decomposition which stimulates root development, thereby increasing nutrient uptake and reducing possibility of water stress. Deep flooding will tend to reduce available oxygen and perhaps cause soil toxicity (Robertson 1975). The reduced oxygen is said to account for less than optimum germination and seedling establishment in water-saturated and flooded soils (Turner, Cy-Chain Chen and McCauley 1981). With deep water, temperatures may be lower, leading to slower decomposition of organic matter and reduced tillering rates. In deep water, plants will be taller and lodge more easily and photosynthesis may be affected (Robertson 1975). Many of the above effects depend on the variety, with floating types having stems up to 6 m being able to produce grain provided the upper leaves and panicles float above the water (Robertson 1975).

Even this brief review of the large literature indicates that the question of the water regime is very complex. Studies not cited include Oldeman and Frere (1982), Lomas and Herrera (1985), Novero *et al.* (1985) and various papers in WMO/IRRI (1980). Further aspects are considered in § 8.2. It seems that while water should completely cover the soil, deep water is perhaps unnecessary and probably harmful. More scientific use of water could increase yields with perhaps considerable water saving.

(d) Sorghum

Sorghum is well adapted to arid and semi-arid areas too dry for maize. It is intermediate between maize and pearl millet in tolerance of a water deficit (Norman *et al.* 1984). Salter and Goode (1967) concluded that its resistance to drought is probably because of its well-developed root system and effective internal control over transpiration. Deficits have a marked impact on rooting characteristics (Norman *et al.* 1984). Despite resistance to drought, the crop seems to respond well to plentiful water supply during the booting and heading stages and drought at these times reduces yields. Lawes (1966) in northern Nigeria indicates that despite drought tolerance, rainfall conservation is still important to give an even distribution of moisture, build up reserves and thereby protect sorghum against minor droughts.

(e) Millet

While there is comparatively little information, there are indications that response to water supply at heading and flowering may be similar to those of the more important grain crops, i.e. that these are moisture-sensitive periods. For pearl millet in India, Mahalakshmi and Bidinger (1985) found that water stress prior to panicle initiation did not affect grain yield of the main shoot but increased yield on the tillers, resulting in increased total yield. Stress during panicle development did reduce grain yield on the main shoot but this was compensated by increased number of tiller panicles producing grain. However, stress during flowering and grain filling reduced both grain components. Norman *et al.* (1984) state that a deficit from anthesis onwards means that auxiliary tillers fail to develop grain and reduced evaporation increases temperature thereby reducing the grain filling period and hence grain size is small. Squire *et al.* (1984) compared irrigated and non-irrigated pearl millet under post-monsoon conditions in India. Marked differences in dry matter production and transpiration occurred. Rapid and deep root penetration account for the drought tolerance of pearl millet (Norman *et al.* 1984)

Salter and Goode (1967) summarise conditions for cereals, including those dealt with above. They appear to have a marked sensitivity to water conditions during formation of the reproductive organs and during flowering, so that shortage at these times tends to reduce yields. When flowers start to develop, root growth is often much reduced and this is also a time of high water demand. Unless soil moisture is high to permit rapid water movement to the roots, this can lead to a marked reduction in plant water, a matter of great significance since this is a moisture-sensitive period. Irrigation, therefore, has its maximum beneficial effect on yield during the development of floral organs (late shooting, heading and flowering stages).

(f) Cotton

Because of its very great economic significance in various parts of the world, the effects of water on cotton yields have received considerable attention, particularly in terms of irrigation. The moisture requirements differ according to variety (Ochse *et al.* 1961; Wrigley 1969) but soil characteristics also play a great part. Gerard and Namken (1966), for example, found that because of the greater water availability on medium-textured soil, summer rainfall in the Lower Rio Grande supplied enough water to give 80–90 per cent of the yield potential but only 50 per cent on fine-textured soil. They demonstrate the complexity of water–yield relations, stressing not only the impact of soil type but also rooting habits and related moisture extraction patterns and the different irrigation needs on different soil types. Chang (1968) suggests that for maximum yields, it is necessary for the soil-moisture level to be kept above 50 per cent of available water.

Experiments to determine moisture-sensitive stages are not all in agreement, but many indicate that adequate water is needed before flowering to ensure good growth (Salter and Goode 1967; Marani and Horwitz 1963). At the start of flowering and during boll development, moisture supply is critical in influencing boll growth as well as further vegetative growth. Guinn and Mauney (1984a) refer to conflicting evidence on the effects of a water deficit on fruiting, some reports indicating increased flowering, others a decrease. In an experiment in Arizona they indicate the complexity of the situation including the impact of lygus bugs which were more plentiful in irrigated plots. They found no evidence that water stress provided a physiological stimulation of flowering. A water deficit decreases boll retention (Kramer 1983; Guinn and Mauney 1984b), as does excessive rain (Berger 1969). As in the case of cereals, at the time of flowering and fruit development, the roots may have temporarily stopped growing, creating water-supply problems at a moisture-sensitive stage.

The effects of rainfall and its time of occurrence have been examined by a number of workers. Jowett and Eriaku (1966) have pointed out the problems of such analyses, and in particular the question of correlation between the various factors. For example, a negative correlation between cotton yield and rainfall may be the result of a negative correlation between rainfall and sunshine hours, the latter having a positive effect on yield. A number of workers have found positive correlations between yield and rainfall at particular times of year (e.g. King 1957, in northern Nigeria; Manning and Kibukamusoke 1960, in Uganda; Crowther 1944, in the Sudan Gezira). For Namulonge in Uganda, Manning and Kibu-kamusoke (1960) stress the benefits of early planting or even dry planting to offset the effects of rainfall variability, and of rainfall conservation. Berger (1969) points out that heavy rain can injure young seedlings and even harm fully grown plants and that a drier period is necessary to allow bolls to ripen and be picked. Berger (1969) suggests that a minimum of 500 mm of rain per year is required, with 175–200 mm well distributed during the growing season. Wrigley (1969) quotes Crowther (1944) who found that high total rainfall in the year prior to sowing decreased yields, this being attributed to weed growth on fallows utilising nitrogen and soil moisture.

Marani and Amirav (1971), in examining the effect of moisture stress on two cotton varieties in Israel, found that lint yield ranged from 270 kg/ha on a non-irrigated treatment to 2000 kg/ha on a fully irri-gated treatment. Another Israel experiment in the Negev illustrates the complexity of the impact of irrigation on yield (Amir and Bielorai 1969). They found, for example, that by applying the first irrigation at the beginning of flowering, maximum yields of 1700–2100 kg/ha could be obtained with three irrigations. With the three driest treatments when the plants suffered from moisture stress, lint quality was not affected.

Bennett, Ashley and Doss (1966) found that both a black plastic mulch and irrigation increased cotton yields – in the case of irrigation this increase being more than 1790 kg/ha with or without mulch. From experiments in Uganda, Rijks and Harrop (1969) concluded that the most important function of irrigation in the area was to ensure the germination and establishment of the crop at a date which was optimum in respect of the temperature and sunshine regimes in the early stages of growth and at picking.

Since irrigation water is often in short supply, determination of the minimum requirements for optimum results is of great importance. Accepting previous experience that the beginning of flowering was the best time to irrigate, Marani and Fuchs (1964) found that the amount applied in a single irrigation at that time to produce optimum yield was that needed to wet the soil to a depth of 90 cm. More water did not increase yields but less water reduced them.

(g) Sunflower
Kramer (1983) states that water stress increases the time from seed germination to flowering but does not affect development of flower primordia. Water is critical in the formation of oil, particularly during the period between the formation of flower heads and the ripening of the seeds (Salter and Goode 1967). A shortage of water at this stage seems to result in a much reduced seed yield and a reduction in the oil content of the seed.

(h) Groundnuts
Again, the evidence of various experiments is conflicting. An important factor may be the very great differences in the reaction of different varieties of the crop to water (Wessling 1966). Water stress reduces germinability of seeds of some varieties (Kramer 1983). Salter and Goode (1967) conclude from limited information that for maximum yields an adequate supply of soil moisture is essential during flowering and seed development, and that drought at this time would seriously reduce production. One experiment quoted by them indicated that yields may not be affected by wide variations in soil moisture between germination and flowering. However, work by Holford (1971) in Fiji suggests that growth and yield are very sensitive to moisture stress at all stages and at no time are they resistant to drought, which reduces vegetative growth, flowering and peg formation. Holford found that growth was most responsive to readily available moisture during the establishment, rapid vegetative growth and flowering phases up to the time of maximum flowering midway through the growth cycle. Rainfall received in the first half of the growing season accounted for 82 per cent of the yield

variability. There did not appear to be a relation with rainfall in the second half of the season except for a suggestion that excessive rainfall might depress yields. Nageswara Rao *et al.* (1985) examined impacts of three irrigation procedures at four stages for groundnuts in India. Kernel yield was most reduced by water stress during seed filling and the lowest pod yields occurred when severe stress was applied from emergence to maturity. A decrease in irrigation in the early phases increased pod yield compared with full irrigation. Norman *et al.* (1984) point out that although there are many reported yield responses to irrigation, they have rarely been related to soil water potential.

Ochse *et al.* (1961) suggest that rainfall at frequent intervals during the period of vegetative growth is beneficial but likely to be harmful while the pods are developing and ripening. Williams and Chew (1980) indicate the need for dry conditions during harvest. Chang (1968) states that groundnuts do not respond with an increase in yield if soil moisture is increased above 50 per cent. He suggests also that excessive water application can reduce yields because of leaching and restriction of root development. Ishag *et al.* (1985) compared two plant densities in the Sudan Gezira, finding that plants in the sparse population were more sensitive to stress. Hence, for the dense population, watering interval could be lengthened without serious yield reduction.

(i) Soyabeans

The high protein value of soyabeans means that there is considerable interest in their expansion, particularly in the humid tropics (da Mota 1978). While Norman *et al.* (1984) suggest the need for relatively high soil moisture for germination, da Mota (1978) states that excessive moisture at that time is injurious. Sensitivity of any one component of vegetative growth to stress is offset in part by compensation among yield components, although the ability to compensate diminishes during grain filling (Norman *et al.* 1984). Hence end of season deficits reduce all yield components.

The long flowering period and extensive root system allow the plant to survive short periods of drought stress, but shortage of water during pod filling reduces yields much more than any other stage, including flowering (da Mota 1978; Kramer 1983). Deficits for 2–4 weeks immediately after flower bud differentiation reduce growth and cause heavy flower and pod dropping. Although soyabeans can tolerate short periods of waterlogged soils better than corn, excessive moisture after bud differentiation reduces yields (da Mota 1978). Of particular importance to the introduction of this crop in humid areas is the need for a dry period at harvesting.

(j) Broad and field beans

The moisture-sensitive period is that of flowering (Salter and Goode

1967; El Nadi 1969b; Norman *et al.* 1984) and irrigation at this time is particularly useful. El Nadi (1970) examined the effects of varying the number of irrigations between six and nine in the Sudan. Increasing the number of irrigations increased the yields, but the significance varied depending on when water was saved. Hearn and Wood (1964) in Malawi, during experiments on a number of dry season crops, found that beans were the only crop to show a loss in yield if the full water requirements were not met by irrigation. Norman *et al.* (1984) point out that beans are not particularly tolerant of water stress despite its common occurrence in major bean areas. They also indicate that heavy rain or irrigation produces a microclimate conducive to diseases.

Perennial crops

(a) Cocoa

Work on this crop indicates that it is extremely difficult to separate the effects of water from those of other factors. Ali (1969), in Ghana, found a positive correlation between yield and rainfall at some times of year and a negative correlation at others. He concluded that this was a result of other factors and in particular that a positive correlation was obtained with a high level of soil nitrate and negative when the nitrate level was low. Other factors such as insect occurrence were thought to have an effect.

Salter and Goode (1967) point out that the interrelationships of factors such as rainfall, temperature, humidity and light make interpretation difficult. There is not necessarily a relation between yield and total rainfall amount, but Ochse *et al.* (1961) suggest that an annual total of 1500–2000 mm is needed without irrigation. Williams and Chew (1980) suggest 100–200 mm per month as best, with annual totals over 2000 mm perhaps depressing yields, particularly on heavy, poorly drained soils since roots are sensitive to poor aeration. Cocoa has little resistance to drought and a water deficit produces rapid stomata closure and resultant severe effects on growth and development (Alvim 1960). Alvim suggests that soil moisture needs to be kept above 50–60 per cent. Growth seems to be favoured by the absence of a dry season and Ochse *et al.* (1961) suggest that some rain is needed in every month with as even a distribution as possible.

Pod production is continuous with no dry season, as in the areas where cocoa originates (WMO 1982; Williams and Chew 1980). In areas with a short dry season a period of no bearing occurs about five months later although annual yields may be rather heavier than with continuous production (Williams and Chew 1980). Young cocoa is seriously set back by even two weeks of drought. Since cocoa is sensitive to a dry atmosphere, even with irrigation, set-back occurs with a long dry season

(Williams and Chew 1980). However, excess humidity ($> 90\%$) also affects yield. Growth in shade reduces impacts of moisture conditions (Williams and Chew 1980; WMO 1982). Bean quality seems to be determined by the sum of rainfall amounts in the three months preceding harvest (WMO 1982).

Smith (1964) analysed the effects of three different moisture regimes on cocoa establishment. He found that irrigation increased growth rate and flower production, and flower production occurred earlier. In Ghana, however, mature cocoa growth took place in the drier months and stopped in the wet season (Salter and Goode 1967). In this case, temperature seemed to be the key factor. While adequate water seems essential to good growth there is some evidence that surplus soil water has bad effects (Salter and Goode 1967; Williams and Chew 1980).

(b) Coffee

Moisture requirements at certain times are complex and there is some disagreement. While under natural conditions, growth follows the cycle of the rains, there does not seem to be any evidence to support the idea that a resting period in the dry season is essential. No harm appears to result from maintaining soil moisture at a sufficiently high level to make growth continuous (Wrigley 1969). Salter and Goode (1967) suggest that the vigour of vegetative growth can be increased by providing water during dry periods, particularly during crop development, and that supplementary irrigation is necessary for maximum yield.

Wrigley (1969) states that adequate moisture appears to be particularly important to stimulate flowering, to sustain bean growth about three months later and to keep the crop in good condition for the following season. Rainfall during the period 10–17 weeks before flowering may be important in determining yield (WMO 1982). There is, however, some disagreement about the best conditions for flowering, a very complex sequence being required to produce open flowers. Evidence suggests that an internal water deficit is needed to break bud dormancy (Kramer 1983), but Salter and Goode (1967) point out that when transpiration rates are high, the internal water supply to the bud is sufficiently restricted to promote an internal water deficit regardless of any soil-moisture conditions. Sanders (1964), however, suggests that soil moisture be kept low prior to flowering to assist flower emergence. Therefore, the picture at this time is rather complex – particularly in terms of the possible impact of other factors such as temperature.

The time of blossoming is influenced by the occurrence of showers or irrigation once the buds mature and this may enable some control of the time of flowering (Wrigley 1969). The avoidance of serious water stress favours berry setting and enlargement (Salter and Goode 1967). While it seems that moisture stress at all stages is to be avoided, Sanders

(1964) suggests that it is important to encourage young trees to develop deep and extensive roots by occasional irrigation during the dry period rather than regular irrigation.

A wide range of rainfall amounts is suggested, e.g. 1800–3000 mm per year (Williams and Chew 1980), 750–3000 mm (Opeke 1980). Williams and Chew (1980) suggest that Arabica and Robusta grow well with one rainfall peak but with adequate amounts in most months and a 1–2 month dry season. While Liberica coffee is adapted to uniform humid conditions, a short dry season can be tolerated. Particularly in lower rainfall areas such as East Africa with perhaps only 1000 mm per year, mulch is important (Williams and Chew 1980).

(c) Tea

Tea yields appear to be proportional to solar radiation (Ellis 1967 – quoted by Wrigley 1969) in central Africa, and since there will to a certain extent be an inverse correlation between radiation and rainfall, analysis of the link between the latter and yields is difficult. There is therefore considerable variation in the requirements suggested by different writers. Evenly distributed rain (Williams and Chew 1980), high rainfall and humidity for most of the year (Opeke 1982) and about 100 mm in each month (Wrigley 1969) indicate the need for a reasonable moisture supply in all months. Salter and Goode (1967) and Eden (1976) state that rainfall in the dry period is critical and if monthly averages are less than 50 mm over a period of several months production suffers severely.

Ochse et al. (1961), while agreeing that the crop will suffer during a sharply defined dry season, point out that if there is enough rainfall during the dry season, a higher quality of tea is obtained than in a continuously moist climate. They suggest that this is because of greater sunshine, pointing out that in a prolonged wet season with less sunshine, growth and quality will be impaired. If irrigation water is provided to allow growth during the dry period, since this is when sunshine is high, both yield and quality of tea can be increased considerably.

Carr (1972) suggests that a minimum annual rainfall of 1150–1400 mm is necessary but that seasonal distribution in relation to evaporative demand is important, the latter point also being made by Eden (1976). Minimum annual totals of 1000–1400 mm (Williams and Chew 1980), 1100 mm (Opeke 1982) and 1270 mm (Eden 1976) with no upper limit provided sunshine is not greatly reduced (Williams and Chew 1980; Eden 1976) are suggested. Factors such as altitude and overcast weather modify the minimum requirements (Eden 1976). The success of tea at higher altitudes is in part because of adequate rainfall but also due to the presence of mist as a moisture source and in raising humidity (Williams and Chew 1980). Carr (1972) cites work indicating the influence of rainfall on yields pointing out that this is not always a simple relationship. The complexity is also suggested by Salter and Goode (1967) who point out, for example, that the effects of a drought may not be felt for some time. The overall importance of water has been illustrated by Fordham (1971) in comparing different irrigation regimes. Plant growth and survival were significantly better under the wettest compared with the driest regimes.

(d) Coconuts and oil palms

An even distribution of rainfall throughout the year is beneficial and if there is a well-defined dry season, trees grow less rapidly and yield less (Ochse et al. 1961). The long period of development (of the order of 29–33 months from initiation to flowering) makes it difficult to analyse water–yield relationships, but it seems that moisture conditions in the dry season are important. For both coconuts and oil palms, yields are correlated with dry season rainfall in the previous 2 years (Williams and Joseph 1970). Smith (1966) also found that dry season rainfall and soil-moisture deficits were important. This latter study is discussed further in § 8.1. Rainfall and irrigation at this time have a great effect upon yields, and in the case of young nursery plants it is important to be able to apply water then. Minimum annual rainfalls, evenly distributed, of 1500 mm for oil palm and 1250 mm for coconuts are suggested by Opeke (1982).

(e) Sugar cane

Salter and Goode (1967) conclude that moisture requirements have been largely resolved and that during periods of rapid vegetative growth, cane will respond best – in terms of final yield of high-quality juice – to relatively low soil-moisture tensions. Koehler et al. (1982) found that a single drought stress period during vegetative growth resulted in a serious reduction in stalk elongation prior to a reduction in the rate of soil water depletion. Williams and Chew (1980) state that even a slight water deficit during the establishment phase can reduce growth. Where there is too much rain, especially in the absence of a sharply defined dry season, the cane will produce a great deal of green matter with low sucrose content and a few dry months are essential for proper maturation (Ochse et al. 1961). Near harvest time and during the ripening phase, sucrose accumulation and quality are favoured by water stress due to a slowly imposed drought (Salter and Goode 1967).

Sugar yield depends not only on tonnage but also sugar content and cane quality, and the effects of moisture on all of these must be considered. The great influence of moisture on yields has been demonstrated by a considerable number of workers. In Natal, Thompson and Wood (1967) found that in two successive 12-month seasons, mean yields per plot were 11.33 and 45.8 t/ha. The great reduction in the first season was ascribed to poor rainfall conditions causing a severe moisture shortage. Oguntoyinbo (1966) found a significant correlation between yields and water deficits in Barbados. Deficits seemed particularly critical in the early growth stages.

Chang (1968) also confirms that a drought at the stage of active growth will reduce yields more than one at a later stage. Irrigation is of great importance in areas where rainfall is insufficient or occurs in only a few months. Thus Thompson, Gosnell and De Robillard (1967) and Thompson and De Robillard (1968) found a significant linear increase in yield with quantity of irrigation water supplied. They found, however, that soil type was of considerable importance and concluded that where irrigation water is limited, heavy soils should be irrigated in preference to sandy ones.

In reviewing studies of the influence of meteorological factors in India, Barbados and Mauritius, WMO (1982) concluded that relationships with yield were complex and that there are dangers in transferring results from one area to another. Rainfall (either too little or too much) is important, but its influence at one period depends on moisture conditions at earlier stages. Williams and Chew (1980) suggest minimum rainfall of 1000–1200 mm/yr with a dry season to aid maturation. Recent studies include work in India by Rupar Kumar (1984) and Venkataramana, Shunmugasundaram and Naidu (1984), the latter concerned with drought resistance of different varieties.

(f) Bananas

This crop does best in areas with no marked dry season. Wrigley (1969) and Williams and Chew (1980) suggest that a minimum of 100 mm is necessary in each month, and there is much evidence to indicate that when grown under conditions of adequate moisture throughout the year plants will be larger with higher yields, larger fruits and improved quality than if they are subject to moisture stress at times. Without irrigation, a dry season delays maturity (Williams and Chew 1980). Suggested annual totals are > 1250 mm/yr and even 2000–2500 mm/yr (Norman *et al.* 1984). Salter and Goode (1967) indicate that each stage of growth is dependent upon previous stages as well as current conditions. The preceeding stage sets an upper limit on what can be achieved and current conditions determine whether this limit will be reached. Harm done in the first three months cannot later be undone. From this, it is clear that irrigation can be of great benefit, particularly during seasons when the price is high. However, there are dangers in over-irrigation. Holder and Gumbs (1983c) indicate that waterlogging adversely affects growth and yield. They also found that over-irrigation reduced the number of hands per bunch, and fingers per hand although bunches

were heavier than with no irrigation because of an increase in finger
length.

Tropical pasture and fodder crops

These are of great importance in many tropical areas, particularly drier
regions (§ 8.3). The large range in environmental conditions and species
makes discussion impossible here. Overviews are presented by
Humphries (1978, 1981). The effects of plant water potential, rooting
density and water use on dry matter production of several tropical
grasses during short periods of water stress in Colombia are discussed
by Jones, Pena and Carabaly (1980). Influence of rainfall patterns on
the production and quality of four Bermuda grasses is examined by
Griffin and Watson (1982). Responses of 16 sub-tropical grass species
to different irrigation and harvest regimes is considered by Mislevy and
Everett (1981).

Water–yield relationships: general comments

A considerable number of crops have not been discussed (e.g. sisal,
cassava, tobacco, pineapples and other tropical fruits), material on them
being contained in references cited earlier in this section. However, the
coverage has been sufficiently wide to indicate general points. Experimental evidence is conflicting in a number of cases, the problems in
interpretation having been indicated at the beginning of the section. The
stage of growth is critical in terms of response to moisture availability
and there are considerable differences between crops. Furthermore, it
appears that for a particular crop, response can differ depending upon
the variety (e.g. groundnuts). The response to water supply and irrigation in some cases depends to a certain extent on both soil type and
nutrient status (especially nitrogen level) of the soil.

Not only yield but quality of the crop is important and the two in some
cases respond differently to moisture conditions. The necessity for drier
conditions at certain times for sugar cane is an example of this. A point
not directly related to yield is that wet conditions in the sugar-harvesting
period seriously impede burning, cutting and transporting to the mills.
Some moisture stress during ripening would seem to have beneficial
effects in the case of rubber and tobacco (Chang 1968; Salter and
Goode 1967). Water use efficiency is of economic importance to irrigation and is discussed by Kramer (1983). He points out that although
in general terms it can be defined as the amount of water used per unit
of plant material produced, the concept can be applied in various ways.
It is most commonly defined by agronomists as units of water used per
unit of dry matter produced. The efficiency varies with crop type, yield

and water used as well as climate (Chang 1968). Aggarwal and Sinha (1983) illustrate how water use efficiency of wheat varies widely depending upon several factors, and the wide range for paddy rice is referred to by Robertson (1975). Often the efficiency seems to increase up to the potential water-use figure, but in some cases (e.g. sugar cane) maximum water-use efficiency seems to be reached at a point considerably below potential evapotranspiration (Chang 1968). Sorghum has a higher water-use efficiency than wheat, giving the former an advantage where water supplies are limited.

Cultivation practices, particularly the use of fertilisers, can alter the efficiency of water use. Proper fertiliser use can markedly increase yields without an appreciable change in water use and in times of water shortage, fertilized crops are more resistant to drought because of a better-developed root system (Chang 1968, quoting Smith 1953). Soil type affects water-use efficiency. Thompson, Gosnell and De Robillard (1967), for sugar cane in Natal, found that on a sandy soil there was little difference in cane yield per unit of water for a range of soil-moisture deficiencies. On clay soils, however, the return per unit of water did depend upon the deficit they allowed to develop in the soil.

Irrigation water is often limited and costly. Excessive irrigation can reduce the uptake of nutrients by the plant because of dilution and can also cause leaching (see discussion of groundnuts). If soils become saturated, lack of oxygen is a problem. For these reasons, over-application can create problems, underlining the value of careful assessment of water needs and efficiency of use. Experiments for a number of crops, including maize, wheat, groundnuts and cocoa, indicated a yield reduction with high irrigation.

It has been suggested that observed variable impacts of growth regulators on yields may be influenced by environmental stress. Regehr (1982) examined the impact of two such regulators (Dinoseb and Triacontanol) on maize under irrigated and non-irrigated conditions and concluded that the inconsistent results reported in the literature should not be attributed to environmental factors.

Discussion of water–yield relationships raises the question of modelling, for example in yield prediction. This includes water-balance models (§ 8.1). Weather-based models for estimating development, ripening and yields are discussed by Baier (1977), WMO (1982) and Robertson (1983). Yao (1981) discusses a three-part system for crop–climate modelling using meteorological and satellite data and comprising: (1) crop potential assessment to assess suitability of crops for an environment; (2) crop condition assessment to assess growth and for use with yield models to adjust model estimates; (3) crop yield assessment models using meteorological and agricultural data. Five approaches to crop yield models are suggested: (a) crop growth simulation; (b) crop/weather analysis; (c) empirical–statistical models; (d) analogue models; (e) the yield index. Norman *et al.* (1984) suggest that the first of these (crop growth simulation or the physiological approach), which analyses environmental effects at each development stage, involves large-scale computer simulation. They suggest it is rather inappropriate at present due to ignorance of particular tropical environments. Lomas and Herrera (1984) provide an example of an agrometeorological model for maize in Costa Rica.

CHAPTER 8
WATER AND AGRICULTURE IN THE TROPICS

8.1 WATER-BALANCE STUDIES – THEIR APPLICATION TO AGRICULTURE

The water balance of an area depends on meteorological factors influencing precipitation and evaporative loss from plants and the soil as well as other factors influencing surface and sub-surface water movement, soil infiltration, percolation and water storage capacity characteristics, crop species and development stage and agricultural practices. The complexity of these different elements, indicated in previous chapters and in the next section as far as agricultural practices are concerned, presents enormous problems and means that water-balance studies are inevitably a simplification. The degree of simplification will be related to the purpose of the study, the available data and the analytical facilities available. Some of the major applications of water-balance studies are as follows:

1. To provide a general overview of the water conditions in an area. Here, precipitation, potential and actual evaporation, soil moisture, soil-moisture change and drainage are considered. This basic, widely used water-balance approach was introduced in § 5.3 and application in climatic classification referred to in § 5.4. While useful, the simplifications involved can be misleading for agriculture. For example, the use of a potential evaporation value rather than a crop-water requirement is a handicap.
2. To form part of a model for investigating rainfall–runoff relationships and streamflow prediction from climatic data.
3. To assess the suitability of an area for a particular crop and vice versa (i.e. analysing the extent to which the water requirement of the crop will be met). Related to this is the assessment of favourable planting and harvesting dates.
4. Following on from (3), to assess irrigation requirement – both quantity and interval. Alternatively, in the case where there is a surplus, a water balance will indicate its magnitude and hence drainage requirements.

5. To examine water–yield relationships. Due to the complexity of the system (§ 6.3), a simple investigation of rainfall–yield relationships may be inappropriate. A water-balance approach allows yields to be related to effective rather than total rainfall. The question of modelling water–yield relationships was introduced at the end of § 7.3.
6. To assess water use by a particular vegetation or crop type.
7. To assess human impact on the system. The effect of irrigation, changing land use and management practices can be analysed, either by mathematical simulation of the water balance or catchment experiments (Ch. 9).

The water-balance equation as given in § 5.4 is

$$R = E_t + \triangle S + \triangle G + Q + L$$

In practice, assumptions and simplifications are adopted. A common assumption is that all the precipitation infiltrates into the soil and therefore, strictly speaking, there is no 'surface' runoff, a water surplus occurring only when the soil-moisture storage capacity has been exceeded. A second frequent assumption is that the rate of evaporation (transpiration) does not change with change in the amount of available water. Both of these aspects were discussed in § 6.3. If the fact that plants may experience difficulty in obtaining water once soil moisture falls below field capacity, with a resultant fall in evapotranspiration, is taken into account, then the calculation is sometimes termed a 'modulated water balance' approach. However, as was indicated in § 6.3, a number of factors influence the relation between the actual water use and soil moisture and hence the decision as to appropriate values of the former presents problems. This aspect is discussed by Chang (1968), Baier (1981) and Giambelluca (1986) for example. The time period considered for the calculations (i.e. monthly, weekly, daily basis) is important. Doorenbos and Pruitt (1977) examine water-balance applications in irrigation scheduling and Baier (1981) considers their use in crop-yield models. Oldeman and Frere (1982) describe methods of calculating the soil water-balance with particular reference to humid regions of South-East Asia. The water balance for rice is considered by Robertson (1975), in this case water storage comprising both water in saturated soil and depth in the flooded paddy. Use of the water balance in assessing irrigation need is considered. In using a water-balance approach to estimate evaporation, McGowan and Williams (1980a) refer to the problem of distinguishing the latter from drainage. They describe a simple graphical approach to overcome the problem. Problems of soil-moisture assessment (§ 6.3) have implications for the use of water-balance approaches to estimate water use by a crop, although McGowan and Williams (1980a) suggest that neutron probe errors can be made small enough for such a purpose. However, it is because there are problems that a water-balance approach to estimate soil moisture is important for various purposes such as those indicated above. To illustrate the application of the water-balance method to agriculture, a number of case studies of different types will be used.

Dagg (1965a) used a simple water-balance approach to test the suitability of a marginal area in Kenya for growth of two varieties of maize. The calculations were carried out using monthly values (Tables 8.1 and 8.2). A reliable rainfall amount which could be expected 7 years out of 10 on average rather than mean monthly rainfall was used. Use was made of E_t/E_o ratios for the two varieties at different stages of their growth and an E_o estimate (using the Penman formula) in order to calculate E_t. It was assumed that all the rain infiltrated into the soil. The results of the calculations supported the observed facts of maize cultivation in the area. In particular, the imported variety of maize, which had a shorter growing season than the local variety, but higher water demand at certain stages of its growth, was found to be more suited to the rainfall season. The local variety, with a longer season, was likely to be subjected to moisture stress towards the end.

In southern Tanzania, yields of cashew nut trees at close spacing of 6 m were found to decrease with age, whereas yields were maintained and increased up to at least 5 years with a spacing of 15 m. It was thought that the decline in yields under close spacing was due to the development of water stress as water demand increased with age and canopy cover. The correctness of this hypothesis was tested by Dagg and Tapley (1967) using a monthly water-balance approach. Three cases were considered:

1. Close spacing, resulting in a completely closed tree canopy.
2. Isolated trees with annual grasses growing beyond the canopy.
3. Isolated trees with bare soil beyond the canopy.

The calculations used average monthly rainfall, a Penman open-water surface evaporation figure (E_o), E_t/E_o ratios taking the proportion of the ground covered by the various types of vegetation or bare soil into account and soil-moisture storage capacities. The calculations indicated that only in case (3), i.e. isolated trees with bare soil, would the cashew trees not experience moisture stress at certain times of the year.

Giambelluca (1986) used a modulated monthly water balance to examine land use impacts on the hydrology of Oahu, Hawaii. Plantation crops (sugar cane and pineapple) had most impact on ground water recharge, urban influence varying with climate.

Table 8.1 Water supply and demand for local maize varieties at Muguga planted on 20 March (after Dagg 1965a)

	Month											
	Jan	Feb	Mar	Apr	May	June	July	Aug	Sept	Oct	Nov	Dec
Penman E_o (mm)	171.7	179.5	183.2	140.1	98.0	98.4	105.1	85.6	147.6	184.1	147.9	116.3
E_t/E_o for maize			From 20th 0.45	0.48	0.63	0.87	1.00	0.92	0.67	To 15th 0.45		
E_t(mm)			30.0	67.2	61.7	85.6	105.1	78.7	98.8	40.1		
Reliable rainfall 7 years in 10 (mm)	50.1	32.9	47.0	185.8	164.0	38.2	16.5	19.3	18.7	44.3	115.8	80.9
Cumulative water storage in soil (mm)			17.0	135.6	238.1	190.7	102.1	42.7	-37.4			

Table 8.2 Water supply and demand for imported short-term maize varieties at Muguga planted on 20 March (after Dagg 1965a)

	Month											
	Jan	Feb	Mar	Apr	May	June	July	Aug	Sept	Oct	Nov	Dec
Penman E_o (mm)	171.7	179.5	183.2	140.1	98.0	98.4	105.1	85.6	147.6	184.1	147.9	116.3
E_t/E_o for maize			From 20th 0.45	0.49	0.70	0.94	1.00	0.75	To 20th 0.47			
E_t (mm)			30.0	68.7	68.7	92.5	105.1	64.3	46.2			
Reliable rainfall 7 years in 10 (mm)	50.1	32.9	47.0	185.8	164.2	38.2	16.5	19.3	18.7	44.3	115.8	80.9
Cumulative water storage in soil (mm)			17.0	134.1	229.6	175.3	86.7	41.7	14.2			

Since plant–water relations over short periods are important, a more detailed analysis than monthly conditions is often necessary, particularly for irrigation. In § 6.2, it was pointed out that some plants suffered a decline in photosynthetic rate with even slight water stress and therefore would require frequent irrigation. The complexities of water availability in relation to soil type, rooting habits and transpiration rate were discussed in § 6.3. This will influence the level of moisture which must be maintained in the soil and hence the frequency of irrigation. Therefore, irrigation practices will need water-balance calculations over different time periods depending on crop type and environmental conditions. Crops reacting to a slight water stress may need daily or 5-day calculations, while more tolerant crops would only need calculations on, say, a weekly or fortnightly basis.

A water-balance approach was adopted by Munro and Wood (1964) in analysing water requirements and yield of maize in Malawi. The basin system of irrigation was used (§ 8.2) and irrigation treatments were based on readings of an evaporation pan (122 cm diameter, 42 cm deep used in East and Central Africa) multiplied by a series of 'irrigation factors' ranging from 0.6 to 2.4. A running water balance was kept for each treatment (irrigation factor) and 50 mm of water were applied when the cumulated soil-water deficit reached or exceeded this amount.

The yield differences between treatments were barely significant, but soil-moisture measurements showed that those with the lower irrigation rate (i.e. with a low irrigation factor) were drawing on soil-moisture reserves. In the case of soil with a lower moisture capacity than those in this experiment, yield differences could have been substantial. A considerable difference in water use with stage of growth of the crop was found (see § 7.2), and while an irrigation factor of 0.6 was sufficient up to tasselling, one of 1.2 was needed in the month after tasselling.

Hearn and Wood (1964), again in Malawi, compared two approaches to assessing irrigation need:

1. A water-balance approach.
2. Actual soil moisture measurements.

A wide range of crops was studied including maize, wheat, beans and onions. As in the previous case, various irrigation factors from 0.3 to 1.8 were applied to pan evaporation. The frequency of irrigation was varied by allowing different soil-moisture deficits to develop. A reasonable agreement was found between the water-balance approach and soil-moisture measurements.

In Uganda, Rijks and Harrop (1969) used a water-balance approach in an experiment concerning irrigation and fertiliser use on cotton. For irrigation control, the water balance, including rainfall, irrigation and crop-water requirement was used to determine the occurrence of a moisture deficit. The crop-water use was determined by applying a series of crop factors ranging from 0.33 to 1.20 at different stages of growth to a Penman open-water surface evaporation figure. Details of the factors are contained in Table 7.6 (§ 7.2). An important conclusion was that the use of a water-balance approach reduced water expenditure, lowered labour cost and gave similar yields to a pre-set pattern of irrigation at regular intervals.

In Kenya, Wallis (1963) examined the use of water-balance methods to estimate soil-water deficit and crop-water use. He assessed the impact of water deficits on yield and also related this to irrigation needs. Mean daily evaporation rates derived from this experiment were quoted in Table 7.3 (§ 7.2). Hudson (1969) examined the water-storage capacities of various soils in Barbados. He then used these in irrigation and cultivation decision-making and as the basis for an ecological grouping of sugar estates according to their probable water balances.

A modulated water-balance approach was used by Smith (1966) for coco-nut in Trinidad. He assumed that the ratio of the actual evaporation to the potential evaporation (water freely available) varies *linearly* with the soil-moisture deficit. The water-balance approach was used on a weekly basis to determine soil-moisture deficit which was then correlated against yield of dry copra. This relation was compared with the relation between rainfall and yield for various periods, and in almost every case the soil-moisture deficit was more closely related to yield than was rainfall. A regression relation between yield of dry copra and an integrated soil-moisture deficit over the 29 months before harvesting was derived. This is an example of a fairly advanced application of the water-balance method.

In Colombia, Jones, Pena and Carabaly (1980) used a soil water balance to investigate water relations of several tropical grasses. In both wet and dry seasons, dry matter production was closely related to transpiration as estimated by the water balance model. Reddy (1983c) used a water balance to estimate soil-moisture loss under cropped and fallow situations in India. Estimated evaporation and soil moisture compared favourably with observed values. Hari Krishna (1982) discusses a water-balance model for small agricultural watersheds in India to evaluate hydrological response to traditional and improved land management techniques.

Stewart and Hash (1982) used a water-balance approach for maize in Machakos District, Kenya. They evaluated the suitability of a given crop for any location. Individual seasons were categorised according to date of rain onset and adequacy of rainfall for maize. This allowed recommendations on seed and fertiliser rates and thinning out of plants to be made and prediction of yields for planning purposes. Porter (1984) used a daily balance with separate scheduling of soil evaporation and crop transpiration in Kenya. Runoff and drainage when soils reached field capacity was made soil specific and there was careful specification of rooting depth, soil depth, moisture availability and crop transpiration coefficients. Effects on soil moisture of soils with farmyard manure were allowed for.

In southern Brazil, daily modulated soil-moisture budgets for corn, soyabean and upland rice were examined by da Mota (1983) and da Mota *et al.* (1984). A daily water stress index was calculated, based on the assumption that yield reduction is proportional to reduction in evaporation below the maximum. Indices were summed and models developed for yield prediction.

The above examples indicate the range of application of water-balance studies as well as the differing levels of complexity which can be used. In many cases they were, to a large extent, concerned with experiments, but the technique has routine application as well, particularly in assessing irrigation requirements. The conclusion reached by Rijks and Harrop (1969) (see above) about the benefits compared with pre-set irrigation patterns is important here.

8.2 THE ROLE OF AGRICULTURAL TECHNIQUES IN CROP–WATER RELATIONS

Techniques will be considered under three interrelated headings.

1. Water- and soil-conservation measures.
2. Irrigation practices.
3. Adapting crop type and variety, timing of planting and harvesting to the water regime.

Water- and soil-conservation measures

Soil erosion by water results from surface or sub-surface runoff. Methods of land use, agronomic and mechanical measures for reduction of soil erosion are aimed at directing and controlling runoff or reducing it by increasing the proportion of the rainfall entering the soil, the latter also conserving water. Wind erosion occurs when the soil is dry and water conservation to raise the moisture content of the soil also helps to alleviate this problem. It is also necessary to avoid excessive leaching. While soil and water conservation in many ways are interdependent, Hudson (1981) points out that it can sometimes be a mistake to design protection works to do both. The result may be a system which does neither effectively.

Factors influencing the proportion of rainfall contributing to runoff have already been discussed in previous chapters (e.g. §§ 4.1, 7.1). They are:

(a) The amount, intensity, duration, occurrence interval and seasonal distribution of rainfall.
(b) The slope of the land.
(c) The vegetation cover.
(d) The type and condition of the soil.
(e) Land use and management practices, the distinction between the two having been made in § 3.4.

It is the character of rainfall (Ch. 4) which makes soil and water conservation of tremendous significance in the tropics. Understandably, therefore, this matter has received considerable attention in recent years. Examples of recent general coverage of the topic are shown in Table 8.3. These and other more specific analyses are referred to below.

As well as influencing the proportion of rainfall infiltrating into the soil (§ 6.3) the slope of the land affects the velocity of runoff and hence erosive power of the water. Length of slope influences amount, velocity and depth of surface runoff. Moreover, the energy and hence erosive capacity increase at a faster rate than does the velocity. This point is illustrated by Wrigley (1969):

> If the rate of flow is doubled, it has four times the scouring capacity, thirty-two times the carrying capacity and can carry particles sixty-four times as large.

Table 8.3　Selected list of material with general coverage of soil and water conservation (specifically tropical or with substantial tropical content)

Author(s)	Date	Title/ general topic
Lal and Russel	1981	Tropical Agricultural Hydrology
Krantz	1981b	Water Conservation Management and Utilization in Semi-Arid Lands
Hudson	1975	Field Engineering for Agricultural Development
Hudson	1981	Soil Conservation
Morgan	1981	Soil Conservation
El-Swaify, Dangler and Armstrong	1982	Soil Erosion by Water in the Tropics
Gumbs	1982	Soil and Water Management in Trinidad and Guyana
Arakeri and Donahue	1984	Soil Conservation and Water Management
Goodland, Watson and Ledee	1984	Environmental Management in Tropical Agriculture
Walling, Foster and Wurzel	1984	Soil Erosion/Conservation in Africa
Morton	1985	Plants for Conservation of Soil and Water in Arid Ecosystems
El-Swaify, Moldenhauer and Lo (eds)	1985	Soil Erosion and Conservation

Any obstacles such as natural irregularities, stones, plants, ridges, etc., will decrease the rate of flow and give more time for infiltration.

A vegetation cover protects the soil against compaction (capping) and slows down surface runoff, allowing water time to infiltrate (§§ 6.3, 7.1). Density of the cover is important, especially in protection against rain-drop impact. While closed forest is generally thought of as very effective, in § 7.1 it was indicated that findings of Wiersum (1985) and others suggest the situation is more complex than at first sight. Drip from the forest canopy may increase potential erosion hazard, the litter layer being perhaps of most significance. In § 7.1 the findings of Ruangpanit (1985) concerning relationships between percentage crown cover and soil and water losses were discussed. Forest litter reduces raindrop impact, improves soil structure and filters out fine particles of soil which could block soil pores. Litter also slows down movement of surface water. Root holes aid deeper percolation. A dense grass cover is almost as good as forest if the sward is kept in good condition. Overgrazing or indiscriminate burning which exposes bare soil, excessive trampling by stock, destroying soil structure and hence lowering infiltration rate would all reduce the effectiveness of a grass cover (Ch. 9).

The importance of a vegetation cover is stressed by many authorities. El-Swaify, Dangler and Armstrong (1982) refer to it as the key and that accelerated erosion would be no problem if all soils had permanent cover of mature forest or grass sward. They list plant species useful for erosion control and discuss covers under non-cultivated and cultivated situations. Stocking (1985) states that the only long-term method of erosion control in Africa is by vegetation protection and fertility-enhancing farming systems rather than by mechanical protection (see below). Re-vegetation of degraded land is discussed by Morton (1985).

For crops planted in rows, soil and water loss will be reduced if the rows run across the slope rather than down, but the effectiveness of this will depend on plant spacing. Tree crops, for example, will usually be spaced so far apart that the influence of row direction is lost. The proportion of the ground covered by crops is a major factor, and particularly for perennials a distinction must be made between mature and immature stages. Mature cocoa and rubber, for example, with a complete or nearly complete, canopy, provide good protection. In its establishment phase, cocoa is usually interplanted with other crops to provide shade and ground cover and is thus still effective in conservation, but the incomplete canopy of immature rubber is not. Unless good cover crops and mechanical conservation works are used, erosion occurs and with the resultant deterioration in soil-infiltration capacity, this may persist into the mature stage. These aspects are considered by Goodland, Watson and Ledee (1984) who also discuss oil palm cover. The general spacing of about 9 m, although providing soil protection, is not as closed a canopy as mature rubber, but has a more protective ground layer. With lower density oil palm planting, intercropping with bananas, coffee and cacao reduces erosion and economic dependence on monoculture (Goodland *et al*. 1984). Other perennials such as citrus crops and coconut, even in the mature stage, do not provide a complete cover and convervation measures are necessary. With the exception of sugar, the same applies to most short-term perennials such as sisal and pineapples.

Annual crops vary greatly in their protective effects, and differences in cultivation practices, particularly spacing, are of considerable significance. The influence of spacing is generally considered more important in protection against raindrop impact than in impeding surface runoff. However, Bhardwaj *et al*. (1985), for corn in India, examined the influence of changing inter- and intra-row spacing of plants with a constant density. They found that runoff and soil loss were considerably reduced by increasing inter-row spacing and decreasing intra-row spacing. Increasing the number of plants in a row increased concentration times and infiltration and reduced peak flow. Hence, even with a fixed density of planting, crop geometry can influence soil and water loss. Soil and water losses under intercultivated row crops such as maize or cotton are usually considerable on sloping land unless conservation measures are adopted. This can be reduced by high-density planting and mulching (see below). While there are fears that high-density planting may exhaust soil-moisture reserves, Hudson (1971) indicates that any increase in moisture use will be more than offset by increased water infiltration. David (1973) points out that monoculture such as cotton in Mali, Chad and north Cameroon can increase the erosion risk compared with traditional patterns of interplanting several species which draw their nutrients from different soil horizons and provide a denser vegetation cover. Measurements indicate decisively that cover is far better from intercropping than on monocropped fields (Stocking 1985).

Closely spaced row or broadcast crops that receive little attention after planting (e.g. wheat, finger millet, creeping varieties of groundnut) form a reasonably good cover. While established fodder grasses, in general, give good protection, tufted varieties only do so if grown in rows along the contour. Creeping legumes give better protection than erect ones but only if planted early before early rains attack the bare soil.

Surface runoff and soil erosion are influenced by soil type and condition. Factors influencing the rate of infiltration and movement of water away from the surface layer (the latter influencing the former – § 6.3) are texture, structure, porosity, organic matter and colloidal content as well as the depth and nature of the soil profile. Organic matter and colloids, in binding soil particles together, enhance soil structure by increasing porosity, while at the same time stabilising soil against raindrop impact. While the larger pore spaces of a sandy soil may readily allow infiltration, the limited and unstable aggregates of fine-textured sandy soils are susceptible to erosion. Clay loams and ferruginous clays, whose aggregates break down slowly, are more resistant.

Profile depth and subsoil permeability are very important. Shallow soils over a hardpan become saturated, with a resultant marked decline in infiltration and increase of surface runoff. El Swaify *et al.* (1982) review studies concerned with assessing the erodibility of different soil types. Erosion is a vicious circle. Removal of the well-structured, humic upper layers reduces the infiltration rate, hence leading to increased runoff and erosion. A decline in soil structure and nutrient status due to erosion of the upper layers, resultant lower water holding capacity and reduced infiltration, by reducing protective plant growth and density, further increases erosion risk.

Land-management practices are of prime importance to water and soil conservation. Differences in erosion due to different management of the same soil are far greater than the differences in erosion from different soils under the same management. Traditional farming systems of the tropics included the practice of water- and soil-conservation methods. However, with increased population pressure, introduction of new cash crops and farming systems, in many cases these practices ceased. Planting of annual crops on steep slopes previously under forest, tree crops or permanent pasture led to great water and soil loss. Ridge construction, ploughing down slopes and clean weeding also create problems. Stocking (1985) presents a framework and general guidelines for soil conservation within the context of development projects for small farmers. He points out that conservation is not simply a technical exercise, but needs consideration of social, political and economic factors as well as environmental and technological aspects.

By increasing the proportion of rainfall lost as surface runoff, problems of flooding and downstream erosion arise following rains, and because less water percolates down to provide base flow for streams the dry season flow may be greatly lowered or even reduced to zero. Lowering of the water table can lead to the loss of well water. Decreased soil-moisture supplies result in a progressively poorer vegetation. Removal of the fertile topsoil by erosion will reduce crop yields. Deposition on to low-lying areas of coarse sand, gravel and stones removed from steep slopes decreases the agricultural potential of the former. Sediment deposition in channels and reservoirs resulting from upstream erosion is a major problem. Suspended sediment represents a deterioration in water quality. Fine clay particles need expensive treatment by chemical flocculation and filtration.

Water- and soil-conservation measures to combat tropical rainfall characteristics involve manipulation of soil, vegetation and slope in order to: (a) protect the soil surface from the compacting effects of raindrops; (b) allow as much water to enter the soil as is practicable; (c) ensure drainage of surplus water away while minimising erosion risk.

Three aspects are involved:

1. Appropriate land use (i.e. crop or vegetation type).
2. Mechanical measures such as earthworks.
3. Appropriate agronomic practices.

Detailed consideration of the measures is contained in the references listed in Table 8.3.

Appropriate land use will be examined in Chapter 9 within the general question of human impact on the hydrological cycle. Here attention is focused on (2) and (3) which are related to agricultural practices, i.e. the management aspects (see above and § 3.4). However, there is a close link between what follows and the discussion of land use in Chapter 9. The interrelation between (1), (2) and (3) is highlighted by the rather different grouping adopted by Hudson (1981). El-Swaify *et al.* (1982) divide measures into 'vegetative' and 'mechanical', the latter including tillage practices which are often considered 'agronomic' measures. El-Swaify *et al.* (1982) provide a flow chart of conservation practices (Fig. 30, pp. 122–123) which indicates the complex range of possibilities. Here only a few points can be made.

Water- and soil-conservation measures must take into consideration the whole catchment area. For example, if mechanical measures are applied only to part of a catchment they may be ineffective, especially if higher parts of the catchment are not properly used. Furthermore, all three measures must be considered together. Good agronomic practices, for example, will minimise the cost of mechanical measures but both may be ineffective if an appropriate land use is not adopted.

Choice of land use often dictates the necessary mechanical conservation methods. Agronomic and mechanical measures are complementary, not alternatives. Hudson (1981) points out that if they are needed, mechanical measures must precede crop management. Three phases will usually be involved. First, a survey to assess the suitability of land for various purposes would be necessary. Here the land will be classified into a number of categories, but the conservation of water and soil will usually be only one aspect of a land potential survey. Of particular concern to conservation will be slopes and soil type in relation to infiltration capacity and susceptibility to erosion. Existing vegetation will also be of importance. The second phase will be the construction of appropriate conservation works and the third phase the adoption of appropriate agronomic techniques.

Techniques of land clearing will have impacts upon soil properties and hence influence hydrology and soil erosion. Dias and Nortcliff

(1985) compared effects of mechanical clearance using a bulldozer with traditional slash and burn on the physical properties of an oxisol in the Brazilian Amazon. While the former resulted in considerable topsoil removal and deterioration in soil structure, the latter produced conditions little different from the natural forest. In West Africa, manual clearing and burning caused less soil compaction than three mechanical techniques (IITA 1984).

Mechanical conservation works consist mainly of earth banks, terraces and channels constructed across the slope. These break up the slope, intercept runoff before its volume and velocity become sufficient to cause serious erosion and give more time for infiltration, hence conserving water for agriculture. Water is diverted in the channels down safe gradients to suitable discharge or outlet points which carry away water in such a way as to minimise erosion damage to other land.

The amount of water concentrated in defined channels must be minimised. Careful design of these works is essential and badly planned measures cause more damage than none at all. For example, gullies started at breaks in a weak bank or scouring of too steep a channel are a hazard. Surface runoff from an extensive area is concentrated in a raging torrent if a bank is overtopped or breached. Examples of some works are shown in Fig. 8.1. Apart from measures adopted on the area itself, storm drains or diversion ditches (Fig. 8.2) can be used to protect these areas from runoff on higher uncultivated land. Details of these measures can be found in Hudson (1981), El-Swaify *et al.* (1982) and Webster and Wilson (1966). Dhruvnarayana and Sastry (1985) discuss the beneficial effects of terraces and other mechanical measures such as ditches in India.

Many agronomic measures for water and soil conservation exist. While ploughing, by breaking up the surface, temporarily increases infiltration, in the long run the more cultivation the more the soil aggregates are broken up, gradually decreasing soil permeability as well as increasing erosion susceptibility. In general, therefore, cultivation should be kept to a minimum.

Macartney *et al.* (1971) review some of the major problems of soil-moisture conservation such as capping, and report on the effects of different cultivation techiques at Kongwa, Tanzania. Their results indicate the importance of soil physical properties in determining maize growth and the need to cultivate the Kongwa soil to alleviate a compaction problem. Disc ploughing and harrowing combined with flat planting, although giving good germination and establishment, were ineffective for water and soil conservation. Also, they were expensive and delayed planting, the latter aspect perhaps greatly reducing yields (see below). Ripping prior to the start of the rains was found to be excellent for conserving early rainfall, being good for water infiltration and also

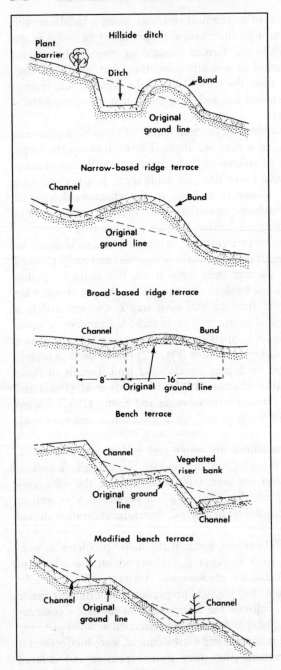

Fig. 8.1 Conservation works (after Webster and Wilson 1966).

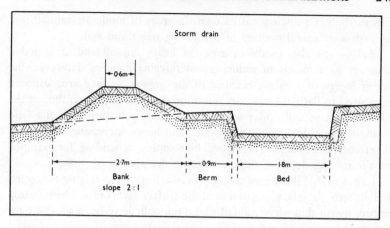

Fig. 8.2   Storm drain (after Webster and Wilson 1966).

assisting root development in the compacted soil. The benefits of a zonal tillage system with reduced surface cultivation, inter-row mulching (see below) and ripping are discussed in this paper.

Considerable attention is now being paid to what is generally termed 'conservation tillage'. The definition of this term is the subject of some debate. Some authorities refer to it as any system with less disturbance (or fewer operations) than 'conventional' (sometimes called 'clean') tillage, or where some crop residues are left as ground cover, others refer to it as 'minimum' tillage. Pierce (1985) accepts the problems of definition and states that all tillage practices which conserve soil and water are 'conservation' tillage. Discussions of the meaning and a variety of practices such as no-till, contour tillage, ridging, strip tillage, non-inversion tillage, sweep tillage, stubble mulching, chemical fallow, are presented by Young (1982), Hayes (1982), D'Itri (1985), El-Swaify et al. (1982), Arakeri and Donahue (1984), Darby (1985). Comparisons of different tillage methods include Simpson and Gumbs (1985), IITA (1984, 1985). Macartney and Northwood (1969) demonstrated that minimal tillage methods combined with the use of herbicides are at least as good as conventional methods and have the additional merit of disturbing the soil as little as possible. Ploughing and cultivating along the contour is a simple and cheap conservation measure although only on gentle slopes is it adequate by itself. Many crops grow well on ridges, and contour ridging can be used either by itself or with other measures. A development of this is the practice of tie-ridging where adjacent ridges are joined at regular intervals by barriers or ties of the same height. The benefits of tie-ridging are discussed by Dagg and Macartney (1968). The basins formed, allow water to infiltrate and prevent runoff except during intense storms. Except on very steep slopes this method is sufficient on its own. In areas of moderate rainfall it is an extremely useful method of conserving water and soil.

Ridges are also useful in areas of heavy rainfall and/or impeded drainage as a means of reducing waterlogging dangers. However, the upper layers of a ridge, because of the greater surface area exposed compared to flat land, may dry out more than flat land even though the lower layers of soil under a ridge have more moisture than flat land owing to the ridges holding up water and hence increasing infiltration. Therefore, the effect of ridging will be complex, depending, for example, on the rooting depth of the crop during dry periods.

Lawes (1961) illustrated the beneficial effects of tie-ridging for cotton in northern Nigeria. Compaction of the surface layers of soil by raindrop impact meant that while rainfall would normally have been adequate for the crop, because 69 per cent constituted surface runoff, even a seasonal total of 1067 mm did not wet the soil to a depth of 75 cm. Tie-ridging led to moisture increase down to the maximum depth examined of 122 cm. Cotton yields were increased from a low figure of 224 kg/ha, to 2242 kg/ha with the application of fertilisers. Three different forms of tie-ridging are examined by IITA (1982) which found that for maize in Upper Volta they reduced the risk of drought stress and substantially increased yields.

Lawes (1966) examined the effects of ridges, cross-ties and mulching on sorghum and groundnut yields in northern Nigeria. He found that these systems were not so important as in the case of cotton discussed above, with no real increase in groundnut kernel yield, and suggested a number of possible reasons for this. Both sorghum and groundnuts are sown at the beginning of the rains and because the crops soon protect the soil from raindrop impact, surface capping is not so severe as when the soil is unprotected until later. Because of earlier sowing and a shorter growing period than cotton, both crops passed the maximum leaf area and growth stages, when moisture stress can decrease yields, before the rains finished. Adventitious roots from the lower nodes of sorghum also help break up the soil-surface cap and the heavy leaf fall from groundnuts provides a mulch. Infiltration in both cases will therefore be better than for cotton. Nevertheless, Lawes suggests that rainfall conservation is still important to give an even distribution of moisture and protection against minor droughts, especially early in the season. Peat and Brown (1960) for Sukumaland, Tanzania, give the following yield increases when tie-ridging is compared to normal contour ridging: bulrush millet 128 per cent, sorghum 57–87 per cent (depending on soil type), groundnuts 59 per cent, cotton 39 per cent, maize 15 per cent.

Contour strip cropping involves the planting along contours of narrow strips of erosion/runoff susceptible crops separated by dense protective crops. Webster and Wilson (1966) quote as an example successive strips of the following:

1. A wide-spaced row-tilled crop such as maize or cotton.
2. A dense untilled crop such as clover or grass.
3. A close-spaced crop receiving little or no cultivation after planting, such as finger millet or a creeping variety of groundnut.

The densely planted strips slow up runoff, allowing more time for water to infiltrate, decreasing erosion risk and causing the deposition of silt. This method, often used with contour terracing, gives protection to the more widely spaced crops. A related method uses permanent contour strips of grass or shrubs, either alone or with mechanical measures. These are termed 'vegetative buffer strips' or 'barriers' and an essential requirement is that cover is dense and continuous. If there are gaps in the cover then channelling of water can lead to gully formation. However, spread of the cover may provide competition for the crops. Deposition of silt behind the barriers will gradually produce a bench-terracing effect. Hudson (1981) and El-Swaify et al. (1982) classify the above methods such as tie-ridging as mechanical protection works, thus highlighting the interrelation between the various methods.

On steep slopes, tree crop plantations usually need a good ground cover even if mechanical conservation measures are adopted. Particularly during the establishment period of the plantation, when the trees provide little protection, cover crops are usually planted. Creeping legumes are often used, since as well as providing protection against raindrop impact they also increase infiltration because of the leaf litter layer, as well as improving the organic content of the soil. Grass also provides a good cover but tends to compete with the trees. The amount of ground cover depends on the slope, soil type, density of the tree canopy and the amount of leaf fall which the latter provides.

Mulching with vegetation trash protects the soil from raindrop impact, slows down water movement and increases soil organic matter content, hence increasing permeability. Permeability is also increased because of the promotion of increased earthworm and termite activity. Lal (1984) reviews the benefits and requirements for erosion control by mulch with no-till systems in the tropics. The mulch rate required is generally less with no-till than with conventional cultivation practices but is also influenced by a range of other factors including slope, soil characteristics and crop characteristics such as the degree of protection afforded the soil.

Thompson and Wood (1967), for sugar cane in Natal, found that a trash layer conserved moisture, this being especially beneficial during the dry season. They found that runoff from both trashed and bare surfaces was negligible and therefore concluded that the main effect was

Table 8.4 Effect of cropping practice on runoff and soil loss on Tatagura clay loam, Rhodesia (6.5% slope) (Hudson 1957, data taken from Webster and Wilson 1966)

	Continuous maize 24 700 plants/ha	Maize trashed 37 000 plants/ha
1954–55		
Soil loss (t/ha)	4.28	0.76
Runoff (mm)	177.8	17.8
Per cent runoff	16.0	2.0
1955–56		
Soil loss (t/ha)	10.58	2.27
Runoff (mm)	198.1	71.1
Per cent runoff	22.0	8.0

in cutting down on evaporation. However, the increased infiltration resulting from mulching is generally considered to be more important than the other benefit of decreasing evaporative loss from the soil.

Stubble mulching or trash farming involves leaving all or part of the crop residues and weeds as a protective cover. It perhaps has potential, especially if accompanied by dense planting. In Table 8.4, Hudson (1957) demonstrated that dense planting and trashing of maize considerably reduced runoff and soil loss, increased soil moisture and eased soil working.

Benefits of various mulch treatments are indicated by Opeke (1982), Raghavulu and Singh (1982), De *et al.* (1983a), Wade and Sanchez (1983), Bonsu (1985) and when combined with no tillage by IITA (1985) or minimum tillage (Suwardjo and Abujamin 1985). Tumuhairwe and Gumbs (1983) found that in Trinidad, necessary soil moisture for good yields of cabbage could be maintained with less than half the amount of water with mulching compared with no mulching. Pla, Florentino and Lobo (1985), for sorghum in Venezuela, found that a relatively low cost asphalt mulch was an effective measure not only for soil and water conservation but in benefiting seed germination, growth and yields. Othieno (1980) found that although the effects on soil moisture of five different mulches varied, all were better than a no-mulch situation. Grass mulches had most effect on water holding capacity but induced a shallow rooting system in tea which made it more susceptible to drought.

Fallow cropping, whereby the land is cropped in alternate seasons to store water, is practised in some dry areas, usually with clean weeding to eliminate water use by weeds. However, in the tropics, with high evaporation rates, the proportion of rainfall stored is low, although the practice may be worth while if it permits crop production not otherwise possible. The problem with clean weeding is that the soil has no protection against compaction and erosion. Stubble mulching or a shallow-rooting, low-water-use crop may be preferable during fallow. In a trial at Kongwa, Tanzania, for example, Pereira *et al.* (1958) found

Table 8.5 Runoff and soil erosion under various slopes, soil and crop types and agronomic practices (from Temple 1972).

1. Soil erosion from three clay plots of different slopes under continuous maize in Rhodesia (m^3/ha) (after Hudson and Jackson 1959)

Slope (degrees)	3.5	2.5	1.5
Average (1953–59)	6.7	3.7	3.1

2. (a) Runoff (mm) and (b) percolation rates (mm/hour) for small plots under various treatments in Namulonge, Uganda (after Hutchinson *et al.* 1958)

	Rain (mm)	Bare soil		Stone mulch		Grass mulch		Grass (Cynodon)	
		a	b	a	b	a	b	a	b
Average (1951–53)	1410	474	25	137	51	65	70	45*	75*

3. Runoff and soil erosion at Mpwapwa, Tanzania (after Staples 1936)

Average 1933–35 seasons

	Bare: uncult.	Bare: flat cult.	Bare: ridge cult.	Bulrush millet or sorghum	Grass
(a) Percentage runoff of total rainfall	50.4	31.5	23.0	26.0	1.9
(b) Soil erosion (m^3/ha)	97.8	85.2	43.3†	52.0	0

4 Runoff and soil erosion from various crops on identical plots at Tengeru, Tanzania, 1958–60; runoff (a) as percentage of annual total; soil loss (b) in m^3/ha.

	Plot treatment	a	b
Coffee:	pruned; 3×3 m spacing; clean weeded continuously; no shade	5.0	22.4
Maize:	stover removed; no conservation measures	3.4	12.0
Maize:	stover and dead grass removed; grass bunds at 3 m vertical interval	2.3	7.2
Maize:	stover and dead grass removed; grass bunds at 2.1 m vertical interval	2.2	5.0
Maize:	stover and trash bunds at 3 m vertical interval	2.1	3.9
Maize:	stover and trash bunds at 2.1 m vertical interval	2.0	1.0
Bananas:	banana trash mulch	1.8	0.5
Grass:	cut for hay (average of eight plots)	1.4	0

* Average of two years.
† Average obtained by doubling 1933–34 value and probably an underestimation.

that shallow-rooting teff grass allowed storage of subsoil water, enabling a yield of 450 kg/ha of groundnut kernels to be obtained in the following season during which only 215 mm of rain occurred. Table 8.5 illustrates the influence of various factors on runoff and soil erosion.

Singh *et al.* (1980) refer to high percolation losses with rice irrigation and investigated the possibility of minimisation by soil compaction with a roller. Water use efficiency was increased, results suggesting that compaction may be a better method of minimising water requirements of rice than other methods such as laying sub-surface barriers.

Irrigation practices

Thornthwaite and Mather (1955) distinguish four kinds of drought:

1. Permanent – characteristic of driest climates.
2. Seasonal – where well-defined wet and dry seasons exist.
3. Contingent – due to rainfall variability. Can occur anywhere including seasonal drought areas but most characteristic of sub-humid and humid areas.
4. Invisible – can occur in any area but most common in more moist regions, even when there is rain on every day. If rain does not meet evaporative needs, a borderline water deficiency exists which limits growth and yield. This condition may be difficult to detect.

In view of these different conditions, the role of irrigation will vary. In some areas it may be a precaution against the inherent variability of rainfall and in others an essential requirement for the production of a crop in every year. However, an important feature in more recent times has been the recognition of the advantages of irrigation even in areas normally considered to have sufficient rainfall for reasonable crop production. Even under such generally favourable conditions, the advantages of supplementary irrigation – either in amount, but perhaps more particularly in terms of timing of water supply in relation to crop needs as a means of obtaining substantial yield increases – is becoming more and more recognised. As has already been indicated (§ 3.1), new high-yielding varieties may also be more sensitive to fluctuations in water supply, making irrigation provision important.

To be successful, irrigation water must be applied in a controlled, carefully calculated manner. The design and operation of an irrigation system demands knowledge of the water requirements of different crops at various stages of growth (Chs 6, 7). It is also necessary that the irrigation water is applied in such a way as to avoid many of the problems associated with natural rainfall discussed previously. Irrigation water lost as surface runoff is wasted and can result in soil erosion. The rate of application should therefore allow the water to infiltrate and not cause either scouring or silting of irrigation channels. The dangers of water-logging and leaching of the soil must also be avoided (§ 7.3). As in the case of natural rainfall conditions, various mechanical and agronomic measures may be necessary to avoid these problems. Over-irrigation can be as harmful as under-irrigation. Goodland, Watson and Ledee (1984) refer to a common tendency to over-irrigate and discuss resulting problems such as waterlogging, soil deterioration and wastage of water.

Waterlogging harms crops, causing stunted growth, delayed crop maturity and shallow rooting. Longer term problems include raised water tables and salinisation (§ 9.6). Thus regulation of quantity and quality (see below) of water as well as means of removing excess water and leaching of salts are important.

Kramer (1983) provides a brief review of irrigation history. While irrigation has been widely practised for a very long time in the tropics (e.g. in parts of Asia and India), this is not the case in Africa south of the Sahara with the exception perhaps of the Kilimanjaro area of Tanzania. The absence of irrigation in southern Africa can perhaps be explained by the general lack of population pressure, but in recent times increasing population and general increased demands on agricultural production have changed this. Many African governments have embarked on schemes. These may take the form of very large-scale, capital-intensive schemes as well as self-help schemes at the village level.

Early irrigation systems, for example in Egypt, often involved the impounding of natural flood waters in basins on flat land bordering rivers, in some cases this being supplemented with water drawn from the river. The latter practice was often adopted for tree crops, a basin being constructed around each tree and flooded at intervals. With these early methods, various precautions such as ridging, terracing, etc., may be necessary, as in the case of rainwater, to decrease surface runoff and soil erosion. The basin system of irrigation was used by Munro and Wood (1964) for maize in Malawi.

In Egypt, the ancient basin system has been replaced by the 'perennial' system, involving the construction of barrages or dams to impound water. Water is led from the storage reservoirs by main irrigation canals into a network of smaller channels and then on to the land. The advantage of the widely used perennial system is in providing all-year-round water rather than only during seasonal flooding.

Vasey et al. (1984) discuss the ancient wetland system of 'raised field' or 'drained field' or 'island bed' agriculture where the field surface is raised above the water table but water remains in surrounding ditches for all or part of the year. Once widespread in the tropics, only scattered remnants of the system remain today. Vasey et al. (1984) examine former and contemporary examples in the Pacific and American tropics.

Subsistence farmers in semi-arid areas since the earliest times have used a system of 'water harvesting' involving the collection and storage of runoff from slopes. This method was used in the Negev Desert of Israel during the Israelite period (c. 950–700 BC) and the Nabatean and Roman–Byzantine periods (c 300 BC to AD 630), where average annual rainfall is only 100 mm. The irrigation works have been restored and high yields obtained from the small irrigated basins (Evanari, Shanan and Tadmor 1968, 1982; Shanan et al. 1970). Ancient techniques of water harvesting and storage in tanks are also being studied and improved upon in India. The ultimate in water harvesting is by sealing the soil surface with silicone resins or asphalt (Myers 1967). In Hawaii, catchments up to 7 ha have been covered by artificial rubber sheeting for water collection.

Branson *et al.* (1981) provide an overview of water harvesting techniques which they consider under four categories: vegetation management, land alteration, chemical treatments and soil covers. Papers on rainfall collection for agriculture in arid and semi-arid areas are presented by Dutt, Hutchinson and Garduno (1981). Techniques of rainfall harvesting for human and livestock consumption in semi-arid areas of Mexico are investigated by Ruiz and Molina (1981). Various catchment types and sealants for the harvest area and storage tanks were investigated, harvesting efficiencies ranging from 45 to 89 per cent. They concluded that better quality water could be collected than was available in dams and reservoirs. Krantz (1981a) investigated rainfall collection as runoff from cropped watersheds in India. Runoff was collected in small reservoirs and used as supplementary irrigation on rainfed crops in the watersheds. The use of a buried membrane collector for harvesting rainfall in sandy areas is discussed by Shanan, Morin and Cohen (1981).

Doorenbos and Pruitt (1977) discuss the identification, design and operation of irrigation projects while Goodland *et al.* (1984) examine impacts and problems associated with irrigation schemes such as siltation of reservoirs, downstream erosion, waterlogging, salinisation, pests, weeds and human disease (§ 9.6). Such problems in arid areas are also considered by Grove (1985).

The wide range of irrigation systems can be classified in various ways. Kramer (1983), for example, groups systems into Basin and Furrow, Sprinkler, Trickle (drip), and Sub-Surface, while Arakeri and Donahue (1984) for India distinguish Wild Flooding, Furrow, Border and Basin, Pot, Trickle and Sprinkler. Basin irrigation, involving construction of a low bund (ridge) around individual trees or plots to enclose water, has already been mentioned. The border system involves construction of bunds 5–15 m apart (Arakeri and Donahue 1984). Furrow irrigation is widely used, water flowing along furrows between rows.

Overhead irrigation by means of portable sprinklers is used particularly for more valuable crops such as fruits and vegetables. Examples are tea in central Africa and India, coffee in India, sugar in central and East Africa, tobacco and oranges in Australia. While capital costs of the system (pumps, pipes, spray equipment) are quite high, if it is used efficiently, it may not be much more expensive over its lifetime (at least 20 years) than surface irrigation which is less flexible and uses part of the land for canals and ditches (Wrigley 1969). An advantage is that foliar nutrients can be sprayed through the system. The 'rain' produced by the sprinklers must not be of sufficient intensity as to cause soil compaction, leaching, waterlogging or crop damage.

A problem with surface and sprinkler irrigation in the tropics is the high evaporation loss. This can amount to 20–30 per cent of the total water applied. In Israel, however, Seginer (1967) found that sprinkler irrigation loss depended on the irrigation interval. The sprinkler system was compared with a special system near the ground which almost completely eliminated spray and interception losses. Two intervals, 3 days and 2 weeks, were tested and it was found that the sprinkler irrigation system used every 3 days used 10 per cent more water than the other treatment. Seginer concluded that regular day-time sprinkling every 2 weeks would not require more than 5 per cent more water than night irrigation or surface irrigation.

A method of underground irrigation developed in the USA using perforated plastic tubes through which water is pumped at high pressure overcomes the evaporation problem, the water being applied directly and precisely to the root zone. While cost (more than twice as much as a sprinkler system) rules out its use in most tropical areas, it could perhaps have application for very high-value crops where water is scarce and evaporation loss is high. While it uses water efficiently and can be combined with fertiliser application, outlets can be blocked. Furthermore, surface salt accumulation can occur since no surplus water is applied to leach out salts (Kramer 1983). In seasonal tropical rainfall conditions where it may be necessary to control the water table, the pipes could be used as drains in the wet season and for irrigation in the dry season.

Trickle (drip) irrigation involving release through small nozzles allows precise placement of water and is used for tree crops and some row crops. Losses through seepage and evaporation are minimised and weed growth is reduced since much of the soil surface is dry (Kramer 1983). In addition to conserving water, salinisation and waterlogging problems are minimised although some water must still pass through the root zone to leach accumulated salts (Goodland, Watson and Ledee 1984). Abrol and Dixit (1971) compared the drip method with conventional basin irrigation in India for onions and lady's finger. They found significant yield and water-use efficiency increases for the drip method which were ascribed to increased availability of soil moisture at low tensions and reduced surface evaporation loss. Yanuka, Leshem and Dovrat (1982) examined forage corn response to several trickle irrigation and fertiliser regimes in Israel, listing good water and nutrient control as advantages of the system.

Although surface irrigation requires less energy than sprinkler or trickle systems to apply the same quantity of water, usually significantly more water is used. For some crops the difference may be 30–60 per cent (Goodland *et al.* 1984). The extra installation and energy costs of sprinkler and trickle systems must be balanced against water saving.

In irrigation schemes, water quality as well as quantity is important. Different crops show different tolerances to salinity for example. Rawlins (1981) discusses the principles of salinity control in irrigation which involve providing water above evapotranspiration requirements to leach salts from the root zone and drainage to remove the leached water. Hussain (1981) describes a simple technique for using highly saline water involving mixing irrigation and drainage water. Goodland *et al.* (1984) discuss the benefits and problems associated with the use of waste water in irrigation. Problems associated with irrigation works such as salinisation, silting and biological hazards such as invasion by plants and disease organisms are considered in § 9.6.

The effectiveness of irrigation varies with soil type. Thompson, Gosnell and De Robillard (1967) compared irrigation effects on sandy and clay soil and found a marked difference in response by sugar cane. The general conclusion arrived at was that where available irrigation water was limited, heavy soils should be irrigated in preference to sandy soils.

Kramer (1983) discusses the possible benefits of high-frequency irrigation and deficit irrigation. The use of tensiometer soil moisture monitoring for irrigation control was examined by Augustin and Snyder (1984) for Bermudagrass. Irrigation water savings of 42–95 per cent were obtained compared with conventional irrigation, benefits being greatest during frequent but unpredictable rainfall since unnecessary irrigation was eliminated. Clawson and Blad (1983) examined the feasibility of scheduling irrigation of maize using canopy temperature differences obtained using a hand-held infra-red thermometer.

Consideration of a few individual crops illustrates many of the points discussed above.

Bananas
Aspects of irrigation of bananas have been examined by Holder and Gumbs (1982, 1983a, b, c) in the West Indies. Land prone to water-logging reduced harvested fruit yields and affected growth and development. Over-irrigation reduced the number of fingers per hand and hands per bunch but bunches were heavier than with no irrigation due to increased finger length (Holder and Gumbs 1983c). Over-irrigation, usually by flooding, in the Canary Islands leads to loss of bases by leaching. Effects of irrigation regimes with soil brought to field capacity when available moisture levels fell to 75, 66 and 50 per cent were compared with a non-irrigated control by Holder and Gumbs (1983a). No significant growth impacts were found but fruit yields increased

significantly for treatments irrigated at the 66 and 75 per cent levels. Yields were increased with increasing irrigation and nitrogen fertiliser levels (Holder and Gumbs 1983b). However, yields obtained with the highest irrigation and nitrogen levels were only slightly greater than with some of the lower combinations. The effects of irrigation at critical stages of development on growth and yield of bananas are examined by Holder and Gumbs (1982).

Coffee

Only with the high prices after the Second World War was serious attention given to irrigation in the main coffee-growing areas. Many important areas in Brazil and East Africa have marginal rainfall for coffee or have long dry seasons. These are the areas where most irrigation development has taken place, yield increases usually of 30–50 per cent occurring. In exceptional cases, however, yields have been doubled and the quality made one or two grades higher (Wrigley 1969). Sprinkler systems or watering in basins round the tree are used.

Irrigation in south India has increased yields two to three times by application during a critical period of a few days to supplement rainfall (Wrigley 1969). In addition to this critical period, irrigation is also used during the dry season, there being no evidence to support the idea that a resting period coinciding with the dry season is essential (§ 7.3). Nitrogenous fertiliser is applied through the sprinkler system in south India.

Cotton

Since cotton thrives under hot, sunny conditions, water is often a limiting factor for yields and quality, therefore irrigation is often of great significance. Furrow irrigation is the main method, although basin and flat irrigation are also common. Unless conditions are unfavourable (e.g. the occurrence of a hardpan) the roots extend to a depth of 2 m or more, but as cotton can take up plenty of water from the top 60–100 cm of the soil where the maximum lateral root spread occurs, there is no need to irrigate to a greater depth. For example, Marani and Fuchs (1964) found that for optimum yields a single irrigation wetting the soil to a depth of 90 cm was sufficient.

It was once common to withhold irrigation water in the early growth stages to force root development, but now it is considered that the crop should never be subjected to water stress. In Egypt, cotton is watered every 18 days in the cooler regions, increasing to every 12–18 days depending upon the weather. Too long an irrigation encourages an excess of vegetative growth. In Israel, Marani and Amirav (1971) found that the optimum time for the first irrigation was 2 or 3 weeks before flowering. Six or seven irrigations, if timed to prevent moisture stress, were sufficient to give high yields and good lint quality. Lint yield ranged from 270 kg/ha in non-irrigated to 2000 kg/ha in the fully irrigated treatments.

Guinn, Mauney and Fry (1981) examined the effect of delaying the first irrigation on growth, fruiting and yield in Arizona. Although a water deficit during the bloom stage hastened cutout, delaying the first irrigation had little effect on earliness. Yield and evapotranspiration under various irrigation schedules in New Mexico were examined by Sammis (1981b). The linear relationship between yield and evapotranspiration for two sites differed from that for a similar Californian study.

Maize

While maize is widely grown in the tropics, its market price is not usually sufficient to make irrigation and fertiliser application economic. In the USA where irrigation is practised, the plants are usually grown on ridges, the irrigation water running down furrows or being applied by sprinklers. Since maize rapidly shows signs of wilting, frequent irrigation is necessary. Fuehring *et al.* (1966) found that in the Lebanon, maize yields were considerably reduced if the irrigation interval exceeded 1 week and increasing moisture stress tended to delay maturity. If the irrigation interval was increased from 1 week to 2, then maize yield was reduced by about 47 per cent. De *et al.* (1983b) found that maize yield was increased by using mulch and antitranspirants (§ 9.2). With these evapotranspiration controls, yields with 2 or 4 irrigations were as high as with 4 or 6 irrigations without the controls and water-use efficiency was increased.

Rice

This is a crop one always associates with irrigation, as many as three crops a year sometimes being grown. This cultivation supports a large proportion of the world's population and yields have not declined even though little fertiliser is applied. One important function of water in rice paddies is to control weeds which would otherwise choke the crop (§ 7.3).

Cultivation and irrigation practices vary in different parts of the world and are often complex. Methods are discussed in a wide variety of texts including Wrigley (1969), Williams and Joseph (1970), Robertson (1975), Fukui (1979), Williams and Chew (1980), Jaw-Kai Wang and Hagen (1981).

Material on deepwater rice is published by IRRI (1979a) where it is indicated that 30–40 per cent of Asia's rice lands are subject to annual monsoon floods. Water depths of 0.5–1 m are common, with even deeper levels in the so-called 'floating' rice areas (see below). In addition to deepwater rice, two other general 'rainfed' systems may be defined (IRRI 1979b): upland (unbunded) and rainfed lowland (generally bunded with water depths less than 1 m). The latter is estimated to occupy about one-third of the world's rice area (IRRI, 1979b).

Krishnamurthy (1979) defines rainfed rice as grown either directly or indirectly with local rainfall, including the use of irrigation where water is derived from such rainfall (i.e. within the same catchment area). One characteristic of rainfed areas is the lack of control over amount and timing of water, with water deficits and excesses occurring. Krishnamurthy (1979) defines lowland areas as bunded areas allowing standing water up to about 1 m, deepwater areas having greater depths. Rainfed areas of all types occupy some 75 per cent of rice lands, the remaining 25 per cent being irrigated (IRRI 1979b). However, as Fukui (1979) points out, the water condition of rice areas is so diverse and complex that it is not really possible to divide land into 'irrigated' and 'rainfed'. Alternatively, depth of water may be used to differentiate between different systems but different authors appear to use different limits.

Seeds will not germinate under water and therefore dry seeds when drilled or broadcast must only have enough irrigation water to ensure germination. In most areas, including Sri Lanka and Thailand, where rice is sown broadcast, it is pre-germinated and sown on the puddled surface or even on standing water (Wrigley 1969). Alternatively, the rice plants are raised in nurseries and transplanted into puddled fields. Williams and Joseph (1970) discuss the advantages of transplanting, listing nine benefits compared to broadcasting. A disadvantage of transplanting is the greatly increased labour requirement. From the irrigation point of view a major benefit of transplanting is the less exacting water control compared to broadcasting, row planting or drilling the seed in rows. With the latter, water control is particularly important during the early stages of establishment when excessive flooding can cover the young seedlings. Excessive flooding can cause the seedlings to float, preventing root establishment in the soil. In view of the exacting water control needed, broadcast and row planting are not practicable where irrigation depends on rainfall. When water control is sufficiently good and correct fertilisation procedures followed, all evidence indicates that broadcast and row-planting methods yield at least as well as transplanting (Williams and Joseph 1970).

Other factors such as nutrient availability, rice variety and mechanisation make comparison between cultivation methods difficult. If weedkillers are used, then it is unnecessary to use as much water for weed control and the water saved can be used to extend the irrigated area or for double cropping. The application of water after establishment of the crop varies from area to area, depending largely on its availability.

In areas subject to seasonal flooding such as the river estuaries of

Sierra Leone, 'floating' rice varieties are grown. The rice grows to a height of 2 m or more, its growth keeping pace with the rising flood. One-third of the rice area of the central plain of Thailand is floating rice (Wrigley 1969). In riverine areas of Bangladesh and Burma, flood waters may rise to 10 m or more and floating varieties, with root systems adapted to floating and deriving all their nutrients from flood water, are cultivated (Williams and Joseph 1970).

Sugar cane

Since irrigation is less efficient than rainfall, double the amount of water is required compared to rainfall (Wrigley 1969). Before the cane canopy closes, the evaporation loss of irrigation water may be as high as 25 per cent. If the land is flat and water plentiful, surface irrigation is often the simplest and cheapest, the cane being planted on ridges or in furrows.

The effect of irrigation in increasing yields has been illustrated by Thompson, Gosnell and De Robillard (1967), and Thompson and De Robillard (1968) in Natal, although soil type influenced response (§ 7.3). Overhead irrigation is preferable when water is not so plentiful, the land is undulating and subject to erosion. Fertilisers may be applied at the same time. Water usage can be cut down by 25 per cent compared to surface irrigation. The effects of different water management systems and planting on raised, level and depressed beds are discussed by Gumbs and Simpson (1981) and Camp (1982).

Tea

In many tea areas there is a loss of water by runoff and drainage in the wet season and a shortage in the dry season. With high-yielding varieties, irrigation, usually overhead, is profitable. As an example, Wrigley (1969) quotes a tea area in Uganda with an annual rainfall of about 965 mm per year with two dry seasons where yields were raised from 1345–1680 kg/ha to 2020 kg/ha. The production was more evenly spread over the year giving a better utilisation of the factory. Sprinkler irrigation can be used to apply nitrogen and correct deficiency diseases such as magnesium which occurs with high production.

In § 7.3, the relation between yields and solar radiation was indicated. Ellis (1967) points out that in Malawi, half the solar radiation occurs in the sunny dry season when soil moisture is limiting. He suggests, therefore, that if this solar radiation could be fully utilised by dry season irrigation then an increase of 80–100 per cent in yield could be possible with a better distribution of the crop throughout the year and an improvement in quality. Other areas, such as those in Sri Lanka which are sunny but dry, might also profit by the growth of high-yielding teas under irrigation.

Tobacco

While irrigation usually only slightly increases yields, the high value of the crop makes it of considerable significance. In Queensland it is desirable to harvest the crop before the start of the wet season when high winds are common. This necessitates dry-season planting and therefore 90 per cent of the tobacco area is irrigated. Apart from increasing yields, irrigation has the advantage of reliable production in spite of rainfall variability. While furrow (surface) irrigation is normal, some sprinkler irrigation is used. Tobacco is very susceptible to waterlogging and overwatering, especially with young plants, is dangerous. Water quality is important; too high a chloride content lowering leaf quality. Since tobacco gets most of its water from the top 30–45 cm of soil, does not have a high water requirement and recovers well from drought, frequent light waterings are suitable.

Adaptations to the water regime

Here, ways by which plant growth can be fitted into the existing rainfall conditions and/or irrigation supplies are examined. This involves questions of differences in water requirements of different crop varieties and consideration of planting and harvesting times in relation to rainfall occurrence.

Cropping systems research is concerned with increasing production from available physical resources which are not readily changed (Zandastra 1977). Although partly concerned with increasing yields, it is particularly concerned with increasing the number of crops per year (IRRI 1977). The cropping pattern is considered a variable, defined as the spatial and temporal combination of cultivars in a plot (Zandastra 1977). The place of a wide range of tropical crops in cropping systems is discussed by Norman *et al.* (1984). Systems in the semi-arid tropics are discussed by Mattei (1979) and those in the 'humid' tropics (including seasonally wet and dry regions) by Fukui (1979). Mattei (1979) indicates how local varieties of crops such as millet, cowpea and sorghum have life cycles adapted to length of the period of water availability and also how adaptation to climate sometimes helps them to escape unfavourable conditions such as insect pests.

An example of the difference between varieties of a crop was discussed in § 8.1 where work by Dagg (1965a) illustrated that an imported short-term variety of maize, despite its higher water requirement at some stages of growth, was more suited to the rather short rainy season of part of Kenya than a local variety with a longer growing season. Varietal differences in drought tolerance of wheat have been examined by Derera, Marshall and Balaam (1969). Pande and Panjab Singh (1970) illustrated the effects of different life-cycle lengths of two rice varieties and Wessling (1966) found very great differences between groundnut varieties. In considering the suitability of a particular variety, not only must its water requirements be considered but also yield and quality. A quick-growing variety may be uneconomic in an area because of the latter factors.

The time of planting of the crop in relation to soil moisture and therefore rainfall is often critical, particularly in areas where the wet season is short relative to the growth period of the crop. To the meteorologist or climatologist the start of the wet season may be recognised by the presence of a particular atmospheric circulation pattern, but in agricultural terms this is of limited value. It is the build-up of soil-moisture reserves which is important, but in practice this is recognised and forecast in terms of rainfall occurrence. However, since rainfall is only an indirect measure, the problem is to decide what conditions are significant, for example the first day on which a certain amount occurs or a spell of a number of days with a given amount – a number of ways could be used to define the 'start' of the wet season. The day-to-day variability of rainfall at the beginning of a wet season is considerable (Ch. 3) making the problem of definition more difficult.

Critical conditions will vary depending upon soil type and the crop considered. The key factor could be the date on which peasant farmers can work the hard soil. Soil-moisture reserves can be built up to a reasonable level by the early occurrence of rain, but then a dry period before the main season begins creates difficulties. What constitutes the beginning of the wet season will vary from year to year since the pattern of storms will vary from year to year. Perhaps the only really satisfactory treatment is in terms of soil-moisture and water-balance studies. However, in the absence of such comparatively sophisticated methods (certainly to the peasant farmer) and particularly in terms of forecasting, rainfall must be used.

In § 3.5, when discussing short-period variability, a number of rainfall analyses relevant to the question of start, length and end of the rainy season were referred to. In the Lilongwe Plain, Malawi, the season is marginal in terms of water requirements of maize and it is therefore essential to make full use of the season. Hay (1981) discusses two planting rules using standard agrometeorological procedures to predict the earliest safe planting date. Particularly in areas where rainfall seasons are comparatively short, forecasting the onset of the rains is of great importance, especially since planting is often done – or should be done – before the rains begin (see below). Forecasting is a problem all over the world, although in recent years there have been immense strides, for example in the use of satellite data. In the tropics, however, the problems are probably greater than elsewhere (Ch. 2) and forecasting lags behind that of higher latitudes. Of necessity, forecasting must often be general and macro-scale in character. In view of the marked areal variations in rainfall conditions (§ 4.2) this limits its value.

Local knowledge is of great importance. Farmers over the centuries in tropical areas have built up an immense store of local experience. Jackson (1982) suggests the possible value of traditional knowledge in improved forecasting and increasing understanding of local weather. Awareness of such knowledge helps understanding of local agricultural decision-making processes and therefore has a role in planning. Jackson illustrates the wide range of indicators used for forecasting in Tanzania, including birds, animals, plants and meteorological factors, together with the fact that 'traditional' concepts of rainy seasons in agricultural terms may not match those based on rainfall amounts. Techniques used by the Incas in pre-Columbian Peru have been studied by de Mayolo (1984) based on observations of present-day farmers. Again, a range of indicators is used and de Mayolo feels that they have considerable value and need to be used in conjunction with agroclimatic models.

Apart from forecasting using synoptic experience, statistical and analogue methods are possible. Statistical techniques, particularly related to variability, were discussed in Chapter 3. Variability in the incidence of rain in the early part of a season is a major problem facing agriculture in the tropics. Where the wet season is short and particularly when there is only one wet season, early cultivation and planting is advantageous. One obvious reason is that plants will benefit from a full season's rain (important when the season is short) but there are a number of other factors involved, not all of them concerned with water availability. Weed growth is absent or slight during the dry season, but with the onset of rain there is rapid development which is difficult to control. Therefore, it is desirable to prepare the land before the rains to get a clean seedbed and to plant early to reduce competition from weeds in the early growth stages of the crop. Soil nitrate content is high at the beginning of the rains, but later it is reduced because of the slower rate of humus decomposition as well as by leaching. The significance of early planting varies with variety, as illustrated by Kittock, Henneberry and Bariola (1981) for cotton and Parker, Marchant and Mullinix (1981) for soyabean. Khalifa (1981) found that planting date impacts on yield and growth varied between rainfed and supplementary irrigation sites for sunflower in the Sudan.

Mattei (1979) suggests that the reduction in yield of many tropical crops with delayed planting may be linked with photoperiodic mechanisms. Fakorede (1985) speculated that reduction in maize yields in Nigeria with delayed planting resulted from increased overcast skies (i.e. reduced solar radiation) as the season progressed. Early sowing dates for okra in southern Nigeria led to poor seedling emergence but growth was more vigorous than for later planting (Iremiren and Okiy 1986).

Plants attained 50 per cent flowering earlier and had a longer harvest duration. Early planting increased pod numbers per plant, pod weight and pod yield per hectare compared with later planting. Lower yields with later sowing were in part attributed to poor soil aeration due to increased rainfall in their growth period. Sowing date did not affect the percentage moisture, oil or protein in the pods. The complex interaction between maize yield, meteorological factors and planting dates is examined by Fakorede and Opeke (1985) and effects of planting date and mulch on cowpea by Kamara (1981). Many other experiments in various parts of the tropics with a variety of crops indicate the benefits of early planting (e.g. Macartney et al. 1971; Semb and Garberg 1969; Gray 1970; Fisher 1980).

Despite the above facts, peasant cultivators generally plant after the optimum date. There are a number of valid reasons for this, based on long experience. Particularly if seed is scarce, because of the variability and unreliability of the rains, there is a natural tendency to delay planting until the farmer is certain that the wet season has really begun. If planting is too early then a delay in the onset of the rain will result in loss of the seed.

Mattei (1979) refers to the fact that rainfall variability obliges farmers in the semi-arid tropics to make repeated sowings if there are unusually long dry spells after the initial rains. In some areas, traditional systems may involve storing different millet varieties for use depending on time of sowing (Mattei 1979). Wangati (1984) refers to work in Kenya relating rain onset date to total seasonal rainfall expectation and early season intensities. This may assist farmers to decide early what type of season is likely and therefore on crop type, plant population and fertiliser input.

A further problem, mentioned previously, is that the farmer may have to wait for the rains before he is able to work the soil. This is the case to a large extent even with ox-ploughs. Since preparation of the land, for example by hoe, is slow, then usually the land is not prepared early enough for the best results. Tractor cultivation towards the end of the dry season to break up the land would allow earlier preparation at the start of the rains using lighter implements. However, such mechanisation has important economic and social implications. A problem with delayed cultivation is that when the rains do start they are often so heavy and continuous as to again make soil working difficult.

A further factor causing delay in planting is when the land is already being cropped and harvesting is not complete before the optimum planting time of the succeeding crop. Changing the cropping pattern can increase this problem as demonstrated in East Africa if finger millet is replaced by maize. The maize harvest is about 6 weeks later than finger millet and is well after the optimum planting date for the cotton crop which follows (Wrigley 1969).

In the monsoon rain regions, where a large proportion of the cultivated rice area occurs, cultivation is synchronised with the rains.
Planting begins with the first rains, which are usually light, so that plants
will reach a reasonable height to survive the flooding which occurs at
a later date. This early planting, without supplementary irrigation, results
in high crop risk due to uncertainties in early rains. Early planting is
necessary because too much rain or too high a rise of rivers later in the
season in flood-plain culture can cause crop losses if the crop is not
well established. Considerable year-to-year fluctuations in production
are found in all monsoon areas because of these problems. The supply
of even a small quantity of irrigation water over a critical period in the
early part of the season could considerably increase the chance of good
production (Williams and Joseph 1970). Where the wet season is longer
or there are two seasons, early planting is not so critical and for some
crops there may be a considerable latitude in time of sowing.

8.3 RAINFALL REGIMES AND CROPS: A GENERAL SURVEY

In previous chapters, the complexities of the atmosphere–soil–plant–water
system and plant–water relations were indicated. Despite these complexities and the danger of oversimplification, crop types and agricultural
systems can be related in a general way to the rainfall regimes indicated
in § 3.2. Many of the references cited in this and previous chapters
contain relevant material. Aspects of tropical food crops are considered
by Norman et al. (1984) and tree crops by Opeke (1982). Detailed
discussion of semi-arid areas is found in Mattei (1979) and Hall,
Cannell and Lawton (1979) while cropping systems, including problems
and prospects, in the more humid tropics (including seasonal rainfall
areas) are covered by Williams and Joseph (1970) and Fukui (1979). The
latter two references include consideration of rice areas, a topic also
covered by Oldeman and Frere (1982) within a general analysis of the
humid tropics of South-East Asia. Other references on rice are
contained in § 8.2 and, in addition, the meteorological aspects of rice
production in different tropical areas are covered by papers in
WMO/IRRI (1980). Tropical pastures and fodder crops are covered
by Humphries (1978, 1981) and these together with animal production
are considered by Mahadevan (1982) and Mannetje (1982). Webster
and Wilson (1966) present an overall coverage of tropical agriculture.

1. Humid tropics

1a. No dry season. Annual totals usually > 2000 mm and all months
at least 100 mm. In general, lack of water is at no time a handicap to
agriculture. However, the large seasonal and annual totals evident in

most areas, together with high intensities and large storm totals, can cause considerable problems of the kinds discussed in previous chapters such as flooding, soil erosion and leaching. The climax vegetation is usually luxuriant evergreen rainforest. Since neither lack of rainfall nor low temperatures limit crop production, there are no marked agricultural seasons. Crops flourishing under continuous hot, wet conditions and not needing a pronounced dry season for harvesting are grown. Grain ripening and harvesting are difficult (Williams and Joseph 1970).

Commercial crops are mainly perennials such as rubber, oil palm, bananas, liberica coffee and to a lesser extent coconut and cocoa. Other perennials, however, while needing high rainfall, also need a short, drier and preferably cooler season for satisfactory flowering and fruiting. These include mango and citrus which are not found in the humid tropics. Root crops, such as yams and cassava, are often the most important subsistence crops. The absence of a dry season limits the significance of rice. Varieties of maize adapted to conditions in the humid tropics are grown. Because weed growth is continuous and very rapid, its control, together with efficient cultivation, is difficult. Intense leaching often limits fertility, and in perennial crops trace-element deficiencies occur quite frequently as well as those of major elements.

Because there are no extensive natural grasslands and also in some areas because of disease, livestock are unimportant except perhaps for pigs and poultry. While the climate is suitable for cultivated pastures and fodder crops, research into suitable grasses and legumes and management methods is relatively recent. These aspects are discussed by Humphries (1981), Mannetje (1982) and Mahadevan (1982). Mannetje (1982) refers to cattle production under tree crops such as coco-nuts and possibilities for grazing during the initial years under oil palms in humid areas. Mahadevan (1982) refers to the need for more effective use of non-conventional feed resources and development of better forage in smallholder crop rotations in Asia.

1b.   No pronounced dry season. Annual totals <2000 mm but a drier period with a few months having <100 mm.

On the equatorial side of this transitional zone, crops resemble those of category 1a and the climax vegetation is luxuriant evergreen forest. On the poleward margins with a few drier months, a proportion of deciduous trees may occur. The drier period makes rice more important than in category 1a, although the main rice areas are those with a pronounced dry season. Especially in low-lying coastal areas, the problem may be effective drainage during the 'drier' period for the production of dry-land crops. In such areas, double cropping of rice is probably the best way of fully using the land, since crop losses can occur with dry-land species due to inadequate drainage (Williams and

Joseph 1970). Other crops include maize, yams, sugar cane, cocoa, coconut and nutmeg.

The presence of a drier season represents below-optimum conditions for crops such as rubber and oil palms which ideally need high rainfall throughout the year. Climate alone, however, would not preclude their cultivation (Webster and Wilson 1966). Perennials such as coffee and citrus, which thrive best under lower rainfall and a short dry season with lower temperatures, are commercially profitable despite below-optimum conditions. On windward (eastern) coasts, the often marked local variations in amount and seasonality of rain (§ 3.2) increase the range of crops.

Natural grasslands are either limited in extent or of low productivity. However, developments in improved permanent pastures and the breeding of locally adapted cattle are taking place (see, e.g. Mannetje 1982; Mahadevan 1982). Goats, pigs and poultry are extensively kept.

2. Wet and dry tropics

These cover very large areas of great agricultural significance and are very varied in terms of rainfall characteristics (§ 3.2). The various types, in reality, show transitions from one to another and this is reflected in the transition from one natural vegetation type to another. Near the humid tropics semi-evergreen forests are found, but as drier areas with shorter wet seasons are encountered the natural vegetation ranges through various types of deciduous forest, open woodlands to thorn woodlands and thickets. The hottest time of the year tends to be towards the end of the dry season, greatly adding to the labour of preparatory cultivation in the hard ground (§ 8.2).

2a. Annual totals 1000–2000 mm. Two rainy seasons with short dry seasons or months with lower rainfall, usually a few months < 50 mm.

Because of the favourable rainfall conditions (in parts of Uganda for example) perennial crops such as coffee, tea and bananas are grown. For annual crops, two cropping seasons per year are possible, the first often being a food crop and the second a crop such as cotton. In the southern parts of Ghana and Nigeria, the main perennial crops are oil palms and cocoa and the annuals mainly maize, pulses, yams and cassava.

Rice is the chief annual crop in South-East Asia. The high rainfall and only short dry seasons often allow the continuous cultivation of wet rice without excessively costly irrigation (Williams and Joseph 1970). Where the dry seasons are very limited (i.e. akin to region **1b**) double cropping is again probably the best way of utilising the land. For reasons similar to those in the humid tropics, livestock are at present relatively unimportant. Perhaps understandably, research on tropical pasture improvement has been considerable in northern Australia. Great improvements, especially in the drier wet and dry areas of tropical Australia, have been produced by the spreading of Townsville stylo. Pasture improvement in tropical and sub-tropical woodlands of Australia is discussed by Moore (1973) and general aspects of pasture and live-stock development discussed in Humphries (1981), Mannetje (1982) and Mahadevan (1982).

2b. Annual totals 650–1500 mm. Two short rainy seasons separated by a pronounced dry season (a few months < 25 mm) and a shorter, drier season. The less favourable rainfall conditions than the previous category mean that in most areas only drought-resistant perennials such as sisal and cashew can grow successfully. In some more favoured highland areas, however, if moisture conservation is practised, high-yielding perennials can be grown commercially. The large coffee industry mainly at altitudes of 1750–1850 m in the East Rift area of Kenya is an example of this, where below-optimum moisture conditions are to a considerable extent counteracted by growing without shade and with regular mulching and pruning to regulate crop production (Webster and Wilson 1966).

The amount and reliability of the rainfall determines the types of annual crops, and in view of the considerable variation of the former a wide range is found. The main types are maize, sorghum, finger millet, sweet potatoes, cassava, groundnuts, beans and other pulse crops. The variability in rainfall conditions within this group means that in the wetter areas two cropping seasons may be possible, but where the amount and reliability in one of the wet seasons is low, this may not be so.

Open grassland with scattered trees covers wide areas. The presence of the tsetse fly in many areas rules out the possibility of cattle unless the fly is eradicated. Mahadevan (1982) indicates that there are large humid and sub-humid areas in Africa and Latin America which could be highly productive in livestock if animal disease can be overcome in the former and the gap between technology development (e.g. improved legume development) and its application narrowed in the latter. Mannetje (1982) discusses the need to adapt non-tropical breeds to suit tropical conditions, including parasites. In many areas with over 750 mm of rainfall per year, the establishment of temporary grass leys is possible as they can survive the dry season and carry some stock through it (Webster and Wilson 1966). During the rains they are highly productive.

2c. Annual totals 650–1500 mm. One fairly long rainy season (normally 3–5 months > 75 mm) and one long dry season.

These areas are not usually suited to perennial crops unless they are markedly drought resistant (e.g. sisal), or deciduous with a marked dormant period (e.g. tung-oil trees). When rainfall is of the order of

1300 mm, however, slight modifications of climate due to altitude and small falls of rain in the dry season allow commercial growing of perennials such as tea and coffee. However, the long dry season results in lower yields than in more favourable areas. The single long rainy season of 5 or 6 months is generally more suitable for annual crops than the previous category where the rain occurs in two short wet seasons. The range of annual crops is wide, including those mentioned in the previous category as well as certain others such as yams and cotton which require a longer or more reliable rainy season. In the wetter areas, particularly in South-East Asia, rice is of great importance. Mattei (1979) discusses the significance of 'flood retreat' cultivation to traditional agriculture found over small areas along the main African rivers. Various crops such as sorghum, maize and cowpea are maintained by soil moisture along water courses after flooding. The seasonal rainfall pattern over the catchment concerned determines the pattern of land use.

The long dry season, ending the growth of natural grasses, greatly restricts the stock-carrying capacity, only the low numbers which can be supported by poor-quality standing hay being possible. Mannetje (1982) refers to the low nutritive value of pasture during most of the period of active growth but points out that pasture improvement could greatly increase production. However, technical, sociological and economic problems inhibit a rapid increase in the area of improved pasture in developing countries. In the lower rainfall areas of 2c and 2e (see below) cattle on native pastures may lose up to 50 per cent of their weight during the dry season (Mannetje 1982).

2d. Annual totals > 1500 mm. One season of exceptionally heavy rain and one long dry season. This category, of which the south-east Asian monsoon regions are typical, is a more extreme case of category 2c, crops being similar to the wetter areas of the latter. Rice is of great importance. References indicated at the start of this section consider the complex issues of rice cultivation in this and other rainfall regime areas. The synchronisation of rice cultivation with the monsoon has already been discussed in § 8.2. Because of the very marked dry season, irrigation works for wet culture of rice during the off season will be costly, and production of less water-demanding crops may be more profitable. However, double cropping of rice is a declared objective in many countries and a careful consideration of the economics of this, as opposed to an alternating cropping system, is necessary (Williams and Joseph 1970). Alternating cropping, as opposed to rice monoculture, has various advantages, such as decreased risk of pest and disease build-up, soil improvement through annual drainage and sometimes a better market value for the alternate crop. A legume dry-season crop can also increase soil nitrogen. However, Williams and Joseph (1970) indicate that periodic drainage is undesirable on acid sulphate soils. Here, mono-cropping of wet rice seems best if soils are not too acid.

2e.   Annual totals 250–650 mm. One short rainy season (3–4 months > 50 mm) and one long dry season. These are not suitable for perennial crops but short-term and/or drought-resistant annuals can be grown. These include sorghum, bulrush millet, sweet potato, groundnuts and sesame. Mattei (1979) discusses the agricultural systems in these areas. Early season showers are lost due to high evaporation from bare soil and this, together with rainfall variability at the start of the season, creates problems for necessary early sowing. Fallow is important (Mattei 1979). Oliver (1969) points out that in the Sudan, concentration of rainfall in a few months means that cotton can be grown in regions with only 400 mm. Mannetje (1982) refers to the symbiotic relationship between livestock and arable cultivation in semi-arid Africa. Livestock consume crop residues in the dry season and the manure benefits subsequent crops.

Livestock, of considerable significance in these regions, require large areas per animal and are kept by pastoral nomads. Many of the problems of these regions, such as drought, overgrazing and soil erosion were referred to in Chapters 3 and 4 and will be considered in Chapter 9. These problems, also faced in the next category, have received much attention in recent years because of drought conditions (see § 3.4).

3. Dry climates

Annual totals < 250 mm, little rain at any time, but can be concentrated in a very short 'wet' season perhaps of only a few weeks.

The division between these areas and category **2e** is arbitrary, particularly in view of rainfall variability and areal differences in evaporative demand. Oliver (1965, 1969) indicates that in the Sudan, concentration of rain in 3–4 months and in the cooler part of the day helps to explain how nomadic tribes can find pasture and even grow scanty crops of sorghum in areas with a mean annual rainfall of 125–200 mm, and that unexpected densities of acacia occur in such localities. North of Khartoum it is possible for sorghum to be grown in depressions with 100 mm annual rainfall. The rainy season is spread over 3–4 weeks (Oliver pers. comm.). The situation is similar in other parts of northern Africa and south-west Asia, the latter, interestingly enough, being recognised as the origin of cultivation and pastoralism (Walton 1969). Ancient practices of 'water harvesting' (§ 8.2) are used in these areas, techniques being discussed by Pereira (1973) and Walton (1969).

Without irrigation these regions are of limited agricultural significance, but possibilities exist for development in some areas. A survey of the New Valley desert region west of Aswan, Egypt, has shown, for example, that several million hectares of soil are cultivable if water can be provided.

It will be apparent that within each of the above categories there is a range of conditions and overlap between the different zones, particularly within the 'wet and dry' areas. Nevertheless, these broad classes do have associated with them reasonably distinct agricultural systems. An important point, likely to increase in significance in the future, will be the artificial alteration of the water regime of the areas by means of irrigation and water-conservation practices.

CHAPTER 9

HUMAN IMPACT ON THE HYDROLOGICAL CYCLE

9.1 INTRODUCTION

The characteristics of tropical rainfall and evaporation and their implications were considered in previous chapters. In Chapter 8, ways in which people adapt agricultural activities to the characteristics were reviewed, together with methods of modifying their effects by irrigation, water and soil conservation. Here, some further impacts of people on the hydrological cycle, both deliberate and inadvertent, are examined. These include the direct modification of rainfall and evaporation characteristics as well as changes to the factors influencing the disposition and character of surface and ground water by land-use practices. Within a single chapter, only an outline can be presented, attention being mainly focused on those aspects directly related to discussions in earlier chapters.

9.2 MODIFICATION OF CLIMATIC FACTORS

Considerable interest has been shown in recent years in how human activities may influence climate inadvertently as well as the possibility of deliberate modification of climatic factors. Aspects of human influence are considered by Hastenrath (1985). Munn and Machta (1979) discuss three possible ways in which human activity may influence climate: (1) by changing atmospheric composition, including water concentration; (2) by releasing heat into the atmosphere; (3) by altering the physical and biological properties of the surface. They also discuss problems in assessing such impacts, for example due to their multiple nature and interaction. Overgrazing, for example, by changing surface properties has an impact upon energy and water budgets and may also increase atmospheric dust (see below).

Rainfall stimulation

Much effort has been devoted to the possibilities of modifying climatic elements, although for financial and manpower reasons most work has been outside the tropics. Since lack of rain imposes development limitations in many tropical regions, possibilities of rainfall stimulation are attractive. This may be achieved by 'seeding' clouds with particles of dry ice, silver iodide, salt or large water droplets to initiate the precipitation process. Results of experiments are extremely variable and interpretation difficult. A problem is that suitable clouds on the point of producing rain are necessary, and it is difficult to decide whether they would have produced rain even if seeding had not taken place. In an area where some clouds are seeded, others which are not may also produce rain, the problem being to decide whether the former produced significantly larger amounts. Experiments need careful design to overcome these problems.

It must be emphasised that the need for clouds on the point of precipitating rules out seeding in dry or drought-stricken areas. Possible benefits are limited to increasing rainfall in specially favoured areas. This is why the term 'rainfall stimulation' is more appropriate than 'rain-making'. In some areas, increases of rainfall of the order of 10–20 per cent at an acceptable level of statistical significance are recorded, but in others there has been no significant increase and even decreases have been found. As was indicated in § 3.8, increases of rainfall of the order of even 10 per cent can have a more than proportionate influence on, say, streamflow. Examples of experiments are discussed by Simpson and Wiggert (1971) and Simpson et al. (1970). Simpson et al. (1970), for Caribbean studies, report that massive seeding of tropical cumulus clouds can cause enhanced cloud growth and approximately double rainfall. A study of two Florida watersheds showed that sufficient numbers of suitable clouds existed to allow increases in rainfall of as much as 20–30 per cent. Woodley et al. (1982) discuss results of an experiment in the Florida area to assess impacts of seeding on convective rainfall over a target area of $1.3 \times 10^4 \text{ km}^2$. Rain volumes over the total target area and the most intensely treated part were examined, positive results being found. Benjamini and Harpaz (1986) developed a method of assessing impacts of cloud seeding in Israel using a rainfall–runoff approach. Evaluation of 13 years of data indicated only a weak overall increase in runoff, median increments for six target zones ranging from 0 to 14 per cent.

East African experiments (e.g. Bargman, Sansom and England 1955; Brazel and Taylor 1959) indicated positive results and even a slight increase in amounts at critical periods may be important for a crop such as tea. It has been stated, however, that research into this aspect is too expensive for East Africa (Thompson 1965). Research in Australia, a leader in the field, halted several years ago although there are signs of renewed interest. Experiments are also being considered in Indonesia. Economic considerations mean that most tropical studies are being carried out in the Caribbean by the USA. Woodcock and Jones (1970) cite evidence from Queensland and Hawaii that burning of sugar cane, by creating additional condensation nuclei, may influence precipitation. However, the evidence is inconclusive and a complex set of factors is thought to operate. In general, experiments to date are variable in their effects and much is still not understood, particularly as to repercussions in non-seeded areas and evidence that seeding has sometimes led to a decrease in amount.

Hail and hurricane modification

Allied to rainfall stimulation is hail suppression. Large hailstones cause considerable damage to valuable crops in certain areas such as tea in Kenya and tobacco in Zimbabwe. In an area of 388 km^2 of western Kenya, hail is reported on average 54 days/year, causing a mean annual loss of 675 000 kg of made tea (Sansom 1966). Seeding of clouds with silver iodide to prevent the formation of large hailstones has been attempted in various parts of the world but without convincing results, perhaps because insufficient amounts were used. The use of explosive rockets to reduce hail damage has also been attempted, particularly in vineyard areas of northern Italy. The concept behind this method is open to question, a review of experiments and theories being presented by Sansom (1966). Despite theoretical doubts, success is claimed, for example in protection of pineapples in Swaziland.

An experiment in Kenya apparently reduced hail damage to tea, losses averaging 5.6 kg/ha during the experimental period compared to 5.8 kg/ha during a control period (Sansom 1966). This difference could have been because of a change in hailstorm distribution, although nearly all the tea estates reported a slightly higher incidence of storms during the experimental period. The rockets, fired when hail began, did not prevent stones forming, but eyewitnesses' reports suggest that hard, damaging hail was turned into either soft hail or rain.

The great damage caused by hurricanes (Ch. 2) has led to consideration of the possibility of their modification, especially in the Caribbean. Techniques involving seeding of clouds to modify hurricanes are discussed by Simpson and Malkus (1964) and an example of experiments is analysed by Sheets (1973). As in the case of rainfall stimulation, hail suppression and hurricane modification are far from being fully understood and considerable experimental problems exist. Again, economic considerations loom large.

Willheim (1980) points out that predictions of response of hurricanes to seeding are uncertain and largely based upon computer modelling, about which there are reservations. Only a small number of field experiments have been carried out. Only two of four experiments in the

Stormfury project gave some grounds for hope and Willheim (1980) refers to an early study where a mature cyclone reversed its course after seeding, causing extensive damage. Seeding could increase rather than decrease cyclone intensity and/or divert it from its natural path, this uncertainty indicating the need for a proper legal regime for experiments (Willheim 1980).

Modification of atmospheric composition

Possible impacts of gases and particles introduced into the atmosphere as a result of human activity are considered by Munn and Machta (1979), Daniel (1980) and Hastenrath (1985). Pollutants and heat resulting from industrial and urban activity, particularly combustion of fossil fuels, are more significant in non-tropical areas. However, their importance in the tropics will increase and Ofori-Sarpong (1983) discusses ideas put forward by Bryson (1973) on the possibility that pollutants, including carbon dioxide, could influence global circulation patterns which in turn might influence drought occurrence in West Africa. In the tropics, atmospheric particles may result from desert sand-storms and as a result of burning of crop residues, grasslands and forest.

Overgrazing and cultivation may also lead to increased dust levels in the atmosphere as a result of a decline in soil structure and moisture. Evidence suggests that the Rajasthan Desert in the north-west of the Indian sub-continent presents a more hostile environment than in the past. The consensus is that this largely results from forest clearance and overgrazing leading to soil erosion (§ 9.3). Bryson and Baerreis (1967) speculate that resulting dust levels lead to increased atmospheric subsidence (Ch. 2), suppression of convection and reduction in rainfall. They suggest that planting of grass, by reducing dust levels, could reduce subsidence and when combined with the existence of a shallow, moist airflow, this could increase the chance of more rain. In addition, a reduction in atmospheric dust could increase the diurnal temperature range at the surface, perhaps leading to greater dew formation, this being important in semi-arid and arid regions (§ 6.3). Since subsidence in the region partly results from rain elsewhere, it is unlikely that such a change could produce widespread heavy rain; the most that could be hoped for is that the area would become slightly less arid (Lockwood 1974).

Riehl (1979) and Munn and Machta (1979) refer to dust raised in North Africa being transported across the Atlantic to the West Indies. Possible effects of Saharan dust on the energy balance have been modelled by Carlson and Benjamin (1980) and Guedalia, Estournel and Vehil (1984). Carlson and Benjamin (1980) suggested that the dust would have a stabilising effect but stressed the need for further work before the possible climatological significance of the results could be assessed.

The possible effects of human activity on levels of carbon dioxide and resultant climatic impacts have received much attention in recent years. Munn and Machta (1979) discuss this in detail and Grove (1985) and Hastenrath (1985) provide useful summaries of the uncertainties involved. Although burning of fossil fuels is generally cited as the major factor, possible impacts of a reduction of tropical forest on carbon dioxide levels are discussed by Riehl (1979) and Munn and Machta (1979). Both stress that the magnitude of such impacts is uncertain. In addition to such uncertainty, the climatological implications of changes in carbon dioxide levels are the subject of much debate (e.g. Munn and Machta 1979). Ausubel (1983) also discusses the problems in assessing the impacts of a change in climate resulting from a change in carbon dioxide. Variations of carbon dioxide levels on daily and monthly time scales in Hawaii showed no evidence of links with large-scale atmospheric fluctuations (Saddler, Ramage and Hori 1982). Local sugar cane burning did not produce any measurable effect on carbon dioxide levels.

The above discussion of a very complex subject indicates the uncertainties at several levels: first, in terms of the impacts of human activity on atmospheric composition; second, the possible climatological implications (for the energy, water balances and atmospheric stability) of a change in composition; third, of particular concern here, possible resultant impacts on precipitation, and finally, uncertainties of the impacts on people of any change. Unfortunately, the picture is even more complex because human activities of the kinds already indicated (e.g. deforestation, grazing, cultivation) also, by changing the nature of the surface, may have a further set of impacts.

Modification of surface characteristics

Alterations to the surface may influence climate in several ways: (1) by changing albedo; (2) by changing rates of emission of long-wave radiation from the earth's surface; (3) by changing rates of evaporative loss, this having implications for atmospheric moisture and the way in which energy at the earth's surface is partitioned. Changes in albedo, emission of long-wave radiation and energy partitioning will have an impact on the energy balance and atmospheric stability. An alteration in surface roughness has implications for evaporative loss and general turbulence. All these factors have possible significance for precipitation and evaporation which are of concern here. However, the interaction of these factors is complex and the question of scale, both of any surface alteration and also its possible impact, raises a further complication. Understandably, this is an area of considerable uncertainty and debate and a compre-

Table 9.1 Impacts of changes to the earth's surface (after Munn and Machta, 1979)

(a) Causes of large-scale changes in land surface albedo

Processes that cause increases in albedo	Processes that cause decreases in albedo
1 Desertification	1 Overgrazing in regions with moderate to heavy rainfall
2 Overgrazing semi-arid regions	2 Man-made lakes and irrigation (slight)
3 Burning of grassland in semi-arid regions (slight)	3 Construction of towns (slight)
4 Ploughing of fields (slight)	4 Snow removal
5 Clearing of forests	5 Deposition of particles on snow
6 Addition of biological films to water surfaces	

(b) Human processes that change the Bowen ratio

Processes leading to an increase in Bowen ratio	Processes leading to a decrease in Bowen ratio
Desertification	Irrigation
Clearing of forests	Man-made lakes
Drainage of swamps	Urban growth (in dry climates)
Urban growth (in moist climates)	

hensive review is outside the scope of this book. Here, a small sample of ideas and experimental evidence is presented to indicate their nature and uncertainties.

Munn and Machta (1979) present an overview of possible impacts of changes to the earth's surface. Some of the stated implications for albedo and the Bowen ratio (the ratio of heat loss by convection and by evaporative cooling) are presented in Table 9.1. Land cultivation may increase or decrease the Bowen ratio depending upon factors such as crop type, stage of growth and climate regime (Munn and Machta 1979). Impacts of land use changes, including deforestation, are also discussed by Hastenrath (1985).

Both Ofori-Sarpong (1983) and Grove (1985) refer to overgrazing in low-rainfall areas resulting in albedo increase tending to decrease convective rainfall, thereby resulting in further environmental deterioration. However, in a study of albedo and temperature changes associated with desertification in the Sahel zone, Wendler and Eaton (1983) indicate the complexity of the situation but do not relate this to possible impacts on rainfall. Possible climatic impacts of changing albedo associated with human activity, particularly tropical deforestation, are modelled by Henderson-Sellers and Gornitz (1984). They found that despite an increase in albedo, the surface temperature effect of deforestation was negligible, this being ascribed to a reduction in evaporative loss. The model did, however, suggest a local rainfall decrease and a decline in total cloud cover. They concluded that albedo changes due to current levels of tropical deforestation have negligible effects on global climate. However, in an editorial, Sneider (1984) points out that applications of such models are open to much uncertainty. Another example of assessment of the energy balance of tropical forest is provided by Pinker, Thompson and Eck (1980). They measured incoming and outgoing short-wave radiation and long-wave radiation at the top and floor of an evergreen forest in South-East Asia.

Munn and Machta (1979) point out that the rather low emissivity of quartz particles could mean that an increase in desert area could reduce long-wave radiation loss from the earth's surface. They also suggest that on a few days of the year, relatively small changes in surface conditions may be enough to trigger off or suppress convective cloud development and shower activity but the frequencies may be so small that statistically significant differences are difficult to find.

Inadvertent modification of evaporation and precipitation

Impacts of surface alteration on evaporative loss are difficult to assess. While desertification and deforestation may have reduced evaporative loss, other changes may have increased it, Munn and Machta (1979) suggesting an overall increase of about 3.5 per cent. Clearly spatial and temporal variations of such impacts are considerable and effects at different scales are uncertain. If a large proportion of precipitation consisted of locally evaporated water vapour, the possibility might exist of influencing precipitation by altering local evaporation sources, for example by building reservoirs or planting large numbers of trees. Alternatively, by clearing forest or draining ponds or marshes, it might be possible to decrease precipitation in humid areas. However, evidence suggests that the local water vapour contribution to precipitation is usually very small (Lockwood 1974) and any such attempts at control would presumably have to be on a very large scale.

The possibility of an impact such as that just described has been raised by the National Academy of Sciences (1980) in relation to Amazonia. They point out that over half of the region's rainfall is returned to the atmosphere and certainly part of the rainfall is derived from moisture circulating within the region rather than from outside. Since the region serves as a source of much of its own moisture, removal of part of the forest could mean a reduction in total moisture, leading to a steadily desiccating environment. The magnitude of change needing to occur before there was any alteration in rainfall (if at all) is unknown. Gutman (1984) modelled effects of changes in albedo and water availability resulting from processes of desertification, deforestation and irrigation. Results indicated that changes in evapotranspiration rather than in albedo would be predominant in regulating surface temperature. Again, limitations of such numerical models must be remembered. Anthes (1984) hypothesises that planting of bands of vegetation 50–100 km wide in semi-arid areas could, under certain large-scale atmospheric conditions, lead to increases in convective precipitation. Anthes suggests increases could be greater than those associated with uniform vegetation over larger areas. Three mechanisms are discussed and Anthes (1984) states that various studies support the concept.

Creation of large lakes will affect atmospheric conditions, water and energy budgets. For example, water has a lower albedo than many other surfaces and can store large quantities of heat which is released much later. In an arid area, advected energy creates high evaporation rates and losses must be taken into account in planning. However, as indicated in § 5.2, the shape of the lake in relation to airflow will influence losses. In the case of a round lake, Riehl (1979) suggests that atmospheric stability is induced which will limit evaporation. An elongated lake such as Lake Nasser will not be affected in the same way unless the wind direction follows the long axis of the lake and annual evaporation loss has been estimated at 280 cm/year (Riehl 1979).

Evaporation will increase atmospheric moisture downwind of a large lake and there are claims that this could influence precipitation (SMIC 1971). Such claims are difficult to substantiate because of lack of data. A review of the interaction between reservoirs, the atmosphere and hydrometeorological elements is provided by Nemec (1973). He points out that possible effects of Lake Nasser (area 4500 km²) are the subject of controversy, but it is not considered that the resultant increase in atmospheric moisture would produce even clouds, much less precipitation, nor would the lake influence thunderstorm occurrence.

Riehl (1979) agrees that the evidence for possible increase in precipitation resulting from increased atmospheric moisture is all negative. However, as has already been indicated above, energy balance changes occur with the creation of a large lake, and these may result in local circulations which under some circumstances could influence rainfall. In § 2.10, such effects were referred to, Lake Victoria being cited as an example.

For Lake Kariba on the Zambesi River, Hutchinson (1973) found that before-and-after measurements suggested an increase in precipitation on the shore and adjacent land with percentages ranging from very little to nearly 50 per cent. Hutchinson (1973) suggested that this was associated with a lake breeze mechanism (§ 2.10) interacting with nearby mountains to produce convergence and cumulonimbus clouds onshore in the absence of marked stability. Thus it seems that if large reservoir construction does have any impact on precipitation it will be through such effects rather than due to increased evaporation. However, effects will depend on atmospheric structure; stability and subsidence inhibiting such influence (Riehl 1979). Furthermore, sufficient atmospheric moisture is needed, even large lakes not providing enough. Possible effects of large irrigated areas in dry regions are similarly uncertain since they will also influence water and energy budgets (Table 9.1).

The above analysis of possible impacts of human activity is not designed to provide definite answers but merely to indicate the complex nature of the factors and processes which need to be taken into account. Literature cited will hopefully provide a starting point for those wishing to pursue the topic further. The many factors involved mean that local circumstances make generalisation difficult and it is perhaps dangerous to extrapolate results between areas. However, Munn and Machta (1979) point out the need to evaluate possible impacts of major development projects on climate since effects may be difficult to reverse.

Evaporation reduction

High evaporative losses from open water surfaces, soil and plant transpiration in the tropics have led to interest in methods of reduction, particularly in drier areas. A wide variety of approaches including the use of mulches and deficit irrigation (§ 8.2), shelter belts and crop selection are discussed by Krantz (1981b), Davenport and Hagan (1981) and Arakeri and Donahue (1984). Here attention will be focused on two aspects: reduction of evaporation from open water and the use of antitranspirants.

Lake Nasser provides one example of high losses, which are estimated at up to 10 000 million m^3 per year out of a total of 176 000 million m^3 when it is full (Keating 1972). For the Indian sub-continent, losses lie between 1.9 and 2.2 m storage depth/year (Kumaraswamy 1973). The large surface area of storage reservoirs leads to major losses, and in the case of small reservoirs and tanks, which are usually shallow, evaporative loss is frequently more than the amount actually used. Hence increasing depth, thereby reducing ratio of surface area to water volume, is advantageous.

Experiments to cut evaporation loss involve production of films of heavy alcohol applied as a solution, an emulsion, powder or pellets over the water surface. Compounds such as non-toxic cetyl alcohol forming monomolecular films may reduce loss by up to 50 per cent under ideal conditions (Davenport and Hagan 1981). Wax also reduces loss, cracks in the film being partly resealed when the sun re-melts it. However, the effectiveness of all films depends upon windspeed, water surface area and temperature. Over large surfaces, wave action breaks up the film and the saving is greatly reduced. Wave heights of 2.7 m are possible in the Kariba Dam with sustained winds of 38 km/hr over a fetch of 64–80 km (Van der Lingen 1973). Kumaraswamy (1973) found that for a tank of 11.2 ha in Madras State, India, with windspeeds of 2.2–2.4. m/sec and variable direction, extensive areas could be covered with relative ease. For windspeeds <2.2. m/sec practically the whole lake could be covered, but with speeds >9 m/sec it was impossible to maintain a film on the water surface. The efficacy of spraying alcohol decreases with increase in temperature. Kumaraswamy (1973) found that reduction under the tropical conditions of India was only 20 per cent compared to 50–60 per cent for temperate areas of the USA and Australia. Greater success will depend on the development of stronger films. For small reservoirs, experiments using floating hexagonal panels about 1 m² cast from expanded polystyrene have been successful but costs do not yet allow routine use (Pereira 1973). Davenport and Hagan (1981) point out that the use of barriers such as styrofoam blocks and plastic balls still allows water to 'breathe'. However, they point out that all barriers tend to increase water temperature which would tend to increase evaporation from uncovered areas.

In recent years, possibilities of reducing transpiration loss have received attention. Branson *et al.* (1981) define three broad approaches: (1) use of reflecting materials to reduce heat load on leaves; (2) film-forming materials which hinder water vapour loss from leaves; (3) stomatal-closing chemicals. Other discussions of basic principles and problems associated with antitranspirants include Sutcliffe (1979), Krantz (1981b), Davenport and Hagan (1981), Kramer (1983) and Arakeri and Donahue (1984).

Most experimental evidence suggests that a reduction in transpiration is accompanied by reduction in photosynthesis and yield since antitranspirants will also hinder carbon dioxide intake. However, the impact on carbon dioxide intake is generally considered to be less than that on transpiration (§ 6.2). Krantz (1981b) suggests that application to non-agricultural riparian phreatophytic vegetation, such as salt cedar where a possible yield reduction is not important, may help to conserve ground water and maintain streamflow. The high cost of chemicals means they may only be appropriate for high-value crops such as certain fruit trees (Krantz 1981b). Kramer (1983) cites various reviews on the topic and discusses problems. Effectiveness seems to depend on a range of factors including crop type, stage of development and atmospheric conditions. With growing plants, repeated applications of films are needed to cover new leaf surfaces.

While many authors suggest that due to problems of the type indicated above, antitranspirants have only limited practical usefulness, a number of studies have indicated positive results. In a study of mulches and antitranspirants, Raghavulu and Singh (1982) found that of six antitranspirants, only kaolin and atrazine had a marked effect on sorghum in north-west India. Both antitranspirants increased water-use efficiency, grain yield and uptake of nitrogen and phosphorus compared with no suppressants. Babu and Singh (1984a, b) also found that antitranspirants had beneficial impacts on growth, yield, water-use efficiency and nutrient uptake of spring sorghum in north-west India and that savings could be made with irrigation water. Fuehring and Finkner

(1983) found that a single seasonal application of a hydrocarbon film antitranspirant on moisture-stressed corn increased yield. However, the response to spraying was curvilinear, with a yield decrease when rates of application were too high. De *et al.* (1983a, b) found that mulching and antitranspirants had beneficial effects on yield and water-use efficiency of maize and sorghum in India.

9.3 LAND USE AND THE HYDROLOGICAL CYCLE: INTRODUCTION

This can conveniently be considered under three main headings:

1. Forest cover.
2. Grazing.
3. Crop types.

Effects of various cropping systems were reviewed in § 8.2, where it was indicated that land-management practices were of paramount importance. In § 3.4 the distinction was made between land use and land management. Failure to make this important distinction is one of a number of reasons why there is much confusion about impacts of a change in land use on the hydrological cycle and soil erosion. Adverse effects have often resulted from inappropriate management practices rather than problems inherent in the new land use. Here attention focuses on (1) and (2) although there are links with (3), particularly for changes from forest cover to cultivation. The influence of urbanisation is not considered here. This is a complex area in its own right, management, especially during transitional periods when bare soil is exposed being again an important element. Urbanisation is itself a very vague term encompassing everything from low-density housing to industrialisation with virtually 100 per cent impervious surface. Hydrological effects must be considered in terms of whole catchments and furthermore, as with all land uses, the effects will vary depending upon rainfall characteristics, including antecedent conditions. Different parts of an urban area will be at different stages of development and redevelopment. These and other factors mean that the hydrological impacts of urbanisation will vary considerably in space and time. Discussion of tropical urban hydrology is presented by Watkins and Fiddes (1984), Alvarez and Sanchez (1980) providing an example from Brazil.

In the tropics, substantial changes in vegetation and land use have taken place. Much of the natural dry deciduous forests of semi-humid areas in Africa, north-east Brazil, parts of Central America and South-East Asia may have been converted into savanna grasslands by bush fires ignited during the dry season. In India the conversion has been mainly to arable land. These and other changes have had consequences for the heat and water budgets.

The effects of changing land use or management are conveniently studied on a catchment basis. Catchment experiments were referred to in Chapter 5 as a means of assessing water use by different vegetation and land-use types. Apart from this water-balance approach, experiments can serve a variety of other purposes and differ considerably in design and observational detail. They can be concerned with streamflow response to catchment characteristics, including land use and other hydrological elements, particularly rainfall. The impact of a land-use change may or may not be part of an experiment. Examples of catchment studies are considered below. Plot experiments to establish water use (E_t) by vegetation (Ch. 5) also assist in assessing effects of land-use changes. Computer simulation models represent another approach although they involve various simplifications and assumptions.

The water-balance equation for a catchment, together with some of the problems in its use were presented in § 5.5. Leakage (L) from watersheds can only be avoided by careful study of catchment geology. Even then, uncertainties exist but leakage can be detected by methods utilising water-balance data obtained from the catchment. Here, E_t values obtained from the equation or a simplification (§ 5.5) can be related to an estimate of open-water surface evaporation (E_o) and used to compare water use under various land uses in different areas as well as in the same locality.

The physical principles of the effects of land use and management on the hydrological cycle have been covered in previous sections (e.g. §§ 6.3, 8.2). These include: (1) the protection afforded soil against rain-drop impact not only by various vegetation layers but also by the litter layer; (2) effects of vegetation on soil organic matter and hence characteristics including waterholding capacity, infiltration and permeability; (3) water-use characteristics of vegetation including rooting depth and density which influence available water, a feature of considerable importance during dry periods. However, the significance of factors such as those above will vary depending upon a range of environmental factors such as rainfall characteristics, basic soil type and slope.

The aims in hydrological terms of a change in land use will vary. For example, in areas of scarce water supply, the aim may be to maximise water yield (streamflow) but in high rainfall areas the reverse may be true. Minimising stormflow (quickflow, surface runoff – see § 4.4) and soil erosion will normally be important. Factors such as the expense of dam construction may mean that land use and management must play an important part in evening out seasonal flow variations. Although in some cases hydrological aims may dominate requirements of a change in land use, in most cases the change is initiated for other reasons which may well conflict with the hydrological ideal.

Since in many tropical developing countries population pressure and the need to increase production mean that changes in land use are contemplated, the possible consequences of such a change need careful consideration. From the above brief comments and material in previous sections, assessing these impacts presents problems and it is impossible to present a complete discussion here. General coverage of a variety of aspects is given in Goodland, Watson and Ledee (1984), Keller (1983), IAHS (1979, 1980), other material related more specifically to forests and grazing being presented below.

9.4 FOREST COVER AND THE EFFECTS OF CHANGE

General principles

Possible consequences of utilising forests and changing land use in forest areas have received much attention and are the subject of considerable controversy. Useful references dealing with tropical forests include Mergen (1981), Hamilton (1983), Hamilton and King (1983), Lal (1983), Hamilton (1985). Although not specifically tropical, a general coverage of forests, water and soils is presented by Riedl and Zachar (1984), while Bosch and Hewlett (1982) present a review of 94 catchment experiments from various parts of the world. Controversy surrounding the topic has led Ewel (1981) and Hamilton (1985) to refer to 'myths' which have developed. Hamilton (1985) discusses the following concepts which in reality have little, or worse, a countervailing, scientific basis: (a) cutting reduces rainfall; (b) cutting reduces water supplies; (c) cutting causes floods; (d) shifting agriculture causes erosion and reservoir sedimentation; (e) grassland is better than forest; (f) reafforestation or afforestation is a panacea for water problems. In the discussion, Hamilton (1985) provides an eloquent indication of the complexity of the situation and why misunderstanding has arisen.

In many tropical regions, great pressure exists for a change in land use over forested areas, either for cultivation of annual crops or plantation agriculture. Pressure may also exist for utilisation of forest products. For the humid tropics between 10°N and 10°S, Lal (1983) cites an estimate that new arable land is being developed at an annual rate of 11×10^6 ha and that some 10^9 ha of once forested land has been turned into semi-desert during recorded history. One estimate is that some 40 per cent of the remaining forest in the humid tropics will disappear by the year 2000 (Lal 1983). Knowledge of the impacts of such changes is vital and clearly sound management practices are essential. Furthermore, there is a need to adopt measures to restore degraded areas. The disastrous effects of indiscriminate clearing of forest in many areas has led to a general feeling that forests should remain. However, forests exist in a wide variety of environments and this must be considered together with the proposed alternative land use and particularly the management practices. It is perhaps true to say that much of the confusion and conflict stem from a simple focus on the cutting of forest rather than a concern with what happens afterwards – whether there is to be a return to forest, whether it is to be replaced by agriculture or grassland and what type of management is to be used, including cropping systems, cultivation techniques, soil and water conservation practices, stocking densities.

Impacts of a change, whether permanent or temporary, need to be considered in terms of effects on infiltration, water use (evapotranspiration) and soil-moisture storage capacity. If there is a reduction in infiltration due to adverse effects on the soil, such as compaction resulting from decreased protection or a decline in organic content in the surface layers, then stormflow would increase, soil-moisture storage would decline and a fall in the water table might occur. In general streamflow would become more irregular with a decline in dry season flow. However, if removal of forest leads to a decrease in water use this will tend to increase water yield (streamflow). Here, rooting depth, especially in relation to seasonal rainfall regimes must be considered. If deep-rooted trees are replaced by shallow-rooted grasses or crops, then reduced dry season water use will tend to increase dry season flow. Clearly what happens in a particular case will depend upon what happens after removal of forest, both in terms of use and management practices.

It should also be noted that there may be differences in response to any change between temperate and tropical areas due to differences in rainfall characteristics (e.g. intensity, storm size, totals) and evaporative demand. For example, an increase or a decrease in protection of the soil would have greater significance with high tropical rainfall intensities than in the case of temperate areas. High evaporative demand in the tropics means that absolute values of water use could change more than in temperate latitudes, with resultant implications for water yield. Far more experiments on the impacts of a change from forest cover have been carried out in temperate latitudes and inevitably there is a tendency to extrapolate results to the tropics. The differences in meteorological and other parameters such as soil characteristics mean that caution must be exercised in such extrapolation.

As already mentioned, whether or not an increase in water yield is desirable will depend on whether a forested area is a source of supply for areas of marginal rainfall downstream or whether the real problem is an excess of water. If removal of a forest results in a rise in the water table due to reduced evaporative loss then problems may arise if the ground water is saline. Springs from this water can destroy vegetation over a wide area. In high rainfall areas, particularly on steep slopes, minimisation of stormflow and erosion will be critical. Hamilton (1985) points out the importance of tree root shear strength in maintaining stability on steep slopes and that forest cover gives the greatest protection against erosion in such cases. Whether or not regulation of seasonal streamflow is important will depend in part on the existence of storage reservoirs.

In considering land use change, factors other than the hydrological consequences must be examined, particularly nutrient cycling and maintenance of soil fertility within the ecosystem. The deeply leached soils over large parts of the humid tropics depend heavily on the presence of a forest cover for maintenance of fertility. Not only does it modify the leaching and weathering action of high temperatures and rainfall but the deep-rooted vegetation restores minerals to the surface from the deeper soil layers or parent rock. The forest litter decomposes rapidly, bringing minerals quickly back into circulation. Tree clearing for efficient tractor use interferes with the natural regeneration of fertility under bush fallow because of the role of trees in recovering nutrients from the lower soil horizons (David 1973).

A forest cover, composed of many species, is less susceptible to disease than a plantation. Stands of one or two species can be severely affected, leading to destruction of much of the watershed cover. Similarly, where the cover is a single deciduous species, much of the watershed may be unprotected at certain times, this not being the case with natural forest. Removal of biologically diverse forest in favour of another use (e.g. a plantation crop) means a loss of unique genetic material and a habitat for animals, plants and indigenous peoples (Goodland, Watson and Ledee 1984). However, Ewel (1981) points out that removal of forest in most cases is followed by rapid recolonisation. Seldom does forest exploitation *alone* sufficiently degrade a site to become an unproductive environment. An exception would be where large-scale deforestation means that seed sources are devastated (Ewel 1981). The other exception is of course where an alternative land use and management leads to degradation (see below).

Possible impacts on precipitation

One claim sometimes made is that removal of forest leads to a reduction of rainfall. Claims have even extended as far as to suggest that a desert can result from deforestation. Apart from the rapid recolonisation already suggested, the idea that a high rainfall area can be converted to a dry desert is difficult to accept. Certainly by inappropriate land use and management, particularly with some soil profiles, a degraded environment can result (see below). However, this is not due to a reduced rainfall associated with forest removal. The general principles underlying possible human impacts on rainfall were discussed in § 9.2.

Hamilton (1984) makes a case for increased trigger action of convectional rainfall associated with a forest cover based upon greater surface roughness and increased instability of air above it. Despite references cited by Hamilton (1984) in support of forest increasing rainfall, much evidence suggests little or no influence (e.g. Penman 1963; Hamilton 1985). The most uncertain case, already referred to in § 9.2 concerns the possible impact of forest removal in Amazonia. Whatever the uncertainties, compared with other aspects, possible impacts of deforestation on rainfall are of minor significance since at best the effects would concern particular types of rain under particular atmospheric conditions.

One situation where forest removal could have an impact on water input to an area concerns fog drip (§ 6.3), particularly in coastal and elevated locations. Removal may reduce the intercepting surface area in situations where fogs (mists) are prevalent, thereby reducing water inputs to an area. Penman (1963) quotes data from Kisantu (Zaire) where fog drip amounted to 187 mm out of an annual total of 1600 mm precipitation and for the Canary Islands where in one year condensation in a eucalypt canopy trebled the rainfall. Hamilton (1985) cites an extra 760 mm above 2600 mm of rainfall in Hawaii. Other examples are given in § 6.3. Penman (1963) suggests however that the significance of fog drip is limited to forest borders.

Forest infiltration

Reference is often made to the protection a forest cover affords soil against raindrop impact. As already indicated in §§ 6.3 and 7.1, however, Wiersum (1985) suggests that a forest canopy may increase drop size and erosive capacity and that forest litter is a crucial factor in protection. In addition, forests vary in nature and perhaps therefore in the protection they afford. Whatever the degree of protection afforded by forest, that provided by the alternative use must also be taken into account. The idea that a forest cover is always superior to others may stem from comparison with situations following immediately after clearance or where inappropriate use and management practices have been adopted. This does not mean that alternatives are necessarily inferior. Hamilton (1984) refers to the beneficial effects of the litter layers under forest in increasing organic matter in the topsoil, promoting activity of soil organisms and reducing blockage of soil pores by finer soil particles. This will tend to encourage infiltration and permeability. Again, however, this does not necessarily mean that an alternative cover, with appropriate management, will be inferior in this respect.

Soil profile characteristics vary and have an impact upon the significance of forest removal. While shallow soils over impermeable rock can have a high initial infiltration rate, rapid soil saturation occurs, infiltration rates fall and for a prolonged storm surface runoff then is considerable. A forest cover in this case merely delays stormflow but its removal can accelerate it. In the Cherrapunji area (Assam), shallow soils and horizontally bedded sandstones, together with the removal of tree cover, give the surface the appearance of a semi-arid region despite one of the world's record rainfalls (Oliver pers. comm.).

Overland and sub-surface flow

Low rates of overland flow and hence the significance of sub-surface flow under a forest cover are often referred to (e.g. Hamilton 1984; Pathak, Pandey and Singh 1984, 1985). However, in Malaysia, Peh (1980) found that overland flow was a significant hydrological and erosional process. In north-east Queensland, widespread overland flow is commonly found in undisturbed forest due to high rainfall intensities frequently exceeding the saturated hydraulic conductivity below 0.2 m (Bonell, Gilmour and Cassells 1983). This results in rapid saturation of the surface layer and hence overland flow. As a result, Bonell et al. (1983) found no change in runoff hydrology following logging, although suspended sediment levels were doubled during high flows and clearing produced a tenfold sediment increase.

Further evidence of the complexity of the situation is provided by differences between forested sites in the same general region. Even on similar slopes, Peh (1980) found significant variations in total runoff volumes and erosion rates between two study areas in Malaysia. Although surface runoff was small, Pathak et al. (1984) found significant differences in amounts between six forested sites in the Indian central Himalayas as well as differences between rainy season months. In the Philippines, Manubag (1985) found contrasts in interception and surface runoff between two types of forest.

Water yield

Most evidence points to an increase in total water yield following forest removal (e.g. Ewel 1981; Lal 1983; Hamilton 1984; Hamilton 1985; Bosch and Hewlett 1982). This is generally ascribed to high rates of rainfall interception, surface detention and evapotranspiration by forest and extraction of soil moisture from subsoil layers by deep-rooted tree species (Lal 1983). More uncertainty surrounds the question of the impact upon low-season flows almost certainly due to different conditions following forest removal together perhaps with differences between tropical and temperate conditions.

Removal of deep-rooted tree cover will reduce water use during the dry season if replaced by shallow-rooted vegetation (see above). This would lead to an increase in dry period flow. Most experimental evidence does indicate that not only does annual water yield increase following forest removal but that the greatest proportional increase occurs in low-flow months together with an increase in ground water levels (Hamilton 1985). Nevertheless, claims are made that deforestation results in a decline in ground water level and a reduction in dry season flows (e.g. Hamilton 1984). Possible reasons have to be sought in a change in infiltration rates (see above). If a new land use and poor management practices led to a reduction in infiltration (due to surface compaction), then lower water tables and reduced dry season flow could result, together with associated storm runoff increase during the wet season, i.e. a decrease in seasonal control. Characteristics of tropical rainfall, particularly high intensities, could make this aspect more significant than in temperate areas where much of the experimental evidence has been gathered.

Indiscriminate clearing and adoption of inappropriate land use and management practices in the past will certainly have tended to give an overall impression of a more uneven streamflow regime, with reduced dry season flow, and increased wet season flow, particularly peak storm runoff amounts and associated soil erosion problems. Once again, therefore, what happens after clearing is important, appropriate changes having little or no effects, or indeed some positive impacts.

Hamilton (1985) discusses concern about possible effects of deforestation on flooding. Evidence exists that forest provides protection against local flash flooding but Hamilton (1985) provides convincing arguments against the possible significance of deforestation *alone* for large-scale floods in the lower reaches of major rivers.

Soil erosion

The significance of forest removal for soil erosion is, like the hydrological case, a complex issue. Soil losses under forest are generally agreed to be low (e.g. Lal 1983), but whether or not a change in cover does significantly increase erosion will be influenced by what follows removal. Much depends on whether the area is allowed to regenerate itself, or what cropping system, agronomic and mechanical conservation practices, grazing practices, etc., are adopted (§ 8.2). As an example, Hamilton (1985) contrasts 'traditional' with 'new' types of shifting agriculture. The former was a stable, ecologically sound system, presenting minimum disturbance and time for regeneration. Population pressure and a shift to cash cropping have led to new forms, with clearing of large areas, lack of adequate soil conservation measures and cultivation until land becomes degraded before moving to a new area. The impacts of the two systems in hydrological and soil erosion terms will be very different.

Effects of grazing are considered below. Here it is relevant to point out that research has shown that conversion of forest to a grass cover has usually increased yield, produced higher ground water levels in deep soils but with only small or zero increase in stormflow volume and peak flow (Hamilton 1985). In addition, established grass shows no difference in erosion rates. However, Hamilton (1985) cautions against using such evidence to advocate conversion to grassland since most experiments were conducted on moderate slopes and without grazing. Most grass-lands will be grazed with periodic burning, and depending upon management practices, this may or may not have serious impacts (see below).

Afforestation and reafforestation

In general, effects are the reverse of deforestation and are summarised by Hamilton (1985) as follows: (1) only in the case of fog drip is an increase in water input likely; (2) in most experiments, reafforestation leads to a lowering of water tables and reduced streamflow, particularly in the dry season; (3) although there may be some reduction in storm-flow, most experiments show only small effects; (4) only in the case of large, badly degraded areas with resultant large volumes of stormflow and sediment load is large-scale planting likely to have much impact on flooding in the lower reaches of streams. Forest in such cases might slow and reduce surface runoff to a point where flooding is reduced but not eliminated. Despite the utility of the above statements, points already made, particularly the importance of the pre-existing land use and management, mean that these generalisations must be treated with some caution.

Examples of forest cover and change

To illustrate some of the above points, examples of assessment of land use and management impacts are briefly indicated below. Low and Goh (1972) used data from a number of Malaysian catchments to illustrate how a forest cover evaporates a high proportion of rainfall and can reduce streamflow compared with other land uses. They also presented data illustrating limitations of water-balance analyses over shorter time periods. The arbitrary division of the year into calendar months means that rainfall in one month may contribute to runoff in following months. Therefore, the authors found that runoff in one month could exceed the rainfall of that month, hence the difference between the two suggesting a negative water loss value.

For an exceptional storm over Singapore, Chia and Chang (1971) illustrated the differences between a mainly forested catchment and one with a large proportion of paved area. The latter had a lower water loss rate and a higher surface runoff than the forested catchment. However, in both cases the surface runoff as a percentage of rainfall was high (64% for the forest, 74% for the urbanised catchment). This illustrates a point already made – that with exceptional rainfalls, while a forest may perhaps reduce flooding problems it does not eliminate them.

A series of experiments in East Africa illustrates alternative land-use effects. These have been widely reported (e.g. Pereira *et al.* 1962; McCulloch and Dagg 1965; Dagg 1965b; Dagg and Blackie 1965; Pereira 1965, 1967, 1973). One paired catchment experiment in Kenya concerned the effects of replacement of rainforest by tea plantation. One catchment remained under forest as a 'control'. Initial clearing of the catchment to be planted over 50 per cent of its area during a 4-year period gave an 11 per cent reduction in water use when only 30 per cent of the valley was under bare soil at any one time. A heavy storm at the stage of maximum bare soil exposure demonstrated the need for sound conservation measures (§ 8.2). The effectiveness of the forest flood control was demonstrated by it producing a maximum storm-flow of only 0.6 m^3/sec/km^2 while the cleared and cultivated land delivered 27 m^3/sec/km^2. However, the well-constructed soil conservation works coped with this, although if clearing had been carried out on a larger scale, the cumulative effects of high peak flows would have caused problems downstream.

The fraction of annual rainfall lost as stormflow remained low and, after 10 years of tea estate development, had fallen to the original forest level of only 1 per cent. Thus, the experiment demonstrated the hydrological effectiveness of a forest cover in reducing stormflow. However, given high capital inputs and skilled planning and development, with a high level of soil conservation, especially in the critical early stages, the tea estate over the experimental period proved a hydrologically effective substitute.

A second set of plot and catchment experiments concerned replacement of bamboo forest by pine plantations in Kenya. The experiments indicated an increase in water yield in the softwood plantations over the first three or four years of clean weeding when vegetable crops were grown between the young trees (Pereira and Hosegood 1962). However, water use by the pines increased until it equalled that of bamboo. There was some suggestion that older pine plantations might use more water than bamboo in exceptionally dry years because of extensive root development.

Effects of replacing evergreen forest by peasant cultivation were investigated by using paired catchments in south-west Tanzania. During the rainy season, water use by the two catchments was similar. By the end of the dry season the forest with its deep rooting system (+13 m) was extracting deeper water – certainly beyond 3 m. This was not so over the cultivated catchment which therefore used less water and yielded more. For a 10-year period, E_t/E_o ratios were 0.88 for forest and 0.65 for the cultivated catchment. There was, however, some indication of increased soil erosion in the cultivated catchment. Thus the economic, technological and sociological feasibility of soil conservation measures arises. Here there are differences between such a situation and the case of a high-value crop such as tea discussed above.

Stormflow for the forest was again only 1 per cent and was still only 3 per cent for the cultivated catchment. While these values could be 50 per cent underestimates, the latter value is still very low for a steep-sided catchment under such cultivation. This was thought to be due to the high infiltration rates of the ash-derived soil. Such a condition does not exist in many areas, hence illustrating the complexity of the situation referred to in the earlier discussion.

Parfait and Lallmahomed (1980) used hydrograph separation techniques to examine effects of a change from forest to tea in Mauritius. Unlike the Kenyan experiment already discussed, the volume of storm runoff was increased under tea and there was a change in the time distribution of runoff. A paired catchment experiment showed water yield increases following clearcutting of sub-tropical montane forest in Taiwan (Hsia and Koh 1983). Yield increases were 55 per cent and 47 per cent for the first and second wet seasons after clearcutting. Corresponding dry season increases were 108 per cent and 293 per cent but dry season flow accounted for only 10 per cent of total annual yield.

In a Nigerian experiment, Lal (1983) cites an increase in total water yield, direct runoff and interflow following deforestation. The increase in direct runoff was attributed to deterioration of surface soil structure and infiltration. Base flow increase was ascribed to non-utilisation of subsoil moisture by shallow-rooted seasonal crops. Soil-moisture storage capacity was limited by a decrease in organic matter and the relative proportion of detention pores. Experiments in Ibadan, Nigeria, also showed that the effects of deforestation depended on the methods of clearing, certainly initially, but also on crop types, sequences, tillage methods and mulching (Lal 1983). In south Brazilian catchments, Bordas and Canali (1980) reported much increased surface runoff and erosion rates in deforested areas during the initial six months of an experiment. In the Upper Tana Basin, Kenya, Ongweny (1979) reported increased sediment yields from about 20 t km^{-2} yr^{-1} on undisturbed forest areas to about 1000 t km^{-2} yr^{-1} on grazed areas and over 3000 t km^{-2} yr^{-1} on steep cultivated areas.

Concluding remarks

The above discussion and examples indicate a complex situation. Forest areas exist in a wide variety of hydrological situations, slopes, soils and human pressures. Therefore, the possible impacts of alternative land use and management practices must be carefully considered in each case. In some areas, reduction in water loss may be critical but not in others. The economics of alternative land uses compared with indigenous forest must be noted together with the existence of capital and expertise for well-planned development. Illustrations of acceptable hydrological substitutes under the right conditions have been given. Some broad general concepts are indicated in Table 9.2. Above all, hopefully the discussion will have indicated the dangers in making sweeping generalisations and thoughtless extrapolation of the results of investigations in one situation to others.

Table 9.2 Tropical watershed management consideration (after Dasman, Milton and Freeman 1973)

Environmental factors	General characteristics	Management consideration
1. Sub-humid watersheds		
Rainfall	Equal to or slightly less than potential evaporation (PE) on annual average, but higher than PE during rainy season, including intense downpours	Runoff control and regulation measures to maximise both clean runoff to reservoir and availability of soil water for crops and pasture
Soils	Variable in structure, depth and fertility, but sometimes with good potential for agriculture. Erosion a potentially serious problem. Rapid runoff may also cause flooding	Great care required on cultivated land to prevent erosion and rapid runoff, especially on sloping terrain with clay soils
Natural vegetation	Semideciduous forests, savanna 'parklands' (fire climax)	Can include valuable tropical woods. Natural forest should be left on steep catchment areas
Commonly observed land use	Short cycle and perennial crops, grazing, forestry. Conflicts can be expected between agricultural and forest uses of lands. High population densities and intensive land use are likely	Productive agricultural and animal-raising uses possible under resource-conservative management. Vigorous soil and water-conservation programmes needed in high-intensity use, dense population zones. In steep areas natural forests should be conserved, but plantation crops may provide adequate cover
II. Humid watersheds		
Rainfall	Exceeds PE in most months. May characterise middle and upper watersheds, subject to orographic influences. Short dry season if any	Maintenance of natural vegetative cover of paramount importance, in order to absorb rainfall, release clean runoff and stabilise flow

Table 9.2 continued

Environmental factors	General characteristics	Management consideration
II. Humid watersheds		
Soils	Mature soils may be deeply weathered clays with good structure and internal drainage, but low natural fertility. Podsolic soils in cooler zones	Leaching rapidly depletes exposed soils. Ill suited for cultivation except on younger fertile soils. Erosion danger variable but slumping and landslides a danger, especially after deforestation
Natural vegetation	Evergreen forest, including 'rainforest' and 'cloud forest'. Relatively limited commercial value for wood products	Best left in natural state, especially in steep catchment areas, unless economic perennial crop providing equivalent protection is possible
Commonly observed land use	Little permanent farming except on exceptionally fertile soils. Shifting agriculture likely	Soil and forest exploitation technologies not well developed for this environment, except on best soils. Road construction and maintenance costly; fungal diseases serious problem in agriculture

9.5 THE EFFECTS OF GRAZING

Grasslands vary enormously in type, environmental condition and hydrological characteristics. Therefore, it is difficult to generalise about the effects of grazing and here only some basic hydrological aspects are briefly discussed. Although not specifically concerned with tropical conditions, Branson et al. (1981) present a detailed analysis of rangeland hydrology, including runoff processes, management, and the effects of grazing and improvement practices on infiltration, runoff and other components of the cycle. As with forest, impacts of change are influenced by a range of environmental factors. Pastoral areas cover a broad spectrum from the margins of arid areas to those of forest, ranging in type from semi-arid steppe, through natural grasslands to savanna. Despite the great ecological diversity, some of the greatest differences in hydrological characteristics and rangeland productivity are found between well- and poorly-managed areas in the same environmental region. Hydrological characteristics vary with the amount of grazing, trampling and burning. Human activity has had great influence in pastoral areas, both in their creation and in their degradation. Some of the most intensively developed grasslands are in areas cleared of forest or woodland where trees would again develop if management stopped.

Large areas in semi-arid regions originate from natural dry woodland, scrub or savanna where trees and shrubs are held in check. Because of this modification – both by people and their animals and also by wild animals – it may be difficult to judge potential by looking at present conditions. In many areas, productive grasslands have been badly damaged, turned into wasteland and desert because of human activity. The hydrological consequences of this degradation and its reversal are the subject of this section. For a broader discussion of the whole question of tropical pastures and animal production, including a description of native and improved pastures, the main animal species and constraints on the effective utilisation of livestock resources, papers by Mahadevan (1982) and Mannetje (1982) are useful.

Grasses vary considerably in their water use, rooting depth being of special importance (§ 6.3). In semi-arid areas, shallow-rooted species usually use less water but deep-rooted species may be better adapted to survive. Grasslands differ widely in their infiltration rates, erosion and runoff production. Some grasses may use more water but produce great quantities of forage, while others may use less water but be less efficient at producing material. While in many cases grass uses less water than other vegetation, deep-rooted species may transpire more than some trees. The most suitable grass will vary greatly with environment and purpose – whether it is necessary to produce maximum water yield, reduce stormflow, soil erosion or principally to create grazing lands. Often multiple use is planned and needs may be conflicting. For example, a shallow-rooted grass may use comparatively little water, but not encourage infiltration. It may, therefore, produce a high water yield but provide little protection against erosion and stormflow, as well as not being able to withstand drought.

Conversion of an area to grassland may be advantageous or the opposite in hydrological terms depending upon the objectives and the particular environment, including previous land use, soil and rainfall characteristics and management practices. As already suggested, in many cases grass uses less water than other vegetation and hence many experiments indicate that a change from forest to grass increases water yield with a rise in the water table. Hamilton (1985) points out that these experiments often indicate little or no increase in stormflow volume and peakflow, depending on grass density and productivity and that once grass is established, there is no difference in erosion rates and sediment production. However, as already indicated (§ 9.4), the experiments have usually been on only moderate slopes and also have not involved grazing. Although controlled grazing and prescribed burning need not have serious effects, Hamilton (1985) points out that such control is difficult and overgrazing and indiscriminate burning have adverse hydrological and soil erosion effects (see below). Manubag (1985), for example, indicates much greater surface runoff and sediment production from grazed than ungrazed areas in the Philippines.

Problems associated with hydrology in grazing areas cannot be separated from the whole ecosystem and the economic, social and political circumstances. In particular, the way in which changes in the human factors influence the impact of, say, rainfall characteristics was referred to in Chapter 3. Swift (1973) discusses the way in which a nomadic Tuareg tribe protected itself against drought in northern Africa and how this protection has been eroded by changes outside their control such as political restrictions on their movements. Randall Baker (1973) indicates common misconceptions about problems facing pastoral areas. Traditional systems were ecologically well adapted to the environment, but change such as increase in human and animal numbers due to disease control and restriction on tribal warfare, can lead to overgrazing. Provision of extra watering points can lead to overgrazing in their vicinity. Such provision is wrong on its own and must be incorporated within an overall, ecologically sound development plan. Attempts to improve the situation have often failed because of lack of understanding of the people and their traditions, inadequate educational programmes and attempts to gain their support. The complex socio-economic importance of livestock in traditional societies is pointed out by Stocking (1985). For this reason, simple approaches to overgrazing by reducing livestock numbers will fail. Overstocking is merely a symptom of broader problems. Solutions must use a sensitive, incremental approach to development, together with simple range management techniques and development of marketing facilities to increase quality and reduce grazing pressure (Stocking 1985). Mahadevan (1982) stresses that solutions must be based on a combination of educational, technical, economic and social considerations. Such considerations are outside the scope of this book, but – together with a need to have an overall understanding of rangeland ecology and management principles – are fundamental to any scheme incorporating water-resource development.

Overgrazing leads to the destruction of vegetation, compaction of the soil surface due to trampling and lack of protection against raindrops, and destruction of the soil structure. An example of vegetation destruction can be found in southern Tunisia near Nefta where coverage of vegetation inside an area fenced off is 85 per cent in contrast to a 5 per cent coverage outside it. Here, the original dry steppe has changed to semi-desert without any appreciable variation in rainfall (SMIC 1971). Oliver (1968) indicates how overgrazing and soil erosion around water points result from the excavation of *hafirs* (storage tanks or reservoirs) in the central Sudan. Again, however, the picture is complex. This is indicated by a study of grazing and trampling effects on vegetation cover, soil deterioration, sealing, compaction, wind and water erosion around water holes in the Sahel (Valentin 1985). Marked differences in response occurred depending upon environmental factors, particularly contrasts in underlying soil types. In addition, some processes such as crusting resulted more from the combination of several natural factors than from grazing impacts. Surprisingly, it was outside the zone affected by grazing that sealing most often occurred (Valentin 1985). Suppression of grasses by overgrazing in some areas leads to a rapid encroachment of woody shrubs. The question of vegetation destruction due to over-grazing must include consideration of wildlife, particularly in national parks where there is a danger of overstocking.

Trampling by excess animal numbers and raindrop impact on the unprotected soil surface leads to compaction (Chs 7, 8) with a resultant decrease in infiltration capacity. The latter will prevent the build-up of soil moisture reserves for use in dry periods, the recharge of ground water supplies and at the same time lead to an increase in overland flow. The consequence will be a very irregular streamflow, alternating between spate flow with resultant flooding danger and little or no dry season flow since underground water reserves which maintain it are not recharged. The latter will also lead to boreholes and wells drying up.

Overland flow will increase erosion, particularly since overgrazing means that the soil is no longer protected by a vegetation cover. The bare soil will also be subject to wind erosion during dry periods. Erosion means an increase in the sediment load of streams, with resulting down-stream problems of silting up of reservoirs and other water-supply works, rivers and harbours. The overgrazed Bermejo catchment of South America, forming only 4 per cent of the Parana River (which becomes the Paraguay River), contributes 82 out of 102 million tonnes of sedi-ment transported by the latter per year (Pereira 1972). Deposition of this sediment creates great problems for the port of Buenos Aires and for river transport in the Argentine. Trampling and erosion will lead to a deterioration in soil structure, in turn making it more susceptible to erosion, decreasing its water-holding capacity and this, together with a decrease in infiltration, further reduces its ability to maintain vegetation. A consequent further decline in vegetation affords less protection for the soil against raindrop impact and, as indicated in § 8.2, a vicious cycle of environmental deterioration is set up. These and other factors which may tend to increase evaporative demand leading to deterio-ration and desertification are discussed by Hashemi (1979). All of these problems will tend to be most severe around water points.

Fire influences infiltration by changing variables which affect the latter such as litter, soil characteristics and the solute concentration of water. Branson et al. (1981) cite a range of experiments showing either an increase, decrease, or no change in infiltration after burning of grass-land. Effects appear to depend upon a range of factors including soil characteristics, normal infiltration rates, litter depth and number of burns. A number of factors also seem to influence effects of cryptogams (e.g. algae and lichens) on runoff in areas denuded of macro-vegetation and the effects of water repellent substances in soils (Branson *et al.* 1981).

Improvement of grazing areas involves consideration of several elements. In particular, water development must take place within an integrated programme involving the whole ecosystem, including human factors. Without land reform, grazing control and co-operation from the people, water development leads to overgrazing and environmental deterioration. Control of livestock numbers is essential but, in addition, control of their distribution is important. Damage may occur not because of gross overstocking but because of uncontrolled distribution, allowing animals to concentrate in the wrong place at the wrong time. Rotational use of watering points, closing them down at the first signs of over-use, is important. Problems associated with rotation are discussed by Fosbrooke (1973). Usually, improvement initially involves removal of all livestock for a period, or at least holding numbers below carrying capacity. This allows control of erosion and restoration of vegetation. Without such control, measures such as increasing the water supply are wasted and in fact harmful.

Since the ability of plants and soil to withstand drought depends partly on their condition at the end of the wet season (Dasman, Milton and Freeman 1973), one way to improve their condition, both in respect of plant stamina and soil structure, is always to leave a reserve of standing forage and litter at the end of the growing season. Fire can be used to control woody vegetation and undesirable plants, but its application needs skill and it is not always effective. Goats can be of help in controlling such vegetation (Fosbrooke 1973). Randall Baker (1973) suggests a possible framework for the development of pastoral areas. Adjustment on the part of the people must occur but always within the traditional framework. Valentin (1985) stresses the need to assess carrying capacity and to gain co-operation of pastoral communities in any project. He also points out the value of establishing tree shelterbelts within 1–2 km around waterholes.

All the above points merely introduce this extremely complex situation, a detailed consideration of which is outside the scope of this book. Pereira (1973) points out that improvement of grasslands involves consideration of two opposing hydrological features. Control of grazing to prevent exposure of the soil and trampling is essential to curb soil erosion and reduce stormflow. However, improving the density and productivity of grassland, because of high water use, decreases total water yield. As the intensity of grazing increases, water yields increase but are accompanied by more uneven streamflow and the risk of soil erosion.

Some of the basic hydrological aspects associated with overgrazing and the improvement of pastoral areas may be illustrated by an experiment from Uganda (Pereira *et al.* 1962; Pereira 1973). In the Karamoja District of northern Uganda, the annual rainfall of 500–750 mm, while much less than potential evaporation, is adequate to maintain an open woodland cover with a rich flora of grasses. However, persistent misuse by overgrazing reduced large areas to thorn-scrub with desert grasses. Runoff from bare, trampled soil was large, bridges being damaged by peak flows. Construction of extra water points merely led to a concentration of animals in new areas being opened up, resulting in more overgrazing and extension of the areas of destruction. This practice was stopped pending the development of grazing control.

In 1956 at Atumatak, an experiment to study the effects of restoration was set up. In the area, for 9 months of the year, the average bare soil exposure was 40 per cent. There was no permanent streamflow and, typically, a rainstorm of 20 mm produced violent stormflow lasting for only 30 min. For a period of 4 years, two valleys were maintained under heavy grazing and it was found that stormflow amounted to 40 per cent of all rainfall. Penetration of rain into the trampled soil was slight, in many cases to less than 0.5 m.

Calculations of stored soil moisture indicated that at least one-third of the rain became runoff, agreeing well with streamflow measurements. While excessive grazing continued on one catchment, the adjacent one was treated by bush clearing with controlled grazing protected by fencing. For the first season, cattle were excluded and although much of the topsoil had been lost, no reseeding was necessary. Penetration of rain into the soil increased from 0.5 to 1.25 m, peak flows were reduced and a rich flora of grasses developed. The carrying capacity was estimated to be one beast per 2.43 ha, nearly double that before treatment. Thus, the application of simple pasture-management techniques was found simultaneously to improve water infiltration and grazing conditions.

9.6 MODIFICATIONS TO SURFACE WATER AND GROUND WATER

People have had considerable influence on surface water and ground water and here only a selection of aspects can be presented. The aim is to indicate the complexity of human influence.

The construction of a dam to provide irrigation water, power, flood control, fisheries or transport has many consequences, including those of a social and economic nature. Possible impacts upon energy and water balances with resultant climatic effects have been discussed in § 9.2.

It is impossible to discuss all the many other effects of dam construction here. Ackerman, White and Worthington (1973) and Goodland, Watson and Ledee (1984) indicate their range, including loss of land through inundation and implications for water table levels. Silting behind dams reduces downstream sediment loads. Grove (1985) points out that sediment yields are often greatest in semi-arid areas and, furthermore, dams in such areas require large catchments. Hence such reservoirs decline in capacity rapidly, at the same time as their surface area for evaporation changes only slightly. Arakeri and Donahue (1984) present data on rates of silting in Indian reservoirs. Loss of sediment in reservoirs has downstream implications. Vegetation, agriculture and fisheries may not only have been adapted to periodic flooding before dam construction but also to the deposition of sediment, including important nutrients. In central Africa, trapping of silt by Lake Kariba resulted in impoverishment of the alluvial flood plain of the Mana Pools area some 80–100 km below the dam wall (Van der Lingen 1973). Remedies do exist for such effects, for example fertiliser use, but this involves economic and social considerations.

Since water leaving a reservoir has lost much of its sediment load, rapid downstream erosion may result. A further problem is that a dam may have been constructed to cope with a natural hydrological regime which is later altered by human activity. For example, Shalash (1980a) points out that since construction of the Owen Falls Dam at the outlet of Lake Victoria, stages and outflows have increased for this reason. A new reservoir may become depleted of oxygen because of decomposition of flooded biomass and this can also cause downstream problems (Goodland et al. 1984).

Oyebande et al. (1980) discuss how the Kainje Dam on the River Niger has greatly reduced peak flows and tripled low flows but also caused flood plain contraction and altered channel geometry. Effects of the Aswan High Dam (Lake Nasser) have received much attention. Kashef (1981b) provides a background and history of the control of the River Nile together with a discussion of potential projects. Kashef (1981a) discusses hydrological and structural aspects of the High Dam, including data on capacity, flows, hydro-electric power and irrigation before discussing side effects and potential remedies. Problems include silting of Lake Nasser, downstream loss of fertile silt and erosion, and socio-economic impacts including migration of entire villages which were inundated, fishing industry losses and deterioration of summer resorts along the Mediterranean. Proposals to remedy these problems are put forward.

Also for the Aswan High Dam, problems in assessing downstream degradation are discussed by Shalash (1980b) and effects on the hydro-chemical regime in Shalash (1980c). It is all too easy to ignore the benefits of such schemes when confronted by the problems. Despite the latter, the Aswan High Dam has been described as 'one of the success stories of the 1960s' (*New Internationalist* 1975). Other problems discussed later in this section include those associated with human health and water weeds.

Impacts and problems associated with irrigation are discussed by Goodland *et al.* (1984), arid land irrigation aspects being focused on by Worthington (1977). Only some can be considered here. In any irrigation project, evaporation losses from reservoirs, canals and furrows must be considered, particularly in arid areas. Other losses, such as seepage from unlined canals, are important.

Other problems associated with irrigation include waterlogging, salinisation and contamination. Over-irrigation is a common failing and this, together with seepage from unlined canals causes the ground water table to rise if drainage is inadequate, leading to waterlogging and salinisation. The undesirable effects of waterlogging were discussed in Chapter 8. Yaron (1981) states that the essence of the salinity problem is that nearly all irrigation water contains some salts and these tend to concentrate in the soil unless leached out by rain and/or irrigation water in excess of evaporation. Saline irrigation water, low soil permeability, inadequate drainage, low rainfall and poor irrigation management all contribute to salt accumulation. If the water table is within about 1.5 m of the surface the capillary fringe reaches the zone where water evaporates, leaving behind dissolved salts which concentrate near the surface. Apart from the toxic effects of the salts themselves on plants, saline water has less dissolved oxygen which can have adverse effects on aeration in the root zone and on desirable organisms in the soil and canals receiving runoff. Salts in soil water increase osmotic pressure and therefore reduce the ability of plant roots to absorb water (Arakeri and Donahue 1984).

Grove (1985) discusses the problem of waterlogging and salinity in arid areas, pointing out that as much land is going out of cultivation because of the problem as is being introduced by irrigation. Tanwar (1979) discusses effects of irrigation on ground water in the semi-arid area of Haryana, India. More than 60 per cent of the area within the state has been classified as marginal to saline, the depth of the water table varying between 3 and 50 m below the surface but being close to the surface along the major unlined canals and in the flood plains. Yaron (1981) indicates that in India, 6.1 million hectares have been lost to agriculture because of salinity and that about one-third of the world's irrigated land in humid, arid and semi-arid areas is affected.

Aspects of salinity control are discussed by Yaron (1981), Rawlins

(1981), Goodland *et al.* (1984), Arakeri and Donahue (1984), van Alpen (1984). Essential elements are sound irrigation water control and prevention of seepage from unlined canals, provision of an increment of irrigation water in addition to evapotranspiration requirement to leach salts from the root zone and means of removing excess water by provision of adequate drainage. The latter is a problem in flat areas such as the Indus Valley, Pakistan. The solution to a severe waterlogging and salinity problem in this case is to sink large numbers of tube wells. Water pumped from them causes the water table to fall. Since the ground water has a low salt content it can then be used for irrigation, the irrigation water at the same time leaching the salts from the surface layers. Since much of the irrigation water will evaporate, there will be a net loss to the air and the water table should fall as long as pumping continues (Simons 1967).

In areas affected by salinisation the fate of irrigation water used to flush out salts is important. If it percolates into the ground water, the salinity of the latter increases. If the water is drained off from the area then its salinity can create problems downstream. Water which is repeatedly diverted for irrigation and then returned to streams becomes increasingly saline due to leaching of minerals from the soil and the fact that water volume is decreased.

A relatively recent complication is the contamination of irrigation water by fertilisers and pesticides. This is a special problem where canal and field water are used, for example, not only for irrigation but also for domestic use and raising fish. Contaminated water can destroy fisheries. The insecticide lindane, used to control the rice stem borer, has proved lethal to *Tilapia*, a species of fish commonly raised in south-east Asian rice fields (Dasman, Milton and Freeman 1973). Nutrients from heavily fertilised soils entering streams can cause eutrophication (Goodland *et al.* 1984). Nitrogen fertilisers used on Queensland sugar-cane areas could possibly affect offshore marine life (Oliver pers. comm.). Contaminated water may also seep downwards to pollute ground water.

In the dry season without irrigation, problems from insect pests, plant pathogens and weed growth may be suspended (Goodland *et al.* 1984). However, provision of irrigation water in the dry season means that this does not happen. Goodland *et al.* (1984) discuss ways of overcoming such problems, for example by turning off irrigation water at key periods.

Aquifers can be used as underground reservoirs. The advantages of this system are that water is stored without evaporative loss, without vast capital costs, maintenance and gradual sedimentation as in the case of surface reservoirs. A major problem is the salinity of many rock strata. Excessive withdrawal of ground water leads to a lowering of the water table. Thomas and de Bouvrie (1973) point out that in the Sahel, a combination of inadequate recharge during dry years and continuous over-use has in a rapidly increasing number of cases caused wells to run dry. Excessive withdrawal of fresh water may allow the inflow of saline sea water. General discussions of ground water include Balek (1983), Goodland *et al.* (1984) and material on the African situation in Walling, Foster and Wurzel (1984).

Evaporation losses from marshes of freely transpiring vegetation are large. For example, it is estimated that half of the volume of water of the White Nile is lost in an immense papyrus marsh called the Sudd. This led to proposals to build a bypass canal (the Jonglei Canal) to reduce this loss.

Drainage of swamps usually reduces the water lost by transpiration. However, since swamps tend to regulate streamflow, this can have adverse effects, particularly when the swamp is in the upper part of a catchment. Streamflow can become less regular and stormflow increase. Removal of vegetation to replace a swamp by an open-water surface was once believed to reduce evaporation. Advection of energy through tall reeds together with turbulence (§ 5.2) could produce higher rates of water loss than from open water. As was pointed out in § 5.2 however, although this may apply to a limited body of water, Cooley and Idso (1980) concluded that the introduction of any kind of vegetation cover over an extensive water body would reduce evaporative loss.

A non-climatic effect of swamps is that they can act as a filter; lower water velocities than in rivers lead to deposition of suspended sediment. Drainage of swamps can therefore have serious effects downstream. However, where clean water enters a marsh, decaying organic matter can affect both colour and taste. Drainage of swamps can eliminate breeding-grounds of vectors of disease transmission such as the anopheline mosquito.

Creation of reservoirs and irrigation systems has beneficial health effects due to provision of drinking water, improvements in personal hygiene and nutritional benefits associated with irrigation agriculture. However, they also create health problems. Water can carry disease and acts as a habitat for disease vectors such as mosquitoes, snails and tsetse flies. Chemically polluted water (see above) can also affect health. Malaria is a debilitating disease which can, certainly indirectly, cause death. Today it is largely confined to the tropics but remains the most widespread disease in the world (Worthington 1977). It is transmitted by mosquito vectors, most of which require water in which to breed. In 'humid' areas, suitable breeding conditions can exist all the year round, but elsewhere malaria infection fluctuations are likely to correspond to rainfall seasonality. In low-rainfall areas, with a long dry season, breeding-grounds are very limited. Hence, the situation can be radically altered by irrigation or other water-development projects. Unless adequate preventive measures are taken, such projects can lead to the

300   HUMAN IMPACT ON THE HYDROLOGICAL CYCLE

development or increase of malaria incidence. Migrant workers can introduce malaria into new areas where mosquitoes exist. A discussion of these aspects is presented by Prothero (1965). Goodland *et al.* (1984) point out that heavy biocide spraying associated with cotton can produce biocide resistance in mosquitoes. This makes malarial control difficult or even impossible and recent outbreaks in formerly free areas have been attributed to this. Encephalitis and yellow fever are also mosquito borne.

Bilharzia probably ranks second only to malaria as a cause of human suffering and economic loss. Extensive treatment is not practicable due to the side effects and toxicity of available drugs and cure is not always complete. The disease is caused by a minute parasite worm, which spends part of its life cycle in a species of tropical water snail occurring in fresh water. People contract the disease in bathing, fishing, washing clothes and drawing water from infected areas. As perennial irrigation spread in the Nile Valley, so did the disease, and in some rice-growing areas where the peasants spend much of their time in waterlogged channels the incidence is as high as 90 per cent (Keating 1972). Obeng (1973) states that the incidence of bilharzia (schistosomiasis) in children under 10 years old has risen to 90 per cent on the western shore of the Volta Lake in West Africa since the latter was formed. On the eastern shore, due to different environmental conditions, the incidence is less. The complexity of the situation is shown by the fact that the creation of Lake Brokopondo in Surinam has not resulted in explosions of diseases such as malaria, filariasis and bilharzia. It is thought that this is due to environmental conditions such as the acidity of the water as well as low population density (Leentvaar 1973).

Onchocerciasis (river blindness) is spread by a black fly depending on sunlit areas of fast flowing, oxygenated water. It has affected entire villages in Upper Volta with rice irrigation schemes (Goodland *et al.* 1984). Other diseases linked with water development and irrigation schemes include cholera and typhoid, as well as gastro-intestinal infections. Discussions of diseases, their control and prevention are found in Worthington (1977) and Goodland *et al.* (1984).

Rapidly spreading weeds in reservoirs and irrigation channels present a number of problems. On the Volta Lake, concern is with their association with public health hazards and parasitic infection problems, some types harbouring snails, mosquitoes and other intermediate parasite hosts (Obeng 1973). They also present a handicap to fishing in the lake. Dasman, Milton and Freeman (1973) list seven losses due to aquatic weeds, including:

1. Fishery losses due to competition for light or energy and for nutrients.
2. Fishery losses because of physical interference of weed cover with fishing processes.

3. Health losses, such as those indicated above for the Volta.
4. Disruption of navigation.
5. Loss or damage from weed invasion of irrigation systems, including blockage, water loss, competition for nutrients, decrease of fish and increase of disease organisms.

Examples cited of weed infestation include the Jebel Auliya (Sudan), Brokopondo (Surinam), Nam Pong (Thailand), Kainji Lake (Nigeria). Water fern threatened to eliminate fishing and navigation of Lake Kariba when it was filling. However, the weed is damaged by wave action and has been unable to invade the major open-water surface although it remains in creeks and harbours. Before the construction of the Bhakra Lake in India, canal water carried silt and was turbid, preventing weed growth. The clear water after construction has resulted in weed growth in unlined channels causing a drastic reduction in their carrying capacity (Rao and Palta 1973).

A wide variety of control methods is discussed by Goodland *et al.* (1984). Chemical spraying produces environmental problems and is costly. For example, chemical control of the water hyacinth in Lake Brokopondo cost US$2.5 million over the period 1964–70 as well as severe environmental problems (Goodland *et al.* 1984). Manual and mechanical control is not feasible for large areas. Biological control is attractive, for example the introduction of a small weevil from Brazil has been successful in control of *Salvinia* in tropical Australia. Use of other methods such as herbivorous fish, ducks, etc., is discussed by Goodland *et al.* (1984). Clearly, introduction of a new species needs careful research because of possible adverse effects on the environment.

Weed infestation is not without potential benefits, for example as a fertiliser, mulch and animal fodder. Morton and Obot (1984) point out that although the spread of a grass (*Echinochloa stagnina* (Retz) P. Beauv) causes problems by displacement of water and increased silting in Lake Kainji, Nigeria, it also has considerable benefits. It provides a valuable spawning and breeding ground for many economically important fish species and cattle farmers regard it as a valuable dry season fodder. They modelled occurrence of the grass, which demonstrated that 28–46 per cent of the lake surface could have been covered between 1972 and 1983. About 75 per cent could be harvested each year on a sustainable basis. Cattle are fed on water hyacinth of the Jebel Auliya Reservoir (Oliver pers. comm.). Water weeds of various types are a source of fibre for pulp and paper production and some are useful food sources (e.g. water spinach, water cress, chinese water chestnut). Water hyacinths can be converted to methane-rich gas. Many weeds absorb inorganic and some organic compounds from effluent. They can, therefore, assist in reducing environmental damage of effluent as well as being harvested. A fuller discussion of the above is presented by Goodland *et al.* (1984).

CHAPTER 10

PROBLEMS AND PRIORITIES

10.1 INTRODUCTION

A wide range of tropical rainfall characteristics exert great influence on water resources and agriculture. Resulting problems are compounded by high evaporative demand. Human responses to these key environmental variables in terms of land use and management practices are varied. Rainfall characteristics and evaporative demand in the tropics mean that the hydrological response to changes in land use and management may be especially marked. Implications for soil erosion are also serious. Population increase and the need to increase agricultural production in many tropical developing countries means that changes in land use and management practices are underway or being contemplated. Hence there is a great need to assess possible impacts on the hydrological cycle and ways of overcoming or minimising resultant problems.

Research needed to increase understanding and find solutions to problems presents a difficult challenge given the limited resources of tropical developing countries. Furthermore, solutions adopted in developed, temperate countries are not necessarily appropriate given the very different physical, economic, social and political circumstances (§ 1.4).

10.2 SCIENTIFIC AND TECHNICAL ASPECTS

Understandably, there is concern about lack of data on rainfall, evaporation and other environmental variables in the tropics. However, of equal or even greater significance is the quality of existing data. Experience from various countries suggests that quality is a matter for serious concern. Therefore, checking existing instrumentation, site and observing procedures and ensuring routine maintenance should be a first priority before considering network expansion and installation of more sophisticated instrumentation. Careful training of staff concerned with routine observation and maintenance of equipment is needed to make them aware of problems. Regular inspections of networks are essential. Sound data checking procedures prior to analysis are necessary. In the case of rainfall, marked spatial variations (§ 4.2) make standard methods of checking utilising nearby stations particularly difficult.

Sophisticated experimentation and equipment, particularly when the latter are not specially designed for tropical conditions, present problems. Inadequate repair and maintenance facilities for such equipment are a handicap, especially since physical conditions such as high temperatures and humidities create difficulties.

Although sophisticated experimentation at, say, a research station can provide valuable information, wider application of the results may not be possible because necessary data do not exist elsewhere. Therefore, an important aspect of such work should be to compare these results with those from relatively simple approaches having wider application. A positive factor may be the availability of often plentiful but perhaps poorly trained or untrained labour. Detailed experimentation may therefore be possible without the need for elaborate recording systems, given adequate training. Such work may be less costly and more reliable than with a more sophisticated experiment where breakdown of recording equipment may occur. The latter factor is particularly important where experiments are in remote locations removed from repair and maintenance facilities.

The need for innovative approaches and instrumentation presents a challenge. A requirement is to develop cheap, simple, reliable equipment able to be maintained and operated by relatively unskilled labour. Unfortunately, there is a tendency to use sophisticated approaches and instrumentation which are successful in developed temperate areas. Such approaches are expected by the scientific community and hence important for individual recognition and advancement. Simpler approaches which may seem more like data collection exercises, or not at a research 'frontier', are not adopted despite perhaps being very valuable. This applies to both local and expatriate staff and the point is addressed later when discussing education and training.

In §§ 2.2 and 4.4, the dangers of applying extra-tropical models and concepts was referred to. Even within the tropics, extrapolation from one area to another may be dangerous. There is a need to examine similarities and differences in rainfall characteristics and other environmental variables between areas. This is important for technology transfer such as agronomic practices and water and soil conservation measures. Furthermore, as was stressed in § 9.5, an integrated approach is needed, involving consideration of the whole ecosystem rather than water alone.

In decision making, social, economic and political factors may well dominate. Therefore, it is important that data and analysis provide information necessary for the exploration of a range of possible decisions rather than only the one which scientific and technological evidence suggests is most appropriate. Information must enable decisions not necessarily scientifically optimal to be as successful as possible. Scientists may not be aware of all the factors involved in decision making. Furthermore, imposition of an 'expatriate' concept of what is appropriate, for example in terms of 'efficiency', may be dangerous, particularly since the latter can be defined in various ways. Existing systems may be dismissed as 'inefficient' without understanding the underlying reasons for their existence. As was suggested in §§ 3.1 and 9.5, the traditional system may be ecologically sound. If it was not, then it would not have lasted for hundreds, perhaps thousands, of years. This is not arguing against change or to suggest that improvements are not possible. It is a plea not to denigrate an existing system out of hand. Furthermore, changes, often introduced from outside, may lead to a breakdown in the system (§§ 3.1, 9.5) and in such cases adjustments must occur. Understanding the traditional system is important, especially if change is contemplated. Traditional knowledge has considerable value (§ 8.3) and there are real dangers that it will be lost in the face of technological development. Issues of this type are addressed by Chambers (1983).

10.3 THE HUMAN ELEMENT

The above comments lead into the question of education and training. Limited facilities in developing tropical countries often lead to reliance on overseas institutions. This poses problems. The different physical and human environments mean that with the best will in the world, the emphasis in individual courses overseas, particularly in relation to field work and practical experience, will to some extent be inappropriate. As already suggested, models and concepts in temperate areas may not be applicable. An emphasis on sophisticated high technology training means that on return to the home country where equipment, adequate maintenance and data are not available, disillusionment is inevitable. Conditioning towards sophisticated approaches may prevent the search for simpler more appropriate ideas.

In developed countries, increased specialisation is common and hence overseas training may emphasise depth rather than breadth. In developing countries the need may be for a broader training since often individuals are faced with making decisions without ready access to specialist views from a variety of areas. At the very least, a broader training improves their ability to communicate with others. Where greater expertise in a particular field is needed, this could come from outside. Increasingly, developed countries see professional and vocational training in post-graduate terms following a first degree. This represents a long period of time, a matter of perhaps little importance in the developed world. However, for developing countries, the costs in financial terms and in losing their best educated people for many years are great. Arguments can be advanced that such a system is necessary. While true in some cases, in others this view merely reflects inertia.

In addition to the scientific aspects, overseas training is often regarded as an enriching experience. However, it can have drawbacks. Exposure to attitudes and living standards in a developed country can mean that people become out of touch with the realities and aims of their own country. An elitist attitude may result, both in terms of personal goals and a belief that colleagues, even those more senior, without overseas training, are inferior. They may not be easily convinced that concepts and models learnt overseas may be inappropriate.

The above reasons mean that in many cases, training in the home country where possible is likely on balance to be more appropriate. If this involves staff from developed countries in running courses they must be flexible in their attitudes, recognise that they are in a different environment and be prepared to tackle problems in an original manner. Unfortunately, such attitudes can take time to develop. However, they do have a chance to develop, which is not the case where the person concerned teaches a course in his or her own country.

Major river catchments may cross international boundaries. The need to consider catchments as entities in water development and to overcome problems such as erosion (§§ 8.2, 9.3–9.6) means that in such cases political considerations loom large.

Reference has been made in § 10.2 to the need to understand traditional systems and their value. Traditional structures of society also have an important role in the implementation of change. Schemes superimposed from above by central authorities, without regard for the traditional system and not utilising it, face problems. Not only must change utilise the traditional social organisation as far as possible but structures must allow effective exchange of information between national government and grass roots levels *in both directions*. In many respects, therefore, political leadership may be of greater significance than the purely technical aspects of problem solving. This is especially true since all changes may adversely affect some groups.

The significance of human factors such as those raised in this chapter is discussed by Blaikie (1985) and Blaikie and Brookfield (1987) in relation to soil erosion and land degradation. Although the focus of the book has been on the physical aspects, as was suggested in § 1.4, it is important not to divorce them from the political, economic and social realities.

REFERENCES

Abrol, I. P. and **Dixit, S. P.** (1971) 'Studies of the drip method of irrigation', *Experimental Agriculture*, **8**, 171–5.

Ackerman, W. C., White, G. F. and **Worthington, E. B.** (1973) '*Man Made Lakes: Their Problems and Environmental Effects*', American Geophysical Union, Geophysical Monograph 17.

Ackerson, R. C. (1980) 'Stomatal response of cotton to water stress and abscisic acid as affected by water stress history', *Plant Physiology*, **65**, 455–9.

Agee, E. M. (1972) 'Note on ITCZ wave disturbances and formation of tropical storm Anna', *Monthly Weather Review*, **100**(10), 733–7.

Aggarwal, P. K. and **Sinha, S. K.** (1983) 'Water stress and water use efficiency in field grown wheat: a comparison of its efficiency with that of C_4 plants', *Agricultural Meteorology*, **29**, 159–67.

Ali, F. M. (1953) 'Prediction of wet periods in Egypt four to six days in advance', *Journal of Meteorology*, **10**, 478.

Ali, F. M. (1969) 'Effects of rainfall on yield of cocoa in Ghana', *Experimental Agriculture*, **5**, 209–13.

Allan, R. J. (1983) 'Monsoon and teleconnection variability over Australasia during the southern hemisphere summers of 1973–77', *Monthly Weather Review*, **111**, 113–42.

Allison, J. C. S. and **Eddowes, M.** (1968) 'Climate and optimum plant density for maize', *Nature*, **220**(1), 343–4.

Alvarez, C. and **Sanchez, J.** (1980) 'Effects of urbanisation on the hydrology of a suburban basin in Porto Alegre, Brazil', in *The Influence of Man on the Hydrological Regime with Special Reference to Representative and Experimental Basins*, IAHS Publication No. 130, 23–8.

Alvim, P. de T. (1960) 'Stomatal opening as a practical indicator of moisture deficiency in cacao', *Phyton*, **15**, 79–89.

American Meteorological Society (1984) *15th Conference on Hurricanes and Tropical Meteorology*, American Meteorological Society, Boston.

Amir, J. and **Bielorai, H.** (1969) 'The influence of various soil moisture regimes on the yield and quality of cotton in an arid zone', *Journal of Agricultural Science*, **73**(3), 425–30.

Ananthakrishnan, R. and **Parthasarathy, B.** (1984) 'Indian rainfall in relation to the sunspot cycle: 1871–1978', *Journal of Climatology*, **4**, 149–69.

Ananthakrishnan, R., Pathan, J. M. and **Aralikatti, S. S.** (1981) 'On the northward advance of the ITCZ and the onset of the southwest monsoon rains over the southeast Bay of Bengal', *Journal of Climatology*, **1**, 153–65.

Ananthakrishnan, R. and **Rajagopalachari, P. J.** (1964) 'Pattern of monsoon rainfall distribution over India and neighbourhood', *Proceedings of the Symposium on Tropical Meteorology*, Rotorua, New Zealand, 192.

Anderson, J. R. (1970) 'Rainfall correlations in the pastoral zone of eastern Australia', *Australian Meteorological Magazine*, 18(2), 94–101.

Angus, J. F. *et al.* (1983) 'The water balance of post-monsoonal dryland crops', *Journal of Agricultural Science*, 101, 699–710.

Anthes, R. A. (1982) *Tropical Cyclones, Their Evolution, Structure and Effects.* Meteorological Monographs, Vol. 19, No. 41, American Meteorological Society.

Anthes, R. A. (1984) 'Enhancement of convective precipitation by mesoscale variations in vegetative covering in semiarid regions', *Journal of Climate and Applied Meteorology*, 23, 541–54.

Arakeri, H. R. and Donahue, R. (1984) *Principles of Soil Conservation and Water Management*, Rowman and Allanheld Publishers, New Jersey, pp. 253.

Arenas, A. D. (1983) 'Tropical storms in central America and the Caribbean: characteristic rainfall and forecasting of flash floods', in *Hydrology of Humid Tropical Regions*, ed. R. Keller, IAHS Publication No. 140, 39–48.

Arkin, P. A. and Webster, P. J. (1985) 'Annual and interannual variability of tropical–extratropical interaction: An empirical study', *Monthly Weather Review*, 113, 1510–23.

Arnoldus, H. M. J. (1977) 'Predicting soil losses due to sheet and rill erosion', in *Guidelines for Watershed Management*, FAO Conservation Guide, 99, 99–124.

Arnoldus, H. M. J. (1980) 'An approximation of the rainfall factor in the universal soil loss equation', in *Assessment of Erosion*, eds M. DeBoodt and D. Gabriels, Wiley, pp. 127–32.

Ashton, F. M. (1956) 'Effects of a series of cycles of alternating low and high soil water contents on the rate of apparent photosynthesis in sugar cane', *Plant Physiology*, 31, 266–74.

Augustin, B. J. and Snyder, G. H. (1984) 'Moisture sensor-controlled irrigation for maintaining Bermudagrass turf', *Agronomy Journal*, 76, 848–50.

Australian Water Resources Commission (1976) *The Representative Basin Concept in Australia*, Department of National Development, Hydrological Series 2.

Ausubel, J. H. (1983) 'Can we assess the impacts of climatic changes?', *Climatic Change*, 5, 7–14.

Ayoade, J. O. and Akintola, F. O. (1982) 'A note on some rainfall characteristics of rainstorms in Ibadan, Nigeria', *Weather*, 37(2), 56–8.

Azam-Ali, S. N. (1983) 'Seasonal estimates of transpiration from a millet crop using a porometer', *Agricultural Meteorology*, 30, 13–24.

Babu, D. B. and Singh, S. P. (1984a) 'Studies of transpiration suppressants on spring sorghum in north-western India in relation to soil moisture regimes. I. Effect on yield and water use efficiency', *Experimental Agriculture*, 20, 151–9.

Babu, D. B. and Singh, S. P. (1984b) 'Studies of transpiration suppressants on spring sorghum in north-western India in relation to soil moisture regimes. II. Effect on growth and nutrient uptake', *Experimental Agriculture*, 20, 161–70.

Bachelor, C. H. and Roberts, J. (1983) 'Evaporation from the irrigation water, foliage and panicles of paddy rice in north-east Sri Lanka', *Agricultural Meteorology*, **29**, 11–26.

Baier, W. (1977) *Crop-Weather Models and Their Use in Yield Assessments*, WMO Technical Note No. 151, WMO Publication No. 458.

Baier, W. (1981) 'Water balance in crop-yield models', in *Applications of Remote Sensing to Agricultural Production Forecasting*, ed. A. Berg, A. A. Balkema, Rotterdam, pp. 119–29.

Bajard, Y., Draper, M. and Viens, P. (1981) 'Rural water supply and related services in developing countries – comparative analysis of several approaches', *Journal of Hydrology*, **51**, 75–88.

Baker-Blocker, A. and Bouwer, S. D. (1984) 'El Nino: evidence for climatic nondeterminism?' *Archiv fur Meteorologie, Geophysik und Bioklimatologie*, Ser. B, **34**, 65–73.

Balek, J. (1983) *Hydrology and Water Resources in the Tropical Regions*, Elsevier, Amsterdam, pp. 271.

Bargman, D. J., Sansom, H. W. and England, C. (1955) 'Report on experiments on artificial stimulation of rainfall at Mityana, Uganda', *East African Meteorological Department*, Memorandum 3(4).

Barry, R. G. and Chorley, R. J. (1968) *Atmosphere, Weather and Climate*.

Barry, R. G. and Chorley, R. J. (1982) *Atmosphere, Weather and Climate* Methuen, London, 4th edition, pp. 407.

Bates, J. R. (1970) 'Dynamics of Disturbances on the Intertropical Convergence Zone', *Quarterly Journal of the Royal Meteorological Society*, **96**, 677–701.

Benacchio, S. *et al.* (1983) 'Agroclimate information and agricultural production in the western Llanos of Venezuela', in *Agroclimate Information for Development*, ed. D. F. Cusak, Westview Press, Boulder, pp. 190–208.

Benjamini, Y. and Harpaz, Y. (1986) 'Observational rainfall–runoff analysis for estimating effects of cloud seeding on water resources in northern Israel', *Journal of Hydrology*, **83**, 299–306.

Bennett, O. L., Ashley, D. A. and Doss, B. D. (1966) 'Cotton responses to black plastic mulch and irrigation', *Agronomy Journal*, **58**, 57–60.

Berger, J. (1969) *The World's Major Fibre Crops: Their Cultivation and Manuring*, Centre d'Etude de L'Azote, Zurich.

Berlage, H. P. (1957) *Fluctuations of the General Atmospheric Circulation of More Than One Year*, Amsterdam.

Betson, R. P. and Marius, J. B. (1969) 'Source areas of storm runoff', *Water Resources Research*, **5**, 574–82.

Bhalme, H. N. and Jadhav, S. K. (1984a) 'The double (Hale) sunspot cycle and floods and droughts in India', *Weather*, **39(4)**, 112–16.

Bhalme, H. N. and Jadhav, S. K. (1984b) 'The southern oscillation and its relation to the monsoon rainfall', *Journal of Climatology*, **4**, 509–20.

Bhalme, H. N., Mooley, D. A. and Jadhav, S. K. (1983) 'Fluctuations in the drought/flood area over India and relationships with the southern oscillation', *Monthly Weather Review*, **111**, 86–94.

Bhardwaj, S. P., Khybri, M. L. Sewa Ram and Prasad, S. N. (1985) 'Crop geometry–a non monetary input for reducing erosion in corn on four percent slope', in *Soil Erosion and Conservation*, eds S. A. El-Swaify, W. C. Moldenhauer and A. Lo, Publication of the Soil Conservation Society of America, pp. 644–8.

Birch, H. F. (1960) 'Soil drying and soil fertility', *Tropical Agriculture*, **37**, 3–10.

Bishay, A. and McGinnies, W. G. (1979) *Advances in Desert and Arid Land Technology and Development*, Harwood Academic Publishers, London, pp. 618.

Bjerknes, J. (1969) 'Atmospheric teleconnections from the equatorial Pacific', *Monthly Weather Review*, **97**, 163.

Blackie, J. R. (1965) 'A comparison of methods of estimating evaporation in East Africa', *Proceedings of the Third Specialist Meeting on Applied Meteorology in East Africa*, EAAFRO, Muguga, Kenya.

Blackie, J. R. and Bjorking, L. (1968) 'Lysimeter study of the water use of sugar cane', *Proceedings of the 4th Specialist Meeting on Applied Meteorology in East Africa*, Nairobi.

Blaikie, P. (1985) *The Political Economy of Soil Erosion in Developing Countries*, Longman, pp. 188.

Blaikie, P. and Brookfield, H. (1987) *Land Degradation and Society*, Methuen, pp. 296.

Blizard, W. E. and Boyer, J. S. (1980) 'Comparative resistance of the soil and the plant to water', *Plant Physiology*, **66**, 809–14.

Blore, T. W. D. (1966) 'Further studies of water use by irrigated and unirrigated arabica coffee in Kenya', *Journal of Agricultural Science*, **67**, 145–54.

Bonell, M. and Gilmour, D. A. (1980) 'Variations in short-term rainfall intensity in relation to synoptic climatological aspect of the humid tropical northeast Queensland coast', *Singapore Journal of Tropical Geography*, **1**(2), 16–30.

Bonell, M., Gilmour, D. A. and Cassells, D. S. (1983) 'Runoff generation in tropical rainforests of northeast Queensland, Australia, and the implications for land use management', in *Hydrology of Humid Tropical Regions*, ed. R. Keller, IAHS Publ. No. 140, 287–97.

Bonsu, M. (1985) 'Organic residues for less soil erosion and more grain in Ghana', in *Soil Erosion and Conservation*, eds S. A. El-Swaify, W. C. Moldenhauer and A. Lo, Publication of the Soil Conservation Society of America, pp. 615–21.

Bordas, M. P. and Canali, G. E. (1980) 'The influence of land use and topography on the hydrological and sedimentological behaviour of basins in the basaltic region of south Brazil', in *The Influence of Man on the Hydrological Regime with Special Reference to Representative and Experimental Basins*, IAHS Publication No. 130, pp. 55–60.

Bosch, J. M. and Hewlett, J. D. (1982) 'A review of catchment experiments to determine the effect of vegetation changes on water yield and evapotranspiration', *Journal of Hydrology*, **55**, 3–23.

Bose, N. K. (1968) 'The Ganga', in B. C. Laws (ed.), *Mountains and Rivers of India*, Chapter XXI. Indian National Committee for Geography, Calcutta, pp. 356–60.

Bowden, B. N. (1964) 'The dry season of inter-tropical Africa', *Journal of Tropical Geography*, **19**, 1–3.

Box, G. E. P. and Jenkins, G. M. (1976) *Time Series Analysis: Forecasting and Control*, Holden-Day, San Francisco, pp. 575.

Branson, F. A., Gifford, G. F., Renard, K. G. and Hadley, R. F. (1981)

Rangeland Hydrology, Kendall Hunt, Dubuque Iowa, 2nd edition, ed. E. H. Reid, pp. 339.

Brazel, J. H. and Taylor, C. M. (1959) 'Artificial stimulation of rainfall in East Africa by means of rockets', *East African Meteorological Department*, Memorandum 3(6).

Brookfield, H. C. and Hart, D. (1966) *Rainfall in the Tropical South-west Pacific*, Australian National University, Canberra.

Brown, J. A. H., Ribeny, F. M. J., Wolanski, E. J. and Codner, G. P. (1981) 'A mathematical model of the hydrologic regime of the Upper Nile Basin', *Journal of Hydrology*, **51**, 97–107.

Bruijnzeel, L. A. (1983) 'Evaluation of runoff sources in a forested basin in a wet monsoonal environment: a combined hydrological and hydrochemical approach', in *Hydrology of Humid Tropical Regions*, ed. R. Keller, IAHS Publication No. 140, pp. 165–74.

Brutsaert, W. H. (1982) *Evaporation into the Atmosphere*, D. Reidel, Dordrecht.

Bryson, R. A. (1973) *Climatic modification by air pollution, II, the Sahelian effect*, Institute for Environmental Studies, University of Wisconsin Report, 9, pp. 12.

Bryson, R. A and Baerreis, D. A. (1967) 'Possibilities of major climatic modifications and their implications: north-west India, a case study', *Bulletin of the American Meteorological Society*, **48** (3), 136.

Budyko, M. I. (1956) *The Heat Balance of the Earth's Surface*, translated by N. I. Stepanova, US Weather Bureau, Washington.

Bultot, F. and Griffiths, J. F. (1972) 'The equatorial wet zone', Chapter 8 in J. F. Griffiths (ed.), *Climates of Africa*, Elsevier, Amsterdam.

Bunting, A. H. (1961) 'Some problems of agricultural climatology in tropical Africa', *Geography*, **46**, 283–94.

Burdon, D. J., Drouhin, G. and Dijon (1973) 'Water and Sahelian zone drought problems in the republic of Senegal', *Food and Agricultural Organisation*, WS/D9635.

Burpee, R. W. (1972) 'The origin and structure of easterly waves in the lower troposphere of North Africa', *Journal of Atmospheric Sciences*, 29 January, 77–90.

Burr, G. O. *et al.* (1957) 'The sugar cane plant', *Annual Review of Plant Physiology*, **8**, 275–308.

Butler, D. R. and Huband, N. D. S. (1985) 'Throughfall and stem-flow in wheat', *Agricultural and Forest Meteorology*, **35**, 329–38.

Camp, C. R. (1982) 'Effect of water management and bed height on sugarcane yield', *Soil Science*, **133** (4), 232–8.

Campbell, W. H., Blechman, J. B., and Bryson, R. A. (1983) 'Long-period tidal forcing of Indian monsoon rainfall: an hypothesis', *Journal of Climatology and Applied Meteorology*, **22**, 287–96.

Cane, M. A. (1983) 'Oceanographic events during El Nino', *Science*, **222**, 1189–195.

Carlson, T. N. and Benjamin, S. G. (1980) 'Radiative heating rates for Saharan dust', *Journal of Atmospheric Science*, **37**(1), 193–213.

Carpenter, T. H. *et al.* (1972) 'Observed relationships between lunar tide cycles and the formation of hurricanes and tropical storms', *Monthly Weather Review*, **100**(6), 451–60.

Carr, M. K. V. (1972) 'The climatic requirements of the tea plant: a review', *Experimental Agriculture*, **8**, 1–14.

Chacon, R. E. and Fernandez, W. (1985) 'Temporal and spatial rainfall variability in the mountainous region of the Reventazon River Basin, Costa Rica', *Journal of Climatology*, **5**, 175–88.

Chaggar, T. S. (1984) 'Reunion sets new rainfall records', *Weather*, **39**(1), 12–14.

Chambers, R. (1983) *Rural Development: Putting the Last First*, Longman, pp. 246.

Chan, J. C. L. (1985) 'Tropical cyclone activity in the northwest Pacific in relation to the El Nino/Southern Oscillation phenomenon', *Monthly Weather Review*, **113**, 599–606.

Chang, C. P. (1970) 'Westward propagating cloud patterns in the tropical Pacific as seen from time-composite satellite photographs', *Journal of Atmospheric Science*, **27**, 133.

Chang, J. H. (1961) 'Micro-climate of sugar cane', *Hawaiian Planters' Record*, **56**, 195–223.

Chang, J. H. (1968) *Climate and Agriculture*, Aldine, Chicago.

Chang, K. K. (1958) 'The relation between long-period fluctuations of the frequency of occurrence of typhoons in the Pacific and long-period fluctuations of the general circulation', *Acta Meteorologica Sinica*, **29**, 135.

Chatfield, C. (1975) *The Analysis of Time Series: Theory and Practice*, Chapman and Hall, pp. 263.

Chen, W. Y. (1983) 'The climate of spring 1983 – a season with persistant global anomalies associated with El Nino', *Monthly Weather Review*, **111**, 2371–84.

Chia, L. S. (1968) 'An analysis of rainfall patterns in Selangor', *Journal of Tropical Geography*, **27**, 1–18.

Chia, L. S. and Chang, K. K. (1971) 'The Record Floods of 10th December 1969 in Singapore', *Journal of Tropical Geography*, **33**, 9–19.

Chung, J. C. (1982) 'Correlations between the tropical Atlantic trade winds and precipitation in northeastern Brazil', *Journal of Climatology*, **2**, 35–46.

Claasen, M. M. and Shaw, R. H. (1970) 'Water deficit effects on corn: (1) vegetative components, (2) grain components', *Agronomy Journal*, **62**, 649–65.

Clawson, K. L. and Blad, B. L. (1983) 'Infrared thermometry for scheduling irrigation of corn', *Agronomy Journal*, **74**, 311–16.

Cochemé, J. (1966) 'FAO/UNESCO/WHO agroclimatology survey of a semi-arid area south of the Sahara', *Nature and Resources*, Bulletin of the International Hydrological Decade, **2**(4), 1–10.

Cochemé, J. and Franquin, P. (1967) 'An agroclimatology survey of a semi-arid area in Africa, south of the Sahara', *World Meteorological Organization*, **86**(210), TP 110, 136.

Cochrane, T. T. and Jones, P. G. (1981) 'Savannas, forests and wet season potential evapotranspiration in tropical South America', *Tropical Agriculture*, **58**, 185–90.

Cooley, K. R. and Idso, S. B. (1980) 'Effects of lily pads on evaporation', *Water Resources Research*, **16**(3), 605–6.

Cooper, C. F. and Jolly, W. C. (1969) *Ecological Effects of Weather Modification:*

a Problem Analysis, report dated 1 May 1968, Contract 14-06-D-6576, School of Natural Resources, University of Michigan.

Cordery, I. and **Pilgrim, D. H.** (1983) 'On the lack of dependence of losses from flood runoff on soil and cover characteristics', in *Hydrology of Humid Tropical Regions*, ed. R. Keller, IAHS Publication No. 140, pp. 187–96.

Coutts, H. H. (1969) 'Rainfall of the Kilimanjaro area', *Weather*, 24(2), 66–9.

Cowan, I. R. (1965) 'Transport of Water in the soil-plant-atmosphere system', *Journal of Applied Ecology*, 2, 221–39.

Crowther, F. (1944) *Report of Plant Physiology Section*, Research Division, Department of Agriculture, Sudan.

Crutcher, H. L. and **Quayle, R. G.** (1974) *Mariners Worldwide Climatic Guide to Tropical Storms at Sea*, US Department of Commerce, NOAA/EDS, National Climate Center.

Curry, L. (1962) 'Climatic change as a random series', *Annals of the Association of American Geographers*, 52(1), 21–31.

Dagg, M. (1965a) 'A rational approach to the selection of crops for areas of marginal rainfall in East Africa', *East African Agricultural and Forestry Journal*, 30(3), 296–300.

Dagg, M. (1965b) 'The use of agro-meteorological methods in catchment area research', *Proceedings of the Third Specialist Meeting on Applied Meteorology in East Africa*, EAAFRO, Muguga, Kenya.

Dagg, M. (1968a) 'Hydrological implications of grass roots studies at Muguga', *Proceedings of the Fourth Specialist Meeting on Applied Meteorology in East Africa*, Nairobi. East African Agricultural and Forestry Research Organisation.

Dagg, M. (1968b) 'Evaporation pans in East Africa', *Proceedings of the Fourth Specialist Meeting on Applied Meteorology in East Africa*, Nairobi.

Dagg M. (1969a) 'Hydrological implications of grass roots studies at a site in East Africa', *Journal of Hydrology*, 9, 438–44.

Dagg, M. (1969b) 'Water requirements of crops', in W. T. W. Morgan (ed.), *East Africa: Its Peoples and Resources*, Oxford U.P. London, pp. 119–25.

Dagg, M. (1970) 'A study of water use in tea in East Africa using a hydraulic lysimeter', *Agricultural Meteorology*, 7, 303–20.

Dagg, M. and **Blackie, J. R.** (1965) 'Studies of the effects of changes in land use on the hydrological cycle in East Africa by means of experimental catchment areas', *Bulletin of the International Association of Scientific Hydrology*, 4, 63–75.

Dagg, M. and **Macartney, J. C.** (1968) 'The agronomic efficiency of the NIAE mechanised tie ridge system of cultivation', *Experimental Agriculture*, 4, 279–94.

Dagg, M. and **Tapley, R. G.** (1967) 'Cashew nut production in southern Tanzania: V-water balance of cashew trees in relation to spacing', *East African Agricultural and Forestry Journal*, 33, 88–94.

Dalby, D. and **Harrison Church, R. J.** (1973) *Report of the 1973 Symposium Drought in Africa*. Centre for African Studies, School of Oriental and African Studies, London.

Dale, W. L. (1959) 'The rainfall of Malaya', *Journal of Tropical Geography*, 13, 23–37.

da Mota, F. S. (1978) *Soya Bean and Weather*, WMO Technical Note No. 160,

WMO Publication No. 498.

da Mota, F. S. (1983) 'Weather-technology models for corn and soyabeans in the south of Brazil', *Agricultural Meteorology*, **28**, 46–64.

da Mota, F.S., de Oliviera Agendes, M. O., da Costa Rosskoff, J. L. and da Silva, J. B. (1984) 'Modelling and forecasting Brazilian crop yields using meteorological data', in *Agroclimate Information for Development*, ed. D. F. Cusak, Westview Press, Boulder, 238–53.

Daniel, H. (1980) *Man and Climatic Variability*, WMO Publication No. 543, pp. 32.

Darby, G. M. (1985) 'Conservation tillage: an important, adaptable tool for soil and water conservation', in *Soil Erosion and Conservation*, eds S. A. El-Swaify, W. C. Moldenhauer and A. Lo, Publication of the Soil Conservation Society of America, pp. 649–58.

Dasman, R. F., Milton, J. P. and Freeman, P. H. (1973) *Ecological Principles for Economic Development*, John Wiley and Sons, New York.

Davenport, D. C. and Hagan, R. M. (1981) 'Concepts for conserving agricultural water', in *Advances in Food Producing Systems for Arid and Semi Arid Lands*, eds J. T. Manassah and E. J. Briskey, Academic Press, New York, pp. 379–87.

David, N. (1973) 'Extensive development of the agricultural sector in the semi-arid and northern savanna zones of West Africa', in Dalby and Harrison Church, *Report of the 1973 Symposium on Drought in Africa*.

Davies, J. A. and Robinson, P. J. (1969) 'A simple energy balance approach to the moisture balance climatology of Africa', *Environment and Land Use in Africa*, Thomas and Whittington, Methuen, London, pp. 23–56.

Davies, N. E. (1981) 'Meteosat looks at the general circulation: III. Tropical – Extratropical interaction', *Weather* **36**(6), 168–73.

Day, A. D. and Suhbawatr Intalap(1970) 'Some effects of soil moisture stress on the growth of wheat', *Agronomy Journal*, **62**, 27–9.

De Boer, Th.A. (1984) 'Interaction of rainfall with production and quality of Sahelian rangeland', *Progress in Biometeorology*, **3**, 187–99.

De Bruin, H. A. R. (1983) 'Evapotranspiration in humid tropical regions', in *Hydrology of Humid Tropical Regions with Particular Reference to the Hydrological Effects of Agriculture and Forestry Practice*, ed. R. Keller, IAHS Publication No. 140, pp. 299–311.

Denevan, W. M. (1973) 'Development and the imminent demise of the Amazon rain forest', *Professional Geographer*, **25**(2), May, 130–5.

Denmead, O. T. and Shaw, R. H. (1962) 'Availability of soil water to plants as affected by soil moisture content and meteorological conditions', *Agronomy Journal*, **45**, 385–90.

De, R., Bheemiah, K., Ramsheshiah, K. and Yogeswara Rao, Y. (1983a), 'Effects of mulches and antitranspirants on the grain yield of sorghum grown under limited irrigations on a deep vertisol', *Journal of Agricultural Science*, **100**, 159–62.

De, R., Rao, Y. Y., Ikramullah, M. and Giri Rao, L. G. (1983b) 'Maize yield as affected by irrigation and evapotranspiration control treatments', *Journal of Agricultural Science*, **100**, 731–4.

de Mayolo, S. E. A. (1984) 'Climate prediction and agriculture in pre-Columbian

Peru', in *Agroclimate Information for Development*, ed. D. F. Cusak, Westview Press, Boulder, pp. 25–33.

Dennett, M. D., Elston, J. and Rodgers, J. A. (1985) 'A reappraisal of rainfall trends in the Sahel', *Journal of Climatology*, 5, 353–61.

Dennett, M. D., Rogers, J. A. and Stern, R. D. (1983) 'Independence of rainfalls through the rainy season and the implications for the estimation of rainfall probabilities', *Journal of Climatology*, 3, 375–84.

de Oliveira, A. S. and Nobre, C. A. (1985) 'Meridional penetration of frontal systems in South America and its relation to organised convection in the Amazon', *16th Conference on Hurricanes and Tropical Meteorology*, American Meteorological Society, Boston, pp. 85–7.

de Oliveira Leite, J. (1985) 'Interflow, overland flow and leaching of natural nutrients on an alfisol slope of Southern Bahia, Brazil', *Journal of Hydrology*, 80, 77–92.

De Oliveira, M. O., da Mota, F. S. and da Silva, J. B. (1980) 'Estimates of potential evapotranspiration (Penman) as a function of geographical factors in Brazil', *Agricultural Meteorology*, 22, 207–15.

Derera, N. F., Marshall, D. R. and Balaam, L. N. (1969) 'Genetic variability in root development in relation to drought tolerance in spring wheats', *Experimental Agriculture*, 5, 327–37.

Dhar, O. N. and Bhattacharya, B. K. (1977) 'Relationships between central rainfall and its areal extent for severemost rainstorms of north Indian plains', *Irrigation and Power Journal*, 34(2), 245–50.

Dhar, O. N., Kulkarni, A. K. and Rakhecha, P. R. (1981) 'Probable maximum point rainfall estimation for the southern half of the Indian peninsula', *Proceedings, Indian Academy of Science (Earth Planet Science)* 90(1), 39–46.

Dhar, O. N. and Rakhecha, P. R. (1979) 'Incidence of heavy rainfall in the Indian desert region', in *The Hydrology of Areas of Low Precipitation*, IAHS Publication No. 128, 33–42.

Dhar, O. N. and Rakhecha, P. R. (1983) 'Foreshadowing northeast monsoon rainfall over Tamil Nadu, India', *Monthly Weather Review*, 111, 109–12.

Dhar, O. N., Rakhecha, P. R. and Kulkarni, A. K. (1982a) 'Fluctuations in northeast monsoon rainfall of Tamil Nadu', *Journal of Climatology*, 2, 339–45.

Dhar, O. N., Rakhecha, P. R. and Kulkarni, A. K. (1982b) 'Estimation of extreme point rainfall over peninsular India', *Proceedings of Conference on Rain Water Cistern Systems*, June 1982, ed. F. N. Fujimura, pp. 61–7.

Dhar, O. N., Rakhecha, P. R. and Mandal, B. N. (1981a) 'Influence of tropical disturbances on monthly monsoon rainfall of India', *Monthly Weather Review*, 109, 188–90.

Dhar, O. N., Rakhecha, P. R. and Mandal, B. N. (1981b) 'Some facts about Indian rainfall – a brief appraisal from hydrological considerations', *Indian Journal of Power and River Valley Development*, 117–25.

Dhar, O. N. and Ramachandran G. (1970) 'Short duration analysis of Calcutta (Dum Dum) rainfall', *Indian Journal of Meteorology and Geophysics*, 21(1), 93–102.

Dhar, O. N., Soman, M. K. and Mulye, S. S. (1984) 'Rainfall over the southern slopes of the Himalayas and the adjoining plains during "breaks" in the monsoon'. *Journal of Climatology*, 4, 671–6.

Dhar, O. N., Rakhecha, P. R., Mandal, B. N. and Ghose, G. C. (1982) 'A brief appraisal of hydrometeorological studies on Indian rainfall', *Proceedings of Seminar Hydrological Investigations during the last 25 years in India*, May 1982, Pune, 95–105.

Dhruvnarayana, V. V, and Sastry, G. (1985) 'Soil conservation in India', in *Soil Erosion and Conservation*, eds S. A. El-Swaify, W. C. Moldenhauer and A. Lo, Publication of the Soil Conservation Society of America, pp. 3–9.

Dias, A. C. C. P. and Nortcliff, S. (1985) 'Effects of two land clearing methods on the physical properties of an Oxisol in the Brazilian Amazon', *Tropical Agriculture*, 62(3), 207–12.

Dick, R. S. (1958) 'Variability of rainfall in Queensland', *Journal of Tropical Geography*, 11, 32–42.

Diehl, B. (1984) 'Interannual variability of surface fields over the Indian Ocean during recent decades', *15th Conference on Hurricanes and Tropical Meteorology*, American Meteorological Society, Boston, 375–8.

Dietrich, G. and Kalle, K. (1957) *Allgemeine Meereskunde: Ein Einfuhrung in die Ozeanographie*, Gebrüder Bomtraeger, Berlin.

D'Itri, F. M. (ed.) (1985) *A Systems Approach to Conservation Tillage*, Lewis Publishers, pp. 384.

Doorenbos, J. and Pruitt, W. O. (1977) *Guidelines for Predicting Crop Water Requirements*, Irrigation and Drainage Paper 24, FAO Rome, pp. 144.

Dorman, C. E. (1982) 'Comments on "Comparison of ocean and island rainfall in the tropical Pacific"', *Journal of Applied Meteorology*, 21, 109–13.

Doss, B. D., Ashley, D. A. and Bennett, O. L. (1964) 'Effects of moisture regime and stage of plant growth on moisture use by cotton', *Soil Science*, 98, 156–61.

Douglas, A. V. and Englehart, P. J. (1981) 'On a statistical relationship between Autumn rainfall in the central equatorial Pacific and subsequent winter precipitation in Florida', *Monthly Weather Review*, 109, 2377–82.

Dow, H. (1955) *East African Royal Commission 1953–5 Report*, HMSO, London.

Downey, L. A. (1971a) 'Water use by maize at 3 plant densities', *Experimental Agriculture*, 7, 161–70.

Downey, L. A. (1971b) 'A visual irrigation guide for maize (*Zea mays* L.)', *Agronomy Journal*, 63, 887–9.

Druyan, L. M. (1981) 'The use of global circulation models in the study of the Indian monsoon', *Journal of Climatology*, 1, 77–92.

Druyan, L. M. (1982a) 'Studies of the Indian Summer monsoon with a coarse-mesh general circulation model, Part I,' *Journal of Climatology*, 2, 127–39.

Druyan, L. M. (1982b) 'Studies of the Indian summer monsoon with a coarse-mesh general circulation model, Part II,' *Journal of Climatology*, 2, 347–55.

Dubriel, P. L. (1985) 'Review of field observations of runoff generation in the tropics', *Journal of Hydrology*, 80, 237–64.

Dumont, R. and Rosier, B. (1969) *The Hungry Future*, Andre Deutsch, London.

Dunne, T. and Black, R. (1970) 'An experimental investigation of runoff production in permeable soils', *Water Resources Research*, 6, 478–90.

Duru, J. O. (1984) 'Blaney–Morin–Nigeria evapotranspiration model', *Journal of Hydrology*, 70, 71–83.

Dutt, G. R., Hutchinson, C. F. and Garduno, M. A. (eds) (1981) *Rainfall*

Collection for Agriculture in Arid and Semiarid Regions, Commonwealth Agricultural Bureau, pp. 97.

Dyer, T. G. J. (1982) 'On the intra-annual variation in rainfall over the subcontinent of southern Africa', *Journal of Climatology*, **2**, 47–64.

Eagleson, P. S. (1970) *Dynamic Hydrology*, McGraw-Hill, pp. 462.

East African Agriculture and Forestry Research Organisation (1971) 'Progress report on Mbeya catchment data analysis', unpublished manuscript.

East African Meteorological Department (1961) *10% and 20% Probability Maps of Annual Rainfall of East Africa*, Nairobi.

Eden, T. (1976) *Tea*, Longman, pp. 236.

Ekern, P. C. (1964) 'Direct interception of cloud water at Lanaihale, Hawaii', *Proceedings of the Soil Science Society of America*, **28**, 419–21.

Ekern, P. C. (1965) 'Evapotranspiration of pineapple in Hawaii', *Plant Physiology*, **40**, 736–9.

Ekern, P. C. (1966) 'Evapotranspiration by Bermuda grass in Hawaii', *Agronomy Journal*, **58**, 387–90.

Ekern, P. C. (1982) 'Measured evaporation in high rainfall areas, Leeward Koolau Range, Oahu, Hawaii', in *International Symposium on Hydrometeorology*, eds Johnson and Clark, American Water Resources Association, pp 85–90.

Ellis, R. T. (1967) 'The prospects for irrigation of tea in Central Africa', *Investor's Guardian*, **209**(6497), 889–91.

El Nadi, A. H. (1969a) 'Efficiency of water use by irrigated wheat in the Sudan', *Journal of Agricultural Science*, **73**(2), 261–6.

El Nadi, A. H. (1969b) 'Water relations of beans (1) effects of water stress on growth and flowering', *Experimental Agriculture*, **5**, 195–207.

El Nadi, A.H. (1970) 'Water relations of beans (2) effects of differential irrigation on yield and seed size of broad beans', *Experimental Agriculture*, **6**, 107–11.

El-Swaify, S. A., Dangler, E. W. and Armstrong, C. L. (1982) *Soil Erosion by Water in the Tropics*, College of Tropical Agriculture and Human Resources, University of Hawaii Research Extension Series No. 24, pp. 173.

El-Swaify, S. A., Moldenhauer, W. C. and Lo, A. (eds) (1985) *Soil Erosion and Conservation*, Publication of the Soil Conservation Society of America, pp. 793.

Elwell, H. A. (1977) *A soil loss estimation system for southern Africa*, Research Bulletin No. 22, Causeway, Rhodesia: Department of Conservation and Extension.

Elwell, H. A. (1981) 'A soil loss estimation technique for southern Africa', *ICSC*, 281–92.

Elwell, H. A. and Stocking, M. A. (1973a) 'Rainfall parameters for soil loss estimation in a tropical climate', *Journal of Agricultural Engineering Research*, **18**, 169–77.

Elwell, H. A. and Stocking, M. A. (1973b) 'Rainfall parameters to predict surface runoff yields and soil losses from selected field-plot studies', *Rhodesian Journal of Agricultural Research*, **11**, 123–9.

Elwell, H. A. and Stocking, M. A. (1982) 'Developing a simple yet practical method of soil-loss estimation', *Tropical Agriculture*, **59**, 43–8.

Enyi, B. A. C. (1963) 'The effect of time of irrigation on the growth and yield of transplanted rice', *Journal of Agricultural Science*, **61**(1), 115–19.

Evans, A. C. (1955) 'A study of crop production in relation to rainfall reliability', *East African Agricultural and Forestry Journal*, **20**, 263.

Evanari, M., Shanan, L. and Tadmor, N. H. (1968) 'Runoff farming in the desert (1 and 2)', *Agronomy Journal*, **60**, 29–32, 33–8.

Evanari, M., Shanan, L. and Tadmor, N. (1982) *The Negev: The Challenge of a Desert*, 2nd edition, Harvard University Press, pp. 437.

Ewel, J. (1981) 'Environmental implications of tropical forest utilization, in *Tropical Forests: Utilization and Conservation*, ed. F. Mergen, Yale School of Forestry and Environmental Studies, pp. 157–67.

Fakorede, M. A. B. (1985) 'Response of maize to planting dates in a tropical rainforest location', *Experimental Agriculture*, **21**, 19–30.

Fakorede, M. A. B. and Opeke, B. O. (1985) 'Weather factors affecting the response of maize to planting dates in a tropical rainforest location', *Experimental Agriculture*, **21**, 31–40.

Falls, R. (1970) 'Some synoptic models – north Australia', *Proceedings of the Symposium on Tropical Meteorology*, Honolulu, American Meteorological Society, World Meteorological Organisation, Hawaiian Institute of Geophysics.

Fang, L. S. and Rao, G. V. (1985). 'A study of the model-derived offshore vortices in the Indian Ocean after the monsoon onset', *16th Conference of Hurricanes and Tropical Meteorology*, American Meteorological Society, Boston, pp. 66–7.

Farbrother, H. G. and Manning, H. L. (1952) *Empire Cotton Growing Corporation Progress Reports from Experimental Stations, Namulonge, 1951–52*. pp. 3–6.

Fermor, J. H. (1971) 'The weather during northers at Kingston, Jamaica', *Journal of Tropical Geography*, **32**, 31–7.

Fijita, T. T., Watanabe, K. and Isawa, T. (1969) 'Formation and structure of equatorial anticyclones caused by large scale cross-equatorial flows determined by ATS-1 photographs', *Journal of Applied Meteorology*, **8**, 649.

Fisher, N. M. (1980) 'The effect of time of planting on four bean (*Phaseolus vulgaris*) genotypes in Kenya', *Journal of Agricultural Science*, **95**, 401–8.

Fitzpatrick, E. A., Hart, D. and Brookfield, H. C. (1966) 'Rainfall seasonality in the tropical south-west Pacific', *Erdkunde*, **20**, 181–94.

Fleming, P. M. (1968) 'Crop water requirements and irrigation', *Agricultural Meteorology*, Proceedings of the WMO Seminar, Melbourne.

Flohn, H. (1960) 'Intertropical convergence zone and meteorological equator', *WMO tropics', *Munitalp/WMO Joint Symposium on Tropical Meteorology in Africa*, Munitalp, Nairobi.

Flohn, H. (1970) 'Climatic effects of local circulations in tropical and subtropical latitudes', *Proceedings of the Symposium on Tropical Meteorology*, Honolulu.

Flohn, S. (1964) 'Intertropical convergence zone and meteorological equator', *WMO Technical Note 64*, 21.

Fordham, R. (1971) 'Effects of soil moisture on growth of young tea in Malawi', *Experimental Agriculture*, **7**, 171–6.

Fosbrooke, H. S. (1973) 'Some sociological factors in the drought situation in

Africa', in Dalby and Harrison Church, *Report of the 1973 Symposium Drought in Africa*.

Fournier, F. (1960) *Climat et Erosion*, Presses Universitaires de France, Paris.

Fournier, F. (1967) 'Research on soil erosion and soil conservation in Africa', *African Soils*, 12(1), 53–96.

Freise, F. (1936) 'Das Binnenklima von Urwaldern im subtropischen Brasilien', *Petermanns Mitteilungen*, 82, 301–4, 346–8.

Frere, M. (1980) 'Report on development of practical climatic classification system for agriculture in the semi-arid tropics, Plenary Session', in *Climatic Classification in the Semi-Arid Tropics*, ICRISAT, 145–7.

Fuehring, H. D. and Finkner, M. D. (1983) 'Effect of Folicote antitranspirant application on field grain yield of moisture-stressed corn', *Agronomy Journal*, 75(4), 579–82.

Fuehring, H. D., Mazaheri, A., Bybordi, M. and Khan, A. K. S. (1966) 'Effect of soil mositure depletion on crop yield and stomatal infiltration', *Agronomy Journal*, 195–8.

Fukui, H. (1979) 'Climatic variability and agriculture in tropical moist regions', in *Proceedings of the World Climate Conference*, Geneva, WMO Publication No. 537, 426–74.

Gaastra, P. (1963) 'Climatic control of photosynthesis and respiration', in L. T. Evans (ed.), *Environment Control of Plant Growth*, Academic Press, New York, 113–40.

Gadgil, S. and Joshi, N. V. (1980) 'Use of principal component analysis in rational classification of climates', in *Climatic Classification in the Semi-Arid Tropics*, ICRISAT, 17–26.

Gadgil, S. and Joshi, N. V. (1983) 'Climatic clusters of the Indian region', *Journal of Climatology*, 3, 47–63.

Gangopadhyaya, M. and Sarker, R. P. (1965) 'Influence of rainfall distribution on the yield of wheat crop', *Agricultural Meteorology*, 2, 331.

Garbutt, D. J., Stern, R. D., Dennett, M. D. and Elston, J. (1981) 'A comparison of the rainfall climate of eleven places in West Africa using a two-part model for daily rainfall', *Archiv fur Meteorologie, Geophysik und Bioklimatologie*, Ser. B, 29, 137–55.

Gerard, C. J. and Namken, L. N. (1966) 'Influence of soil texture and rainfall on the response of cotton to moisture regime', *Agronomy Journal*, 58(1), 39–42.

Ghadiri, H. and Payne, D. (1981) 'Raindrop impact stress', *Journal of Soil Science*, 32, 41–9.

Ghosh, R. K. (1980) 'Modelling infiltration', *Soil Science*, 130(6), 297–302.

Giambelluca, T. W. (1986) 'Land-use effects on the water balance of a tropical island', *National Geographic Research*, 2(2), 125–51.

Giambelluca, T. W., Lau, L. S., Fok, Yu-Si, and Schroeder, T. A. (1984), *Rainfall frequency study for Oahu, State of Hawaii*, Department of Land and Natural Resources, Report R-73, pp. 34, with appendix.

Gichiuya, S. N. (1970) 'Easterly disturbances in the S.E. monsoon', *Proceedings of the Symposium on Tropical Meteorology*, Honolulu. American Meteorological Society, World Meteorological Organisation, Hawaiian Institute of Geophysics.

Glover, J. and Gwynne, M. D (1962) 'Light rainfall and plant survival in East Africa: (1) maize', *Journal of Ecology*, **50**, 111–18.

Glover, J., Robinson, P. and Taylor, C. M. (1955) 'Assessing the reliability of rainfall if monthly falls are not independent', *Journal of Agricultural Science*, **46**, 387.

Goodland, R. J. A., Watson, C. and Ledee, G. (1984) *Environmental Management in Tropical Agriculture*, Westview Press, Boulder, pp. 237.

Gramzow, R. H. and Henry, W. K. (1972)', 'The rainy pentades of Central America', *Journal of Applied Meteorology*, **11**(4), 637–42.

Gray, R. W. (1970) 'The effect on yield of time of planting of maize in S. W. Kenya', *East African Agricultural and Forestry Journal*, **35**(3), 291–8.

Gray, W. M. (1968) 'Global view of the origin of tropical disturbances and storms', *Monthly Weather Review*, **10**, 669.

Gray, W. M. (1984a) 'Atlantic seasonal hurricane frequency, Part I: El Nino and 30 mb quasi-biennial oscillation influences', *Monthly Weather Review*, **112**, 1649–68.

Gray, W. M. (1984b) 'Atlantic seasonal hurricane frequency. Part II: Forecasting its variability', *Monthly Weather Review*, **112**, 1669–83.

Gray, W. M. and Jacobson, R. W. (1977) 'Diurnal variation of deep cumulus convection', *Monthly Weather Review*, **105**, 1171–88.

Gregory, K. J. and Walling, D. E. (1973) *Drainage Basin Form and Process*, Edward Arnold, pp. 456.

Gregory, S. (1965) *Rainfall Over Sierra Leone*, Dept. of Geography, University of Liverpool.

Gregory, S. (1969) Rainfall reliability, *Environment and Land Use in Africa*, Thomas and Whittington, pp. 57–82. Methuen, London.

Gregory, S. (1982) 'Spatial patterns of Sahelian annual rainfall, 1961–1980', *Archiv fur Meteorologie, Geophysik und Bioklimatologie*, Ser.B, **31**, 273–86.

Griffin, J. L. and Watson, V. H. (1982) 'Production and quality of four Bermudagrasses as influenced by rainfall patterns', *Agronomy Journal*, **74**, 1044–7.

Griffiths, J. F. (1959) 'Bioclimatology and the meteorological services', *Proceedings of the Symposium on Tropical Meteorology in Africa*, Munitalp Foundation, and World Meteorological Organisation, Nairobi, Kenya, pp. 282–300.

Griffiths, J. F. (1961) 'Some rainfall relationships in East Africa', *Inter African Conference on Hydrology*, Commission for Technical Co-operation in Africa South of the Sahara, Nairobi, pp. 115–20.

Griffiths, J. F. (1972) *Climates of Africa*, Elsevier, Amsterdam.

Grove, A. T. (1973) 'Desertification in the African environment', in Dalby and Harrison Church, *Report of the 1973 Symposium Drought in Africa*.

Grove, A. T. (1980) 'Climatic classification: concepts for dry tropical environments', in *Climatic Classification in the Semi-Arid Tropics*, ICRISAT, 1–5.

Grove, A. T. (1985) 'The arid environment', in *Plants for Arid Lands*, eds G. E. Wickens, J. R. Goodin and D. V. Field, Allen and Unwin, London, 9–18.

Grundy, F. (1963) 'Precipitation, evaporation and runoff in tropical and subtropical Africa', *Deutsche Afrika Gesellschaft-Wesserwirtschaft in Afrika*, Verlag Deutscher Wirtschaftsdienst, Köln, p. 106.

Guedalia, D., Estournel, C. and Vehil, R. (1984) 'Effects of Sahel dust layers upon nocturnal cooling of the atmosphere (ECLATS Experiment)', *Journal of Climate and Applied Meteorology*, **23**, 644–50.

Guinn, G. and Mauney, J. R. (1984a) 'Fruiting of cotton. I. Effects of moisture status on flowering', *Agronomy Journal*, **76**, 90–4.

Guinn, G. and Mauney, J. R. (1984b) 'Fruiting of cotton. II. Effects of plant moisture status and active boll load on boll retention', *Agronomy Journal*, **76**, 94–8.

Guinn, G., Mauney, J. R. and Fry, K. E. (1981) 'Irrigation scheduling and plant population effects on growth, bloom rates, boll abscission and yield of cotton', *Agronomy Journal*, **73**, 529–34.

Gulhati, N. D. (1968) 'The Indus and its tributaries', in B. C. Laws (ed.), *Mountains and Rivers of India*, Chapter XX, Indian National Committee for Geography, Calcutta, pp. 348–55.

Gumbs, F. A. (1982) 'Soil and water management features in Trinidad and Guyana', *Tropical Agriculture*, **59**, 76–81.

Gumbs, F. A. and Simpson, L. A. (1981) 'Influence of flooding and soil moisture content on elongation of sugar cane in Trinidad', *Experimental Agriculture*, **17**, 403–6.

Gutman, G. (1984) 'Numerical experiments on land surface alterations with a zonal model allowing for interaction between the geobotanic state and climate', *Journal of Atmospheric Sciences*, **41**(18), 2679–85.

Gwynne, M. D. (1964) 'Plant characteristics that make them suited to areas of low and erratic rainfall', *Specialist Meeting on Crops of Low and Erratic Rainfall*, East African Agricultural and Forestry Research Organisation, Muguga, Kenya.

Hall, A. H., Cannell, G. H. and Lawton, H. W. (1979) *Agriculture in Semi-Arid Environments*, Springer-Verlag, Berlin, pp. 340.

Hall, A. H., Foster, K. W. and Waines, J. G. (1979) 'Crop adaptation to semi arid environments', in *Agriculture in Semi Arid Environments*, eds A. H. Hall, G. H. Cannell and H. W. Lawton, Springer-Verlag, Berlin, 148–79.

Halm, A. T. (1967) 'Effect of water regime on the growth and chemical composition of two rice varieties', *Tropical Agriculture*, **44**, 33–7.

Hamilton, C. A. (1984) *Deforestation in Uganda*, Oxford University Press, pp. 92.

Hamilton, L. S. (ed.) (1983) *Forest and Watershed Development and Conservation in Asia and the Pacific*, Westview Press, Boulder, pp. 560.

Hamilton, L. S. (1985) 'Overcoming myths about soil and water impacts of tropical forest land uses', in *Soil Erosion and Conservation*, eds S. A. El-Swaify, W. C. Moldenhauer and A. Lo, Soil Conservation Society of America, pp. 680–90.

Hamilton, L. S. and King, P. N. (1983) *Tropical Forested Watersheds: Hydrologic and Soils Response to Major Uses or Conversions*, Westview Press, Boulder, pp. 168.

Hammer, R. M. (1972) 'Rainfall patterns in the Sudan', *Journal of Tropical Geography*, **35**, 40–50.

Hanks, R. J., Gardner, H. R. and Florian, R. L. (1969) 'Plant growth-evapotranspiration relations for several crops in the central Great Plains', *Agronomy Journal*, **61**, 30–4.

Hardy, N., Shainberg, I., Gal, M. and Keren, R. (1983) 'The effect of water quality and storm sequence upon infiltration rate and crust formation', *Journal of Soil Science*, **34**, 665–76.

Hari Krishna, J. (1982) 'A parametric model to estimate runoff from small agricultural watersheds in the semi-arid tropics', *Journal of Hydrology*, **55**, 43–51.

Harnack, R. P., Lanzante, J. R. and Harnack, J. (1982) 'Associations among the tropical Pacific wind and sea surface temperature fields and higher latitude circulation', *Journal of Climatology*, **2**, 267–90.

Harrison, M. S. J. (1984) 'A generalised classification of South African summer rain-bearing synoptic systems', *Journal of Climatology*, **4**, 547–60.

Harrison Church, R. J. (1973) 'The development of the water resources of the dry zone of West Africa', in Dalby and Harrison Church, *Report of the 1973 Symposium Drought in Africa.*

Hashemi, F. (1979) 'Climatology and fight against desertification', in *WMO Special Environment Report No. 13, Meteorology and the Human Environment*, WMO Publication No. 517, pp. 49.

Hastenrath, S. (1984) 'Interannual variability and annual cycle: Mechanism of circulation and climate in the tropical Atlantic sector?', *Monthly Weather Review*, **112**, 1097–107.

Hastenrath, S. (1985) *Climate and Circulation of the Tropics*, D. Riedel, pp. 455.

Haverkamp, R., Vauclin, M. and Vachaud, G. (1984) 'Error analysis in estimating soil water content from neutron probe measurements: 1. Local standpoint', *Soil Science*. **137**(2), 78–90.

Hay, R. K. M. (1981) 'Timely planting of maize – a case history from the Lilongwe Plain', *Tropical Agriculture*, **58**(2), 147–55.

Hayes, W. A. (1982) *Minimum Tillage Farming*, No Till Farmer Inc., pp. 167.

Hearn, A. B. and Wood, R. A. (1964) 'Irrigation control experiments on dry season crops in Nyasaland', *Empire Journal of Experimental Agriculture*, **32**, 1–17.

Heatherly, L. G., McMichael, B. L. and Ginn, L. H. (1980) 'A weighing lysimeter for use in isolated field areas', *Agronomy Journal*, **72**, 845–7.

Henderson-Sellers, A. and Gornitz, V. (1984) 'Possible climatic impacts of land cover transformations, with particular emphasis on tropical deforestation', *Climatic Change*, **6**, 231–57.

Henry, W. K. (1974) 'The tropical rainstorm', *Monthly Weather Review*, October, **102**, 717–25.

Herwitz, S. R. (1985) 'Interception storage capacities of tropical rainforest canopy trees', *Journal of Hydrology*, **77**, 237–52.

Hewlett, J. D. and Hibbert, A. R. (1965) 'Factors affecting the response of small watersheds to precipitation in humid areas', in *International Symposium on Forest Hydrology*, eds W. E. Sopper and H. W. Lull, Pergamon, pp. 275–90.

Holder, G. D. and Gumbs, F. A. (1982) 'Effects of irrigation at critical stages of ontogeny of the banana c.v. "Robusta" on growth and yield', *Tropical Agriculture*, **59**, 221–6.

Holder, G. D. and Gumbs, F. A. (1983a) 'Effects of irrigation on the growth and yield of banana', *Tropical Agriculture*, **60**, 25–30.

Holder, G. D. and Gumbs, F. A. (1983b) 'Effects of nitrogen and irrigation on the growth and yield of banana', *Tropical Agriculture*, **60**, 179–83.

Holder, G. D. and Gumbs, F. A. (1983c) 'Effects of waterlogging on the growth and yield of banana', *Tropical Agriculture*, **60**, 111–16.

Holford, I. C. R. (1971) 'Effect of rainfall on yield of groundnuts in Fiji', *Tropical Agriculture*, **48**, 171–5.

Holland, G. T., Keenan, T. D. and Guymer A. E. (1984). 'The Australian monsoon: definition and variability'. *15th Conference on Hurricanes and Tropical Meteorology*, American Meteorological Society, Boston, pp. 398–402.

Holton, J. R., Wallace, J. M. and Young, J. A. (1971) 'On boundary layer dynamics and the ITCZ', *Journal of Atmospheric Science*, **28**, 275.

Hopkins, B. (1960) 'Rainfall interception by tropical forest in Uganda', *East African Agricultural and Forestry Journal*, **25**(4), 255–8.

Howell, T. A. Phene, C. J. and Meek, D. W. (1983) 'Evaporation from screened class A pans in a semi-arid climate', *Agricultural Meteorology*, **29**, 111–24.

Hsia, Y. J. and Koh, C. C. (1983) 'Water yield resulting from clearcutting a small hardwood basin in central Taiwan', in *Hydrology of Humid Tropical Regions*, ed. R. Keller, IAHS Publ. No. 140, pp. 215–19.

Hudson, J. C. (1969) 'The available water capacity of Barbados soils', *Experimental Agriculture*, **5**, 167–82.

Hudson, N. W. (1957) 'Erosion control research', *Rhodesian Agricultural Journal*, **54**, 297.

Hudson, N. W. (1971) *Soil Conservation*, Batsford, London.

Hudson, N. W. (1975) *Field Engineering for Agricultural Development*, The Clarendon Press, pp. 225.

Hudson, N. W. (1981) *Soil Conservation*, Batsford, 2nd edition.

Huff, F. A. and Shipp, W. L. (1969) 'Spatial correlations of storm monthly and seasonal precipitation', *Journal of Applied Meteorology* **8**, 542–50.

Huke, R. E. (1966) 'Rainfall in Burma', *Geographical Publications at Dartmouth*, No. 2.

Hulme, M. (1984) '1983: An exceptionally dry year in central Sudan', *Weather*, **39**(9), 281–5.

Humphries, L. R. (1978) *Tropical Pastures and Fodder Crops*, Longman, London, pp. 135.

Humphries, L. R. (1981) *Environmental Adaptation of Tropical Pasture Plants*, Macmillan, London, pp. 261.

Hussain, Z. (1981) 'A simple method of using highly saline water for irrigation', *Journal of Agricultural Science*, **96**, 17–21.

Hutchinson, J., Manning, H. L. and Farbrother, H. G. (1958) 'Crop water requirements of cotton', *Journal of Agricultural Science*, **51**, 177–88.

Hutchinson, P. (1970) 'A contribution to the problem of spacing raingauges in rugged terrain', *Journal of Hydrology*, **12**(1), 1–14.

Hutchinson, P. (1973) 'Increase in rainfall due to Lake Kariba', *Weather*, **28**, 499–504.

Hutchinson, P. (1985) 'Rainfall analysis of the Sahelian drought in the Gambia', *Journal of Climatology*, **5**, 665–72.

Hutchinson, P. and **Sam, J. A.** (1984) 'The unusual start of the wet season in the Gambia', *Weather*, **39**(1), 24–8.

IAHS (1979) *The Hydrology of Areas of Low Precipitation*, IAHS Publication No. 128, pp. 502.

IAHS (1980) *The Influence of Man on the Hydrological Regime with Special Reference to Representative and Experimental Basins*, IAHS Publication No. 130, pp. 483.

ICRISAT (1980) *Climatic Classification: A Consultants Meeting*, International Crop Research Institute for the Semi-Arid Tropics, pp. 153.

IITA (1982) *International Institute for Tropical Agriculture, Research Highlights for 1981*, pp. 72.

IITA (1984) *International Institute for Tropical Agriculture, Research Highlights for 1983*, pp. 123.

IITA (1985) *International Institute for Tropical Agriculture, Research Highlights for 1984*, pp. 114.

Ilesanmi, O. O. (1972) 'An empirical formulation of the onset, advance and retreat of rainfall in Nigeria', *Journal of Tropical Geography*, **34**, 17–24.

Iremiren, G. O. and **Okiy, D. A.** (1986) 'Effects of sowing date on the growth, yield and quality of okra (Abelmoschus esculentus (L.) Moench.) in southern Nigeria', *Journal of Agricultural Science*, **106**, 21–6.

IRRI (1977) *Symposium on Cropping Systems Research and Development for the Asian Rice Farmer*, IRRI, Los Banos, Philippines, pp. 454.

IRRI (1979a) *International Deepwater Rice Workshop*, Los Banos, Philippines, pp. 300.

IRRI (1979b) *Rainfed Lowland Rice*, International Rice Research Institute Conference, Los Banos, Philippines, pp. 341.

IRRI (1983) *International Rice Research Institute, Annual Report for 1982*.

IRRI (1984) *Annual Report for 1983*.

Ishag, H. M. *et al.* (1985) 'Growth and water relations of groundnuts (*Arachis hypogaea*) in two contrasting years in the irrigated Gezira', *Experimental Agriculture*, **21**, 403–8.

Jackson, I. J. (1969a) 'Tropical rainfall variations over a small area', *Journal of Hydrology*, **8**, 99–110.

Jackson, I. J. (1969b) 'The persistence of rainfall gradients over small areas of uniform relief', *East African Geographical Review*, **7**, 37–43.

Jackson, I. J. (1969c) 'Annual rainfall probability and the binomial distribution', *East African Agricultural and Forestry Journal*, **35**(3), 265–72.

Jackson, I. J. (1970) 'Some physical aspects of water resource development in Tanzania', *Geografiska Annaler*, **52A**(3–4), 174–85.

Jackson, I. J. (1971a) 'An experiment on the siting of raingauges in tropical highlands', *The Role of Hydrology and Hydrometeorology in the Economic Development of Africa*, ECA/WMO Conference, Addis Ababa, WMO No. 301, 169–78.

Jackson, I. J. (1971b) 'Problems of throughfall and interception assessment under tropical forest', *Journal of Hydrology*, **12**, 234–54.

Jackson, I. J. (1971c) 'Atmospheric pressure and winds', *Tanzania in Maps*, Berry, pp. 34–35. University of London Press, London.

Jackson, I. J. (1972) 'The spatial correlation of fluctuations in rainfall over Tanzania: a preliminary analysis', *Archiv für Meteorologie Geophysik und Bioklimatologie*, Ser. B, 20, 167–78.

Jackson, I. J. (1974) 'Inter-station rainfall correlation under tropical conditions', *Catena*, 1, 235–56.

Jackson, I. J. (1975) 'Relationships between rainfall parameters and interception by tropical forest', *Journal of Hydrology*, 24, 215–38.

Jackson, I. J. (1978) 'Local differences in the patterns of variability of tropical rainfall: some characteristics and implications', *Journal of Hydrology*, 38, 273–87.

Jackson, I. J. (1981) 'Dependence of wet and dry days in the tropics', *Archiv für Meteorologie, Geophysik und Bioklimatologie*, Ser. B, 29, 167–79.

Jackson, I. J. (1982) 'Traditional forecasting of tropical rainy seasons'. *Agricultural Meteorology*, 26, 167–78.

Jackson, I. J. (1985) 'Tropical rainfall variability as an environmental factor: some considerations', *Singapore Journal of Tropical Geography*, 6(1), 23–34.

Jackson, I. J. (1986a) 'Relationships between raindays, mean daily rainfall intensity and monthly rainfall in the tropics', *Journal of Climatology*, 6, 117–34.

Jackson, I. J. (1986b) 'Tropical rainfall and surface water' in *Developing World Water*, ed. by WEDC, Grosvenor Press International, pp. 42–3.

Jaw-Kai Wang and Hagen, R. E. (1981) *Irrigated Rice Production Systems: Design Procedures*, Westview Tropical Agriculture Series No. 3, pp. 300.

Jenkinson, A. F. (1973) 'A note on variations in May to September rainfall in West African marginal rainfall areas', in Dalby and Harrison Church, *Report of the 1973 Symposium Drought in Africa*.

Johnson, D. H. (1962) 'Rain in East Africa', *Quarterly Journal of the Royal Meteorological Society*, 88, 1–21.

Jones, C. A. (1981) 'Effect of drought stress on percentage filled grains in upland rice', *Tropical Agriculture*, 58, 201–3.

Jones, C. A., Pena, D. and Carabaly, A. (1980) 'Effects of plant water potential, leaf diffusive resistance, rooting density and water use on the dry matter production of several tropical grasses during short periods of drought stress', *Tropical Agriculture*, 57, 211–19.

Jordan, C. L. (1980) 'Diurnal variation of precipitation in the eastern tropical Atlantic', *Monthly Weather Review*, 108, 1065–9.

Jowett, D. and Eriaku, P. O. (1966) 'The relationship between sunshine, rainfall and crop yields at Serere Research Station', *East African Agricultural and Forestry Journal*, 31 (4), 439–40.

Jury, W. A. (1979) 'Water transport through soil, plant and atmosphere', in *Agriculture in Semi-Arid Environments*, eds A. H. Hall, G. H. Cannell and H. W. Lawton, Springer-Verlag, Berlin, pp. 180–97.

Juvik, J. O. and Perreira, D. J. (1974) 'Fog interception on Mauna Loa, Hawaii', *Proceedings, Association of American Geographers*, 6, 22–5.

Kamara, C. S. (1981) 'Effects of planting date and mulching on cowpea in Sierra Leone', *Experimental Agriculture*, 17, 25–31.

Kashef, A-A.I. (1981a) 'Technical and ecological impacts of the High Aswan

Dam', *Journal of Hydrology*, **53**, 73–84.

Kashef, A-A. I. (1981b) 'The Nile – one river and nine countries', *Journal of Hydrology*, **53**, 53–71.

Keating, R. (1972) *The Aswan High Dam and its Effects on the Environment*, Open University Technology Foundation Course Unit 23.

Keen, C. S. and **Tyson, P. D.** (1973) 'Seasonality of South African rainfall: a note on its regional delimitation using spectral analysis', *Archiv für Meteorologie Geophysik und Bioklimatologie, Ser.* B, **21**, 207–14.

Keenan, T. D. (1986) 'Forecasting tropical cyclone motion using a discriminant analysis procedure', *Monthly Weather Review*, **114**, 434–41.

Keller, R. (ed.) (1983) *Hydrology of Humid Tropical Regions*, IAHS Publication No. 140, pp. 468.

Kenworthy, J. M. and **Glover, J.** (1958) 'The reliability of the main rains in Kenya', *East African Agricultural Journal*, **23**, 267–72.

Khalifa, F. M. (1981) 'Some factors influencing the development of sunflower (*Helianthus annuus* L.) under dry-farming systems in Sudan', *Journal of Agricultural Science*, **97**, 45–53.

King, H. E. (1957) 'Cotton yields and weather in Northern Nigeria', *Empire Cotton Growers' Review*, **34**, 153–4.

Kininmonth, W. R. (1983) 'Variability of rainfall over northern Australia', in *Variations in the Global Water Budget*, ed. A. Street Perrot *et al.*, D. Reidel, 265–72.

Kishihara, N. and **Gregory, S.** (1982) 'Probable rainfall estimates and the problems of outliers', *Journal of Hydrology*, **58**, 341–56.

Kittock, D. L., Henneberry, T. J. and **Bariola** (1981) 'Fruiting of upland and pima cotton with different planting dates', *Agronomy Journal*, **73**, 711–15.

Kneebone, W. R. and **Pepper, I. L.** (1982) 'Consumptive water use by subirrigated turfgrasses under desert conditions', *Agronomy Journal*, **74**, 419–23.

Koehler, P. H. *et al.* (1982) 'Response of drip-irrigated sugarcane to drought stress', *Agronomy Journal*, **74**, 906–11.

Kotoda, K. (1986) *Estimation of River Basin Evapotranspiration*, Environmental Research Center Papers, The University of Tsukuba, Ibaraki, Japan, No. 8, pp. 66.

Kousky, V. E. (1980) 'Diurnal rainfall variation in northeast Brazil', *Monthly Weather Review*, **108**, 488–98.

Kousky, V. E. (1985) 'Atmospheric circulation changes associated with rainfall anomalies over tropical Brazil', *Monthly Weather Review*, **113**, 1951–7.

Kramer, P. J. (1963) 'Water stress and plant growth', *Agronomy Journal*, **5**, 31–5.

Kramer, P. J. (1969) *Plant and Soil Water Relationships*, McGraw-Hill, pp. 482.

Kramer, P. J. (1983) *Water Relations of Plants*, Academic Press, pp. 489.

Krantz, B. A. (1981a) 'Rainfall collection and utilization in the semi-arid tropics', in *Rainfall Collection for Agriculture in Arid and Semi Arid Regions*, eds G. R. Dutt, C. F. Hutchinson and M. A. Garduno, Commonwealth Agricultural Bureau, pp. 53–9.

Krantz, B. A. (1981b) 'Water conservation, management and utilization in semi-arid lands', in *Advances in Food Producing Systems for Arid and Semi Arid Lands*, eds J. T. Manassah and E. J. Briskey, Academic Press, New York, pp. 339–78.

Kraus, E. B. (1955) 'Secular changes of tropical rainfall regimes', *Quarterly Journal of the Royal Meteorological Society*, **81**, 198–210.

Kraus, E. B. (1958) 'Recent climatic changes', *Nature*, **181**;(4610), 666–8.

Krishnamurthy. Ch. (1979) 'Rainfed lowland rice – problems and opportunities', in *Rainfed Lowland Rice*, IRRl Conference, Los Banos, Philippines, p. 61.

Krishnamurti. T. N. (1979) 'Tropical meteorology' in *Compendium of meteorology for use by Class I and Class II meteorological personnel'*. Vol. II Part 4. WMO Publication No. 364, pp. 428.

Krishnamurti, T. N. (1981) 'The present status of tropical meteorology', *Scientific lectures presented at the 8th World Meteorological Congress, Geneva, May, 1979*. WMO Publication No. 568, pp. 1–36.

Krishnan, A. (1980) 'Agroclimatic classification methods and their application to India', in *Climatic Classification in the Semi-Arid Tropics*, ICRISAT, pp. 59–88.

Kumaraswamy, Es. P. (1973) 'Retardation of evaporation from open water storages', in Ackerman, White and Worthington, *Man Made Lakes: Their Problems and Environmental Effects*, pp. 278–82. American Geophysical Union, Washington.

Kung, P. (1966) 'Desirable techniques and procedures on water management of rice culture', *Mechanisation and the World's Rice Conference*, Massey-Ferguson/FAO, Stoneleigh, England.

Kyaw Tha Paw, U. and Gueye, M. (1983) 'Theoretical and measured evaporation rates from an exposed Piche atmograph', *Agricultural Meteorology*, **30**, 1–11.

Lahey, J. F. (1958) *On the Origin of the Dry Climate in Northern South America and the Southern Caribbean*, Department of Meteorology Monograph, University of Wisconsin.

Lal, R. (1976) 'Soil erosion on Alfisols in western Nigeria, III. Effects of rainfall characteristics', *Geoderma*, **16**, 389–401.

Lal, R. (1983) 'Soil erosion in the humid tropics with particular reference to agricultural land development and soil management', in *Hydrology of Humid Tropical Regions*, ed. R. Keller, IAHS Publication No. 140, pp, 221–39.

Lal, R. (1984) 'Mulch requirements for erosion control with the no-till system in the tropics: a review', in *Challenges in African Hydrology and Water Resources*, eds D. E. Walling, S. S. D. Foster and P. Wurzel, IAHS Publication No. 144, pp. 475–83.

Lal, R. and Russel, E. W. (eds) (1981) *Tropical Agricultural Hydrology*, John Wiley, New York, pp. 482.

Lamb, H. (1973) 'Some comments on atmospheric pressure variations in the northern hemisphere', in Dalby and Harrison Church, *Report of the 1973 Symposium Drought in Africa*.

Lawes, D. A. (1961) 'Rainfall conservation and the yield of cotton in Northern Nigeria', *Empire Journal of Experimental Agriculture*, **29**, 307.

Lawes, D. A. (1966) 'Rainfall conservation and the yields of sorghum and groundnuts in Northern Nigeria', *Experimental Agriculture*, **2**(2), 139–46.

Laycock, D. H. and Wood, R. A. (1963) 'Some observations on soil moisture use under tea in Nyasaland (III)', *Tropical Agriculture*, **40**, 121–8.

Ledger, D. C. (1964) 'Some hydrological characterisitcs of West African rivers', *Transactions of the Institute of British Geographers*, **35**, 73–90.

Leentvaar, P. (1973) 'Lake Brokopondo', in Ackerman, White and Worthington, *Man-made Lakes: Their Problems and Environmental Effects*, pp. 186.

Lehane, J. J. and Staple, W. J. (1962) 'Effects of soil moisture tensions on growth of wheat', *Canadian Journal of Soil Science*, **42**, 180–8.

Leopold, L. B. (1949) 'The interaction of trade wind and sea breeze, Hawaii', *Journal of Meteorology*, **6**, 312–20.

Levine. G., Oram, P. and Zapata, J. A. (1979) *'Water' Report Prepared for the Conference on Agricultural Production: Research and Development Strategies for the 1980s*, Bonn. Oct. 1979. Rockefeller Foundation, New York, pp. 105.

Lewis, L. A. (1985) 'Assessing soil loss in Kiambu and Muranga'a districts, Kenya', *Geografiska Annaler*, **67a**(3–4), 273–84.

Lighthill, J. and Pearce, R. P. (eds) (1981) *Monsoon Dynamics*, Cambridge University Press, Cambridge, pp. 735.

Lim, C. L. (1969) 'Storm rainfall study of Sungei Klang catchment', unpublished M.Sc. thesis, University of Hull.

Lineham, S. (1967) 'The duration and quality of the rainy season in Rhodesia', *First Rhodesian Scientific Congress, pp. 44–82*, Association of Scientific Studies in Rhodesia, Salisbury.

Linsley, R. K., Kohler, M. A. and Paulhus, J. L. H. (1982) *Hydrology for Engineers*, McGraw-Hill, 3rd edition, pp. 508.

Lloyd, C. R., Shuttleworth, W. J., Gash, J. H. C. and Turner, M. (1984) 'A microprocessor system for eddy-correlation', *Agricultural and Forest Meteorology*, **33**, 67–80.

Lo, A., El-Swaify, S. A., Dangler, E. W. and Shinshiro, L. (1985) 'Effectiveness of EI_{30} as an erosivity index in Hawaii', in *Soil Erosion and Conservation*, eds S. A. El-Swaify, W. C. Moldenhauer and A. Lo, Soil Conservation Society of America, pp. 384–92.

Lockwood, J. G. (1966) '700 mb contour charts for South-East Asia and neighbouring areas', *Weather*, **21**, 325.

Lockwood, J. G. (1967) 'Probable maximum 24 hour precipitation over Malaya by statistical methods', *Meteorological Magazine*, **96**, 11.

Lockwood, J. G. (1968) 'Extreme Rainfalls', *Weather*, **23**(7), 284–9.

Lockwood, J. G. (1974) *World Climatology: An Environmental Approach*, Edward Arnold, London.

Lockwood, J. G. (1984) 'The S. O. and El Nino', *Progress in Physical Geography*, **8**, 102–10.

Lockwood, J. G. (1985) *World Climate Systems*, Edward Arnold, London, pp. 292.

Lockwood, J. G. and Sellers, P. J. (1982) 'Comparisons of interception loss from tropical and temperate vegetation canopies', *Journal of Applied Meteorology*, **21**, 1405–12.

Lomas, J. and Herrera, H. (1984) 'Weather and maize yield relationships in the tropical region of Guanacaste, Costa Rica', *Agricultural and Forest Meteorology*, **31**, 33–45.

Lourensz, R. S. (1981) *Tropical Cyclones in the Australian Region, July 1909 to June 1980*, Bureau of Meteorology, pp. 94.

Love, G. (1985a) 'Cross-equatorial influence of winter hemisphere subtropical cold surges', *Monthly Weather Review*, **113**, 1487–98.

Love, G. (1985b) 'Cross-equatorial interactions during tropical cyclone genesis,' *Monthly Weather Review*, **113**, 1499–509.

Low Kwai Sim and **Goh Kim Chuom** (1972) 'The water balance of five catchments in Selangor, west Malaysia', *Journal of Tropical Geography*, **35**, 60–6.

Lu, A. (1954) *Chinese Climatology. Collected Scientific Papers (Meteorology)*, Academica Sinica (1944), Peking.

Lumb, F. E. (1966) 'Cycles and trends of rainfall over East Africa', *Kenya Coffee*.

Lumb, F. E. (1970) 'Topographic influences on thunderstorm activity near Lake Victoria', *Weather*, **25**(9), 404–10.

Lumb, F. E. (1971) 'Probable maximum precipitation (PMP) in East Africa for durations up to 24 hours', *East African Meteorological Department*. Technical Memorandum No. 16.

Lyons, S. W. (1982) 'Empirical orthogonal function analysis of Hawaiian rainfall', *Journal of Applied Meteorology*, **21**, 1713–29.

Macartney, J. C. and **Northwood, P. J.** (1969) 'Trials with direct drilling of wheat', Ministry of Agriculture Library, Dar es Salaam (unpublished).

Macartney, J. C., Northwood, P. J., Dagg, M. and **Dawson, R.** (1971) 'The effect of different cultivation techniques on soil moisture conservation and the establishment and yield of maize at Kongwa, Central Tanzania', *Tropical Agriculture*, **48**(1), 9–24.

McBride, J. L. (1984) 'Observational studies of the Australian northwest monsoon', *15th Conference on Hurricanes and Tropical Meteorology*, American Meteorological Society, Boston, pp. 406–9.

McBride, J. L. and **Nicholls, N.** (1983) 'Seasonal relationships between Australian rainfall and the southern oscillation', *Monthly Weather Review*, **III**, 1998–2004.

Mahadevan, P. (1982) 'Pastures and animal production', in *Nutritional Limits to Animal Production from Pastures*, ed. J. B. Hacker, Commonwealth Agricultural Bureau, pp. 1–17.

McCulloch, J. S. G. and **Dagg, M.** (1965) 'Hydrological aspects of protection forestry in East Africa', *East African Agricultural and Forestry Journal*, **30**, 390–4.

McGowan, M. and **Williams, J. B.** (1980a) 'The water balance of an agricultural catchment. I. Estimation of evaporation from soil water records', *Journal of Soil Science*, **31**, 217–30.

McGowan, M. and **Williams, J. B.** (1980b) 'The water balance of an agricultural catchment. II. Crop evaporation: Seasonal and soil factors', *Journal of Soil Science*, **31**, 231–44.

McIntosh, D. H. and **Thom, A. S.** (1969) *Essentials of Meteorology*, Wykeham Publications, London, pp. 238.

McMahon, T. A. (1979) 'Hydrological characteristics of arid zones', in *The Hydrology of Areas of Low Precipitation*, IAHS Publication No. 128, pp. 105–24.

McNaughton, D. L. (1971) 'Calendar singularities in Rhodesian precipitation and the implications', *Journal of Applied Meteorology*, **10**(3), June, 498–501.

McQuate, G. T. and **Hayden, B. P.** (1984) 'Determination of intertropical convergence zone rainfall in northeastern Brazil using infrared satellite imagery', *Archiv fur Meteorologic, Geophysik und Bioklimatologie*, Ser. B, **34**, 319–28.

Mahalakshmi, V. and **Bidinger, F. R.** (1985) 'Water stress and time of floral initiation in pearl millet', *Journal of Agricultural Science*, **105**, 437–45.

Mannetje, L't. (1982) 'Problem of animal production from tropical pastures', in *Nutritional Limits to Animal Production from Pastures*, ed. J. B. Hacker, Commonwealth Agricultural Bureau, pp. 67–85.

Manning, H. L. (1956) 'The statistical assessment of rainfall probability and its application in Uganda agriculture', *Proceedings of the Royal Society* **B144**, 460–80.

Manning, H. L. and **Kibukamusoke, D. E. B.** (1960) 'The cotton crop', *Progress Reports from Experimental Stations 1958–59 Uganda, Empire Cotton Growing Corporation*, pp. 6–8. London.

Manubag, J. M. (1985) 'Rainfall interception, surface runoff and sedimentation of dipterocarp stand, mixed forest and grassland in the CMU forest reservation', *CMU Journal of Agriculture, Food and Nutrition*, **7**, 43–54.

Marani, A. and **Amirav, A.** (1971) 'Effects of soil moisture stress on 2 varieties of upland cotton in Israel (III) the Bet-She'an Valley', *Experimental Agriculture*, **7**, 289–301.

Marani, S. and **Fuchs, Y.** (1964) 'Effect of the amount of water applied as a single irrigation on cotton grown on dryland conditions', *Agronomy Journal*, **56**(3), 281–2.

Marani, A. and **Horwitz, M.** (1963) 'Growth and yield of cotton as affected by the time of a single irrigation', *Agronomy Journal*, **55**, 219–22.

Marshall, T. J. (1959) 'Relations between water and soil', Technical Communication No. 50, Commonwealth Bureau of Soils, Harpenden Commonwealth Agricultural Bureau.

Mascarenhas, A. C. (1968) 'Aspects of food shortages in Tanganyika (1925–45)', *Journal of the Geographical Association of Tanzania*, **3**, 37–59.

Mathews, M. A. and **Boyer, J. S.** (1984) 'Acclimation of photosynthesis to low leaf water potentials', *Plant Physiology*, **74**, 161–6.

Matsushima, S. (1966) 'Some experiments and investigation on rice plants in relation to water in Malaysia', *Symposium Service Centre for South-East Asian Studies*, Kyoto University, **3**, 115–23.

Mattei, F. (1979) 'Climatic variability and agriculture in the semi-arid tropics', in *Proceedings of the World Climate Conference*, Geneva, WMO Publication No. 537, pp. 475–509.

Meher-Homji, V. M. (1974) 'Variability and the concept of a probable climatic year in bioclimatology with reference to the Indian sub-continent', *Archiv für Meteorologie, Geophysik und Bioclimatologie*, Ser. B, **22**(1–2), 149–68.

Meher-Homji, V. M. (1980) 'Classification of the semi-arid tropics: climatic and phytogeographic approaches', in *Climatic Classification in the Semi-Arid Tropics*, ICRISAT, pp. 7–16.

Meisner, B. N. (1983) 'Time-extrapolated rainfall normals for central equatorial Pacific islands', *Journal of Climate and Applied Meteorology*, **22**, 440–6.

Meisner, B. N. and Arkin, P. A. (1985) 'Spatial and temporal variability in large scale tropical convective precipitation', *16th Conference on Hurricanes and Tropical Meteorology*, American Meteorological Society, Boston, pp. 88–9.

Melice, J. L. and Wendler, G. (1984) 'Precipitation statistics in southern Tunisia – a contribution to the desertification problem in the Sahel zone', *Archiv fur Meteorologie, Geophysik und Bioklimatologie*, Ser. B, **33**, 331–48.

Mergen, F. (ed.) (1981) *Tropical Forests: Utilization and Conservation*, Yale School of Forestry and Environmental Studies, New Haven, pp. 199.

Merrill, R. T. (1984) 'A comparison of large and small tropical cyclones', *Monthly Weather Review*, **112**, 1408–18.

Meyer, W. S. and Ritchie, J. T. (1980) 'Water status of cotton as related to taproot length', *Agronomy Journal*, **72**(4), 577–80.

Michaels, P. J. (1982) 'The response of the "green revolution" to climatic variability', *Climatic Change*, **4**, 255–71.

Miller, E. C. (1938) *Plant Physiology*. McGraw Hill, New York & London.

Milton, D. (1980) 'The contribution of tropical cyclones to the rainfall of tropical Western Australia', *Singapore Journal of Tropical Geography*, **1**(1), 46–54.

Mink, J. F. (1960) 'Distribution pattern of rainfall in the leeward Koolan Mountains, Oahu, Hawaii', *Journal of Geophysical Research*, **65**, 2869–76.

Mislevy, P. and Everett, P. H. (1981) 'Subtropical grass species response to different irrigation and harvest regimes', *Agronomy Journal*, **73**, 601–4.

Mohr, E. C. J. and Van Baren, F. A. (1959) *Tropical Soils*, Interscience Publishers Ltd under the Auspices of the Royal Tropical Institute, Amsterdam.

Mohr, E. C. J., Van Baren, F. A. and Van Schuylenborgh, J. (1972) *Tropical Soils*, Mouton-Ichtiar Baru-Van Hoeve, pp. 481.

Monteith, J. L. (1973) *Principles of Environmental Physics*, Edward Arnold, pp. 241.

Monteny, B. A., Humbert, J., L'homme, J. P. and Kalms, J. M. (1981) 'Le rayonnement net et l'estimation de l'evapotranspiration en Cote D'Ivoire', *Agricultural Meteorology*, **23**, 45–59.

Mooley, D. A. (1971) 'Independence of monthly and bimonthly rainfall over S.E. Asia during the summer monsoon season', *Monthly Weather Review*, **99**(6), 532–6.

Mooley, D. A. and Appa Rao, G. (1971) 'Distribution function for seasonal and annual rainfall over India', *Monthly Weather Review*, **99**(10), 796–9.

Mooley, D. A. and Parthasarathy, B. (1983) 'Indian summer monsoon and El Nino', *Pageoph*, **121**(2), 339–52.

Mooley, D. A. and Parthasarathy, B. (1984) 'Fluctuations in all-India summer monsoon rainfall during 1871–1978', *Climatic Change*, **6**, 287–301.

Mooley, D. A., Parthasarathy, B. and Sontakke, N. A. (1982) 'An index of summer monsoon rainfall excess over India and its variability', *Archiv fur Meteorologie, Geophysic und Bioklimatologie*, Ser. B, **31**, 301–11.

Mooley, D. A., Parthasarathy, B. and Sontakke, N. A. (1985) 'Relationships between all-India summer monsoon rainfall and the southern oscillation/eastern equatorial Pacific sea surface temperature', *Proc. Indian Academy of Science (Earth Planet. Sc.)* **94**(3), 199–210.

Mooley, D. A., Parthasarathy, B.. Sontakke, N. A. and Munot, A. A. (1981) 'Annual rain-water over India, its variability and impact on the economy'. *Journal of Climatology*, **1**, 167–86.

Moore, R. M. (ed.) (1973) *Australian Grasslands*, Australian National U.P., Canberra.

Moore, P. H. and Osgood, R. V. (1985) 'Assessment of sugarcane crop damage and yield loss caused by high winds of hurricanes', *Agricultural and Forest Meteorology*, **35**, 267–79.

Morel-Seytoux, H. J. *et al.* (eds) (1979) *Modelling Hydrologic Processes*, Water Resources Publications, Fort Collins, Colorado, pp. 818.

Morgan. R. P. C. (1979) *Soil Erosion*, Longman, pp. 113.

Morgan, R. P. C. (ed.) (1981) *Soil Conservation: Problems and Prospects*, J. Wiley and Sons, pp. 576.

Morris, R. A. and Rumbaoa, F. M. (1980) 'Rainfall recurrence analysis for extrapolating rice-based cropping patterns', in WMO/IRRI *Symposium on the Agrometeorology of the Rice Crop*, Los Banos, Philippines, pp. 221–33.

Morth, H. T. (1967) 'Investigations into the meteorological aspects of the variations in the level of Lake Victoria', *East African Meteorological Department Memoirs*, **4**(2).

Morth, H. T. (1970) 'A study of the areal and temporal distribution of rainfall anomalies in East africa', *Proceedings of the Symposium on Tropical Meteorology*, Honolulu. American Meteorological Society, World Meteorological Organisation, Hawaiian Institute of Geophysics.

Morton, A. J. and Obot, E. A. (1984) 'The control of *Echinochloa Stagnina* (Retz.) P. Beauv by harvesting for dry season livestock fodder in Lake Kainji, Nigeria – a modelling approach', *Journal of Applied Ecology*, **21**, 687–94.

Morton, F. I. (1983) 'Operational estimates of areal evapotranspiration and their significance to the science and practice of hydrology', *Journal of Hydrology*, **66**, 1–76.

Morton, H. L. (1985) 'Plants for conservation of soil and water in arid ecosystems', in *Plants for Arid Lands*, eds G. E. Wickens, J. R. Goodwin and D. V. Field, Allen and Unwin, London, pp 203–14.

Motha, R. P., Leduc, S. K., Steyaert, L. T., Sakamoto, C. M. and Strommen, N. D. (1980) 'Precipitation patterns in West Africa', *Monthly Weather Review*, **108**, 1567–78

Munn, R. E. and Machta, L. (1979) 'Human activities that affect climate', in *Proceedings of the World Climate Conference*, WMO, 12–23 Feb. 1979, WMO Publication No. 537, pp. 170–209.

Munro, J. M. and Wood, R. A. (1964) 'Water requirements of irrigated maize in Nyasaland', *Empire Journal of Experimental Agriculture*, **32**(126), 141–52.

Murphy, A. H. and Katz, R. W. (eds) (1985) *Probability, Statistics and Decision Making in the Atmospheric Sciences*, Westview Press, pp. 545.

Myers, L. E. (1967) 'New water supplies from precipitation harvesting', *Proceedings of the International Conference Water for Peace*, Vol. 2.

Nagel, J. F. (1956) 'Fog precipitation on Table Mountain', *Quarterly Journal of the Royal Meteorological Society*, **82**, 452–60.

Nageswara Rao *et al.* (1985) 'Effect of water deficit at different growth phases of peanut. I. Yield responses', *Agronomy Journal*, **77**, 782–6.

Naqvi, S. N. (1958) 'Periodic variations in water balance in an arid region. A preliminary survey of 100 years' rainfall at Karachi', *Climatology and Microclimatology*, Proceedings of the Canberra Symposium, Paris, UNESCO, pp. 326–45.

National Academy of Sciences (1980) *Conversion of Tropical Moist Forests*, National Academy of Sciences, Washington DC, pp. 205.

Nemec, J. (1973) 'Summary: interaction between reservoirs and the atmosphere and its hydrometeorological elements', in Ackerman, White and Worthington, *Man Made Lakes: Their Problems and Environmental Effects*, pp. 398–405.

Newell, R. E., Selkirk, R. and **Ebisuzaki, W.** (1982) 'The southern oscillation: sea surface temperature and wind relationships in a 100-year data set', *Journal of Climatology*, **2**, 357–73.

Newton, O. H. O. and **Riley, J. A.** (1964) 'Dew in the Mississippi Delta in the fall', *Monthly Weather Review*, **92**, 369–73.

Nicholls, N. (1984a) 'A system for predicting the onset of the north Australian wet-season', *Journal of Climatology*, **4**, 425–35.

Nicholls, N. (1984b) 'The southern oscillation and Indonesian sea surface temperature', *Monthly Weather Review*, **112**, 424–32.

Nicholls, N. (1985) 'Predictability of interannual variations of Australian seasonal tropical cyclone activity', *Monthly Weather Review*, **113**, 1144–9.

Nicholls, N. McBride, J. L. and **Ormerod, R. J.** (1982) 'On predicting the onset of the Australian wet season at Darwin', *Monthly Weather Review*, **110**, 14–17.

Nicholson, S. E. (1980) 'The nature of rainfall fluctuations in subtropical West Africa', *Monthly Weather Review*, **108**, 473–87.

Nicholson, S. E. (1981) 'Rainfall and atmospheric circulation during drought periods and wetter years in West Africa', *Monthly Weather Review*, **109**, 2191–208.

Nicholson, S. E. (1983) 'Sub-Saharan rainfall in the years 1976–80: Evidence of continued drought', *Monthly Weather Review*, **111**, 1646–53.

Nicholson, S. E. and **Chervin, R. M.** (1983) 'Recent rainfall fluctuations in Africa – interhemispheric teleconnections', in *Variations in the Global Water Budget*, eds A. Street-Perrott *et al.*, D. Reidel, pp. 221–38.

Nieuwolt, S. (1968) 'Uniformity and variation in an equatorial climate', *Journal of Tropical Geography*, **27**, 23–39.

Nobre, C. A. and **Renno, N. de O.** (1985) 'Droughts and floods in South America due to the 1982–83 El Nino/S.O. episode'. *16th Conference on Hurricanes and Tropical Meteorology*, American Meteorological Society, Boston, pp. 131–3.

Nobre, P., Nobre, C. A. and **Moura, A. D.** (1985) 'Large scale circulation anomalies and prediction of north east Brazil droughts'. *16th Conference on Hurricanes and Tropical Meteorology*, American Meteorological Society, Boston, pp. 128–30.

Norman, M. J. T., Pearson, C. J. and Searle, P. G. E. (1984) *The Ecology of Tropical Food Crops*, Cambridge University Press, pp. 369.

Novero, R. P., O'Toole, J. C., Cruz, R. T. and Garrity, D. P. (1985) 'Leaf water potential, crop growth response and microclimate of dryland rice under line source sprinkler irrigation', *Agricultural and Forest Meteorology*, 35, 71–82.

Obeng, L. E. (1973) 'Volta Lake: physical and biological aspects', in Ackerman, White and Worthington, *Man Made Lakes: Their Problems and Environmental Effects*, pp. 87–97.

Ochse, J. J., Soule, M. J., Dijkman, M. J. and Wehlburg, C. (1961) *Tropical and Sub-Tropical Agriculture*, Vols. I and II, p. 1446. New York, Macmillan.

O'Connell, P. E. (1982) 'Raingauge network design – a review', in *Modelling Components of Hydrologic Cycle*, ed. V. P. Singh, Water Resources Publications, Littleton, Colorado, pp. 13–50.

Odumodu, L. O. (1983) 'Rainfall distribution, variability and probability in Plateau State, Nigeria', *Journal of Climatology*, 3, 385–93.

Oelke, E. A. and Mueller, K. E. (1969) 'Influences of water management and fertility on rice growth and yield', *Agronomy Journal*, 61, 227–30.

Ofori-Sarpong, E. (1983). 'The drought of 1970–77 in Upper Volta', *Singapore Journal of Tropical Geography*, 4(1), 53–61.

Ogallo, L. (1984) 'Variation and change in climate in sub-Saharan Africa', in *Advancing Agricultural Production in Africa*, Commonwealth Agricultural Bureau, pp. 308–12.

Oguntoyinbo, J. S. (1966) 'Evapotranspiration and sugar cane yields in Barbados', *Journal of Tropical Geography*, 22, 38–48.

Ogunkoya, O. O., Adejuwon, J. O. and Jeje, L. K. (1984) 'Runoff response to basin parameters in southwestern Nigeria', *Journal of Hydrology*, 72, 67–84.

Oguntoyinbo, J. S. and Akintola, F. O. (1983) 'Rainstorm characteristics affecting water availability for agriculture', in *Hydrology of Humid Tropical Regions*, ed. R. Keller, IAHS, Publication No. 140, pp. 63–72.

Oguntoyinbo, J. S. and Odingo, R. S. (1979) 'Climatic variability and land use', *Proceedings of the World Climate Conference*, WMO Publication No. 537, pp. 552–80.

Ojo, O. (1983) 'Recent trends in aspects of hydroclimatic characteristics in West Africa', in *Hydrology of Humid Tropical Regions*, ed. R. Keller, IAHS Publication No. 140, pp. 97–104.

Oldeman, L. R. and Frere, M. (1982) *A Study of the Agroclimatology of the Humid Tropics of Southeast Asia*, WMO Technical Note No. 179, WMO Publication No. 597.

Oliver, J. (1965) 'Evaporation losses and rainfall regime in central and northern Sudan', *Weather*, 20, 58–64.

Oliver, J. (1968) 'Problems of the arid lands: the example of the Sudan' in *Land Use and Resources: Studies in Applied Geography*, Institute of British Geographers, Special Publication No. 1, pp. 219–39.

Oliver, J. (1969) 'Problems of determining evapotranspiration in the semi-arid tropics illustrated with reference to the Sudan', *Journal of Tropical Geography*, 28, 64–74.

Omar, M. H. and El-Bakry, M. M. (1981) 'Estimation of evaporation from the lake of the Aswan High Dam (Lake Nasser) based on measurements over the Lake', *Agricultural Meteorology*, 23, 293–308.

Omar, M. H. and **Mehanna, A. M.** (1984) 'Measurements and estimates of evapotranspiration over Egypt', *Agricultural and Forest Meteorology*, 31, 117–29.

Omotosho, J. B. (1985) 'The separate contributions of line squalls, thunderstorms and the monsoon to the total rainfall in Nigeria', *Journal of Climatology*, 5, 543–52.

Ongweny, G. S. (1979) 'Patterns of sediment production within the Upper Tana basin in eastern Kenya', in *The Hydrology of Areas of Low Precipitation*, IAHS Publication No. 128, pp. 447–57.

Opeke, L. K. (1982) *Tropical Tree Crops*, John Wiley and Sons, pp. 312.

Orchard, A. Q. and **Sumner, G. N.** (1970) 'East African rainfall project', Network Report No. 4.

Othieno, C. O. (1980) 'Effects of mulches on soil water content and water status of tea plants in Kenya', *Experimental Agriculture*, 16, 295–302.

O'Toole, J. C. and **Cruz, R. T.** (1980) 'Response of leaf water potential, stomatal resistance and leaf rolling to water stress', *Plant Physiology*, 65, 428–32.

O'Toole, J. C. and **Tomar, V. S.** (1982) 'Transpiration, leaf temperature and water potential of rice and barnyard grass in flooded fields', *Agricultural Meteorology*, 26, 285–96.

Oyebande, L. *et al.* (1980) 'The effect of Kainji Dam on the hydrological regime, water balance and water quality of the River Niger', in *The Influence of Man on the Hydrological Regime with Special Reference to Representative and Experimental Basins*, IAHS, Publ. No.130, pp. 221–8.

Palhaus, J. L. H. (1965) 'Indian Ocean and Taiwan rainfalls set new records', *Monthly Weather Review*, 93, 331–335.

Palmen, E. (1951) 'The role of atmospheric disturbances in the general circulation', *Quarterly Journal of the Royal Meteorological Society*, 77, 337–54.

Pande, H. K. and **Panjab Singh** (1970) 'Water and fertility management of rice varities under low atmospheric evaporative demand', *Journal of Agricultural Science*, 75, 61–7.

Pao-Shin Chu (1983) 'Diagnostic studies of rainfall anomalies in northeast Brazil', *Monthly Weather Review*, 111, 1655–4.

Parfait, J. A. and **Lallmahomed, H.** (1980) 'The effects of change in land use on the hydrological regime of three small basins in Mauritius', in *The Influence of Man on the Hydrological Regime with Special Reference to Representative and Experimental Basins*, IAHS, Publ. No. 130, pp. 351–8.

Parker, M. B., Marchant, W. H. and **Mullinix, B. J.** (1981) 'Date of planting and row spacing effects on four soybean cultivars', *Agronomy Journal*, 73(5), 759–62.

Parthasarathy, B. (1984) 'Interannual and long term variability of Indian summer monsoon rainfall', *Proceedings Indian Academy of Sciences (Earth and Planetary Sciences)*, 93(4), 371–85.

Parthasarathy, B. and **Pant, G. B.** (1985) 'Seasonal relationships between Indian summer monsoon rainfall and the southern oscillation', *Journal of Climatology*, 5, 369–78.

Pasch, R. J. (1984) 'The influence of deep cumulus convection on the onset of the planetary scale monsoon,' *15th Conference on Hurricanes and Tropical Meteorology*, American Meteorological Society, Boston, pp. 367–9.

Pathak, P. C., Pandey, A. N. and **Singh, J. S.** (1984) 'Overland flow, sediment output and nutrient loss from certain forested sites in the central Himalaya, India', *Journal of Hydrology*, **71**, 239–51.

Pathak, P. C., Pandey, A. N. and **Singh J. S.** (1985) 'Apportionment of rainfall in central Himalayan forests (India)', *Journal of Hydrology*, **76**, 319–32.

Patrick, W. H. (1967) 'Effect of continuous submergence versus alternate flooding and drying on growth yield and nitrogen uptake of rice', *Agronomy Journal*, **59**(5), 418–19.

Pazan, S. E. and **Meyers, G.** (1982) 'Interannual fluctuations of the tropical Pacific wind field and the southern oscillation', *Monthly Weather Review*, **110**, 587–600.

Peat, J. E. and **Brown, K. J.** (1960) 'Effect of management on increasing crop yields in the lake province of Tanganyika', *East African Agricultural Journal*, **26**, 103–9.

Pedgley, E. D. (1969) 'Cyclones along the Arabian coast', *Weather*, **24** (11), 456–69.

Peh, C. H. (1980) 'Runoff and sediment transport by overland flow under tropical rainforest conditions', *The Malaysian Forester*, **43**(1), 56–67.

Penman, H. L. (1963) *Vegetation and hydrology*, Commonwealth Bureau of Soils, Harpenden, Technical Communication **53**, 124.

Penman, H. L. *et al.* (1956) 'Discussions of evaporation', *Netherlands Journal of Agricultural Science*, **4**, 87–97.

Penman, H. L. and **Schofield, R. K.** (1951) 'Some physical aspects of assimilation and transpiration', *Symposia of the Society of Experimental Biology*, **5**, 115–29.

Pereira, H. C. (1952) 'Interception of rainfall by cypress plantations', *East African Agricultural Journal*, **18**, 73–6.

Pereira, H. C. (1965) 'Land use and stream flow', *East African Agricultural and Forestry Journal*, **30**, 395–7.

Pereira, H. C. (1967) 'Effects of land use on the water and energy budgets of tropical watersheds', in W. E. Sopper and H. W. Lull (eds), *International Symposium on Forest Hydrology*, Pergamon Press, Oxford, pp. 435–50.

Pereira, H. C. (1972) 'Influence of man on the hydrological cycle: guide to policies for the safe development of land and water resources', *Status and Trends of Research in Hydrology 1965–1974*, UNESCO, Paris.

Pereira, H. C. (1973) *Land Use and Water Resources*, Cambridge University Press, London.

Pereira, H. C. and **Hosegood, P. H.** (1962) 'Comparative water use of softwood plantations and bamboo forest', *Journal of Soil Science*, **13**, 229–314.

Pereira, H. C. *et al.* (1958) 'Water conservation by fallowing in semi-arid East Africa', *Empire Journal of Experimental Agriculture*, **26**, 213–28.

Pereira, H. C. *et al.* (1962) 'Hydrological effects of changes in land use in some East African catchment areas', *East African Agricultural and Forestry Journal*, Special Issue 27.

Petrossiants, M. A. (1981) 'The structure of the intertropical convergence zone in the light of GATE observations', *Scientific Lectures Presented at the 8th World Meteorological Congress*, Geneva, May 1979, WMO Publication No. 568, pp. 37–81.

Philander, S. G. H. (1985) 'El Nino and La Nina', *Journal of Atmospheric Sciences*, **42**(23), 2652–62.

Pierce, F. J. (1985) 'A systems approach to conservation tillage: Introduction', in *A Systems Approach to Conservation Tillage*, ed. F. M. D'Itri, Lewis, pp. 384.

Pinker, R. T., Thompson, O. E. and Eck, T. F. (1980) 'The energy balance of a tropical evergreen forest', *Journal of Applied Meteorology*, **19**(12), 1341–50.

Pittock, A. B. (1972) 'How important are climatic changes?', *Weather*, **27**, 262–71.

Pittock, A. B. (1984) 'On the reality, stability and usefulness of southern hemisphere teleconnections', *Australian Meteorological Magazine*, **32**, 75–82.

Pla, I., Florentino, A. and Lobo, D. (1985) 'Soil and water conservation problems in the Central Plains of Venezuela'. in *Soil Erosion and Conservation*, eds S. A. El-Swaify, W. C. Moldenauer and A. Lo, Publication of the Soil Conservation Association of America, pp. 66–77.

Pope, C. (1971) 'Tropical cyclone eye reformation after a 30-hour movement over the Malagasy Republic', *Monthly Weather Review*, **99**(6), 478–84.

Porter, P. W. (1984) 'Problems of agrometeorological modelling in Kenya', in *Agroclimate Information for Development*, ed. D. F. Cusak, Westview Press, Boulder, pp. 276–90.

Potts, A. S. (1971) 'Application of harmonic analysis to the study of East African rainfall data', *Journal of Tropical Geography*, **33**, 31–42.

Pramanik, S. K. and Jagannathan, P. (1954) 'Climatic changes in India', *IUGG 10th General Assembly*, International Union of Geodesy and Geophysics' Rome, p. 86.

Prasad, T. G. *et al.* (1985) 'Regulation of water loss under moisture stress in sunflower genotypes: stomatal sensitivity in relation to stomatal frequency, diffusive resistances and transpiration rate, at different canopy positions', *Journal of Agricultural Science*, **105**, 673–8.

Prothero, R. M. (1965) *Migrants and Malaria*, Longman, London.

Quansah, C. (1981) 'The effect of soil type, slope, rain intensity and their interactions on splash detachment and transport', *Journal of Soil Science*, **32**, 215–24.

Radin, J. W. and Ackerson, R. C. (1981) 'Water relations of cotton plants under nitrogen deficiency. III. Stomatal conductance, photosynthesis and abscisic acid accumulation', *Plant Physiology*, **67**, 115–19.

Ragab, R. A. (1983) 'The effect of sprinkler intensity and energy of falling drops on soil surface sealing', *Soil Science*, **136**(2), 117–23.

Raghaven, K. (1967) 'Influence of tropical storms on monsoon rainfall in India', *Weather*, **22**(6), 250–5.

Raghavulu, P. and Singh, S. P. (1982) 'Effect of mulches and transpiration suppressants on yield, water use efficiency and uptake of nitrogen and phosphorus by sorghum under dryland conditions of north-western India', *Journal of Agricultural Science*, **98**, 103–8.

Rakhecha, P. R. and Kennedy, M. R. (1985) 'A generalised technique for the estimation of probable maximum precipitation', *Journal of Hydrology*, **78**, 345–59.

Rakhecha, P. R., Mandal, B. N. and Ramana Murthy, K. V. (1985) 'Analysis of hourly rainfall distribution of Karanja catchment', *Transactions of The Institute of Indian Geographers*, **7**(2), 95–103.

Ramage, C. S. (1964) 'Diurnal variation of summer rainfall of Malaya', *Journal of Tropical Geography*, **19**, 62–8.

Ramage, C. S. (1968) 'Problem of a monsoon ocean', *Weather*, **23** (1), 28–36.

Ramage, C. S. (1971) *Monsoon Meteorology*, Academic Press, New York & London.

Ramage, C. S. and Hori, A. M. (1981) 'Meteorological aspects of El Nino', *Monthly Weather Review*, **109** (9), 1827–35.

Randall Baker, R. (1973) 'The need for long term strategies in areas of pastoral nomadism', in Dalby and Harrison Church, *Report of the 1973 Symposium Drought in Africa*.

Rao, K. L. (1968) 'Hydro power in India', in B. C. Laws (ed.), *Mountains and Rivers of India*, Indian National Committee for Geography, Calcutta, pp. 304–18.

Rao, K. L. and Palta, B. R. (1973) 'Great man-made lake of Bhakra, India', in Ackerman, White and Worthington, *Man Made Lakes: Their Problems and Environmental Effects*.

Rao, K. N. (1963) 'Climatic changes in India', *Arid Zone Research*, 20 Changes in Climate, Rome Symposium, pp. 49–66.

Rasmusson, E. M. (1984) 'Meteorological aspects of El Nino/Southern Oscillation'. *15th Conference on Hurricanes and Tropical Meteorology*, American Meteorological Society, Boston, pp. 1–9.

Rasmusson, E. M. and Carpenter, T. H. (1983) 'The relationship between eastern equatorial Pacific sea surface temperatures and rainfall over India and Sri Lanka', *Monthly Weather Review*, **111**, 517–28.

Rasmusson, E. M. and Wallace, J. M. (1983) 'Meteorological aspects of El Nino/Southern Oscillation', *Science*, **222**, 1195–202.

Raudkivi, A. J. (1979) *Hydrology: An Advanced Introduction to Hydrological Processes and Modelling*, Pergamon Press, Oxford, pp. 479.

Rawlins, S. L. (1981) 'Principles of salinity control in irrigated agriculture', in *Advances in Food Producing Systems for Arid and Semi Arid Areas*, eds J. T. Manassah and E. J. Briskey, Academic Press, pp. 391–420.

Reddy, S. J. (1983a) 'Agroclimatic classification of the semi-arid tropics. I. A method for the computation of classificatory variables', *Agricultural Meteorology*, **30**, 185–200.

Reddy, S. J. (1983b) 'Agroclimatic classification of the semi-arid tropics. II. Identification of classificatory variables', *Agricultural Meteorology*, **30**, 201–19.

Reddy, S. J. (1983c) 'A simple method of estimating the soil–water balance', *Agricultural Meteorology*, **28**, 1–17.

Reddy, S. J. (1984a) 'Agroclimatic classification of the semi-arid tropics. III. Characteristics of variables relevant to crop production potential', *Agricultural Meteorology*, **30**, 269–92.

Reddy, S. J. (1984b) 'Agroclimatic classification of the semi-arid tropics. IV. Classification of India, Senegal and Upper Volta', *Agricultural Meteorology*, 30, 293–325.

Reed, R. J. (1970) 'Structure and characteristics of easterly waves in the equatorial western Pacific during July–August 1967', *Proceedings of the Symposium of Tropical Meteorology*, American Meteorological Society, World Meteorological Organisation, Hawaiian Institute of Geophysics, Honolulu.

Reed, R. K. (1980) 'Comparison of ocean and island rainfall in the tropical north Pacific', *Journal of Applied Meteorology*, 19, 877–80.

Reed, R. K. (1982) 'Reply to C. E. Dorman (1982)', *Journal of Applied Meteorology*, 21, 114–15.

Regehr, D. L. (1982) 'Dinoseb and Triacontanol as growth regulators in irrigated and non irrigated field corn', *Agronomy Journal*, 74, 111–15.

Reich, B. M. (1963) 'Short duration rainfall intensity estimates and other design aids for regions of sparse data', *Journal of Hydrology*, 1, 3.

Reiter, E. (1983a) 'Teleconnections with tropical precipitation surges', *Journal of Atmospheric Sciences*, 40, 1631–47.

Reiter, E. R. (1983b) 'Surges of Tropical Pacific rainfall and teleconnections with extratropical circulation patterns', in *Variations in the Global Water Budget*, ed. A. Street-Perrott *et al.*, D. Reidel, pp. 285–99.

Rhodda, J. C. (1967) 'The rainfall measurement problem', *Proceedings of the Berne Assembly, International Association of Scientific Hydrology*, pp. 215–31. International Association of Scientific Hydrology, Gentbrugge.

Richards, L. A. and Wadleigh, C. H. (1952) 'Soil water and plant growth', in B. T. Shaw (ed.), *Soil Physical Conditions and Plant Growth*. Academic Press, New York.

Riedl, O. and Zachar, D. (eds) (1984) *Forest Amelioration*, Elsevier, pp. 623.

Riehl, H. (1949) 'Some aspects of Hawaiian rainfall', *Bulletin of the American Meteorological Society*, 30, 176–87.

Riehl, H. (1954) *Tropical Meteorology*, McGraw-Hill, New York.

Riehl, H. (1965) *Introduction to the Atmosphere*, McGraw Hill, New York.

Riehl, H. (1969) 'On the role of the tropics in the general circulation of the atmosphere', *Weather*, 24, 288.

Riehl, H. (1979) *Climate and Weather in the Tropics*, Academic Press, London, pp. 611.

Rijks, D. A. (1969) 'Evaporation from a papyrus swamp', *Quarterly Journal of the Royal Meteorological Society*, 95, 643–9.

Rijks, D. A. and Harrop, J. F. (1969) 'Irrigation and fertiliser experiments on cotton at Mubuku, Uganda', *Experimental Agriculture*, 5, 17–24.

Riou, Ch. (1984) 'Experimental study of potential evapotranspiration (PET) in central Africa', *Journal of Hydrology*, 72, 275–88.

Robertson, G. W. (1975) *Rice and Weather*, WMO Technical Note No. 144, WMO Publication No. 423.

Robertson, G. W. (1983) *Weather Based Mathematical Models for Estimating Development and Ripening of Crops*, WMO Technical Note No. 180, WMO Publication No. 620.

Robertson, W. K., Hammond, L. C., Johnson, J. T. and Boote, K. J. (1980) 'Effects of plant water stress on root distribution of corn, soybeans and peanuts in sandy soil', *Agronomy Journal*, **72**, 548–50.

Robinson, P. and **Glover, J.** (1954) 'The reliability of rainfall within the growing season', *East African Agricultural Journal*, **19**, 137.

Rose, C. W. (1966) *Agricultural Physics*, Pergamon Press, Oxford.

Roy, G. B. and **Ghosh, R. K.** (1982) 'Infiltration rate at long times', *Soil Science*, **134**(6), 345–7.

Ruangpanit, N. (1985) 'Percent crown cover related to water and soil losses in mountainous forest in Thailand', in *Soil Erosion and Conservation*, eds S. A. El-Swaify, W. C. Moldenhauer and A. Lo, Soil Conservation Society of America, pp. 462–71.

Ruiz, G. C. and **Molina, H. A. V.** (1981) 'Rain harvesting for human and live-stock consumption in the semidesert high plains of Mexico', in *Rainfall Collection for Agriculture in Arid and Semiarid Regions*, eds G. R. Dutt, C. F. Hutchinson and M. A. Garduno, Commonwealth Agricultural Bureau, pp. 77–81.

Runge, E. C. A. and **Odell, R. T.** (1960) 'The relation between precipitation, temperature and the yield of soybeans on the agronomy – South Farm, Urbana, Illinois', *Agronomy Journal*, **52**, 245–7.

Rupar Kumar, K. (1984) 'Yield response of sugarcane to weather variations in northeast Andhra Pradesh, India', *Archiv fur Meteorologie, Geophysik und Bioklimatologie*, Ser. B, **35**, 265–76.

Ruprecht, E. and **Bucher, K.** (1970) 'Examples of organised shower structures in the equatorial Atlantic', *Proceedings of the Symposium on Tropical Meteorology*, American Meteorological Society, World Meteorological Organisation, Hawaiian Institute of Geophysics, Honolulu.

Russell, E. W. (1963) *Soil Conditions and Plant Growth*, Longman, London.

Rutter, A. J. (1963) 'Studies on the water relations of *Pinus sylvestris* in plantation conditions', *Journal of Ecology*, **51**(1), 191–203.

Sadler, J. C. (1970) 'Does the convergence zone migrate between hemispheres?', *Proceedings of the Symposium on Tropical Meteorology*, American Meteorological Society, World Meteorological Organisation, Hawaiian Institute of Geophysics, Honolulu.

Sadler, J. C. (1984) 'The anomalous tropical cyclones in the Pacific during the 1982–83 El Nino', *15th Conference on Hurricanes and Tropical Meteorology*, American Meteorological Society, Boston, pp. 52–5.

Sadler, J. C., Ramage, C. S. and **Hori, A. M.** (1982) 'Carbon dioxide variability and atmospheric circulation', *Journal of Applied Meteorology*, **21**, 793–805.

Salas, J. D., Delleur, J. W., Yevjevich, V. and **Lang, W. L.** (1980) *Applied Modelling of Hydrological Time Series*, Water Resources Publications, Littleton, Colorado, pp. 484.

Salter, P. J and **Goode, J. E.** (1967) 'Crop responses to water at different stages of growth', *Commonwealth Agriculture Bureau*.

Sammis, T. W. (1981a) 'Lysimeter for measuring arid-zone evapotranspiration', *Journal of Hydrology*, **49**, 385–94.

Sammis, T. W. (1981b) 'Yield of alfalfa and cotton as influenced by irrigation', *Agronomy Journal*, **73**, 323–9.

Sanders, F. R. (1964) 'Irrigation of Arabica coffee', in J. B. D. Robinson (ed.),

A Handbook on Arabica Coffee in Tanganyika, Tanganyika Coffee Board, 75–82, Dar es Salaam.

Sansigolo, C. A. and Ferraz, E. S. B. (1982) 'Measurement of transpiration and biomass in a tropical *Pinus Caribaea* plantation with tritiated water', *Agricultural Meteorology*, **26**, 25–33.

Sansom, H. W. (1953) 'The maximum possible rainfall in East Africa', *East African Meteorological Department Technical Memorandum*, **3**, 1–16.

Sansom, H. W. (1966) 'The use of explosive rockets to suppress hail in Kenya', *Weather*, **21**, 86–91.

Sarker, R. P. and Biswas, B. C. (1980) 'Agroclimatic classification for assessment of crop potential and its application to dry farming tracts of India', in *Climatic Classification in the Semi-Arid Tropics*, ICRISAT, pp. 89–107.

Schmidt-Ter Hoopen, K. J. and Schmidt, F. H. (1951) 'On Climatic Variation in Indonesia', *Djawatau Met. dan Geof. Djakarta Verb.*, **41**, Corp of Engineers, Army Department.

Schroeder, T. A., Kilonsky, B. J. and Ramage, C. S. (1978) 'Diurnal rainfall variability over the Hawaiian Islands', *11th Tech. Conference on Hurricanes and Tropical Meteorology*, American Meteorological Society, Boston, pp. 72–7.

Schwarz, F.K. (1963) 'Probable maximum precipitation in the Hawaiian Islands', *Hydrometeorological Report*, **39**, Washington.

Seginer, I. (1967) 'Net losses in sprinkler irrigation', *Agricultural Meteorology*, **4**, 281–92.

Selkirk, R. (1984) 'Seasonally stratified correlations of the 200 mb tropical wind field to the southern oscillation', *Journal of Climatology*, **4**, 365–82.

Sellars, C. D. (1981) 'A floodplain storage model used to determine evaporation losses in the Upper Yobe River, northern Nigeria', *Journal of Hydrology*, **52**, 257–68.

Sellers, W. D. (1965) *Physical Climatology*, University of Chicago Press, pp. 272.

Semb, G. and Garberg, P. K. (1969) 'Some effects of planting date and nitrogen fertilizer on maize', *East African Agricultural and Forestry Journal*, **34**, 371–81.

Sen, S. (1968) 'Bhagirathi-Hooghly Basin', in B. C. Laws (ed.), *Mountains and Rivers of India*, Chapter XXIV, Indian National Committee for Geography, Calcutta, pp. 384–95.

Sevruk, B. (1982) *Methods of correction for systematic error in point precipitation measurement for operational use*, WMO Operational Hydrology Report No. 21, pp. 91.

Shalash, S. (1980a) 'Effect of Owen Falls Dam on the hydrological regime of Lake Victoria', in *The Influence of Man on the Hydrological Regime with Special Reference to Representative and Experimental Basins*, IAHS Publ. No. 130, pp. 239–43.

Shalash, S. (1980b) 'The effect of the High Aswan Dam on the hydrological regime of the River Nile', in *The Influence of Man on the Hydrological Regime with Special Reference to Representative and Experimental Basins*, IAHS Publ. No. 130, pp. 244–50.

Shalash, S. (1980c) 'The effect of the High Aswan Dam on the hydrochemical regime of the River Nile', in *The Influence of Man on the Hydrological Regime with Special Reference to Representative and Experimental Basins*, IAHS Publ. No. 130, pp. 251–6.

Shanan, L., Morin, Y. and Cohen, M. (1981) 'A buried membrane collector for harvesting rainfall in sandy areas', in *Rainfall Collection for Agriculture in Arid and Semiarid Regions*, eds G. R. Dutt, C. F. Hutchinson and M. A. Garduno, Commonwealth Agricultural Bureau, pp. 67–76.

Shanan, L., Tadmor, N.H., Evanari, M. and Reiniger, P. (1970) 'Runoff farming in the desert (3)', *Agronomy Journal*, 62, 445–9.

Sharma, S. K. (1983) 'Diurnal variation of rainfall at Nandi airport, Fiji', *Weather*, 38(8), 231–9.

Sharon, D. (1974), 'The spatial pattern of convective rainfall in Sukumaland, Tanzania – a statistical analysis.' *Archiv für Meteorologie, Geophysik und Bioklimatologie, Ser. B*, 22, 201–18.

Sharon, D. (1981), 'The distribution in space of local rainfall in the Namib Desert', *Journal of Climatology*, 1, 69–75.

Sharon, D. (1972) 'The spottiness of rainfall in a desert area', *Journal of Hydrology*, 17,(3), 161–76.

Sheets, R. C. (1973) 'Analysis of hurricane "Debbie" modification results using the variational optimisation approach', *Monthly Weather Review*, 101(9), 663–84.

Shih, S. F. and Snyder, G. H. (1985) 'Leaf area index and evapotranspiration of taro', *Agronomy Journal*, 77, 554–6.

Shukla. J. and Paolino, D. A. (1983) 'The southern oscillation and long-range forecasting of the summer monsoon rainfall over India', *Monthly Weather Review*, 111, 1830–7.

Sill, B. L., Fowler. J. E. and Lagarenne, W. R. (1984) 'Measurement of evaporation by a vapour budget technique', *Water Resources Research*, 20(1), 147–52.

Simmonds, N. W. (1966) *Bananas*, 2nd edition, Longman, London.

Simons, M. (1967) *Deserts, the Problem of Water in Arid Lands*, Oxford U.P., London.

Simpson, J. and Wiggert, V, (1971) 'Florida cumulus seeding experiment: numerical mode results', *Monthly Weather Review*, 99(2), 87–118.

Simpson, J. et al. (1970) 'Potential application of tropical cumulus seeding', *Proceedings of the Symposium on Tropical Meteorology*, American Meteorological Society, World Meteorological Organisation, Hawaiian Institute of Geophysics, Honolulu.

Simpson, L. A. and Gumbs, F. (1985) 'Comparison of three tillage methods for maize (*Zea mays* L.) and cowpea (*Vigna unguiculata* (L.) Walp.) production on a coastal clay soil in Guyana', *Tropical Agriculture*, 62(1), 25–9.

Simpson, R. H. and Malkus, J. S. (1964) 'Experiments in hurricane modification', *Scientific American*, 211(6), 27–37.

Simpson, R. H. and Riehl, H. (1981) *The Hurricane and its Impact*, Basil Blackwell, Oxford, pp. 398.

Singh, N. T., Patel, M. S., Singh, R. and Vig, A. C. (1980) 'Effect of soil compaction on yield and water use efficiency of rice in a highly permeable soil', *Agronomy Journal*, 72, 499–502.

Singh, S. D. (1981) 'Moisture-sensitive growth stages of dwarf wheat and optimal sequencing of evapotranspiration deficits', *Agronomy Journal*, **73**(3), 387–91.

Singh, S. V. and Kripalani, R. H. (1982) 'Some characteristics of daily summer monsoon rainfall over India, suitable for hydrological purposes', in *International Symposium on Hydrometeorology*, American Water Resources Association, pp. 589–96.

Singh, S. V. and Prasad, K. D. (1976) 'Periodicities in daily summer monsoon rainfall of India', *Proceedings of the Indian Institute of Tropical Meteorology*, Pune, Sept. pp. 427–35.

Singh, S. V., Kripalani, R. H., Priya Shaha, Ismail, P. M. M. and Dahale, S. D. (1981) 'Persistence in daily and 5-day summer monsoon rainfall over India', *Archiv fur Meteorologie, Geophysik und Bioklimatologie*, Ser. A, **30**, 261–77.

Singh, S. V., Kripalani, R. H., Saha, P. and Prasad, K. D. (1984) 'Analysis and prediction of short period droughts during summer monsoon over India', *Mausam*, **35**(3), 361–6.

Singh, V. P. (ed.) (1982) *Modelling Components of the Hydrologic Cycle*, Water Resources Publications, Littleton, Colorado, pp. 590.

Slatyer, R. O. (1962) 'Methodology of a water balance study conducted on a desert woodland (*Acacia aneura* F. Muell) community in central Australia', *UNESCO Arid Zone Research*, **16**, 15–26.

Slatyer, R. O. (1967) *Plant-Water Relationships*, Academic Press, London.

SMIC (1971) 'Report of the study of man's impact on climate', *Inadvertent Climate Modification*.

Smith, G. E. (1953) 'Less water required per bushel of corn with adequate fertility', *Missouri Farmers' Association Bulletin*, No. 583.

Smith, G. W. (1966) 'The relation between rainfall, soil water and yield of copra on a coconut estate in Trinidad', *Journal of Applied Ecology*, (3), 117–25.

Smith, R. W. (1964) 'The establishment of cocoa under different soil moisture regimes', *Empire Journal of Experimental Agriculture*, **32**, 249–56.

Sneider, S. H. (1984) 'Deforestation and climatic modification – an editorial', *Climatic Change*, **6**, 227–9.

Soliman, K. H. (1953) 'Rainfall over Egypt', *Quarterly Journal of the Royal Meteorological Society*, **79**, 389.

Squire, G. R., Black, C. R. and Gregory, P. J. (1981) 'Physical measurements in crop physiology. II. Water relations', *Experimental Agriculture*, **17**, 225–42.

Squire, G. R. *et al.* (1984) 'Control of water use by pearl millet', *Experimental Agriculture*, **20**, 135–49.

Stanhill, G. (1962) 'The control of field irrigation practice from measurements of evaporation', *Israel Journal of Agricultural Research*, **12**, 51–62.

Stanhill, G. and Fuchs, M. (1968) 'The climate of the cotton crop. Physical characteristics and microclimate relationships', *Agricultural Meteorology*, **5**, 183–202.

Stanley, C. D., Kasper, T. C. and Taylor, H. M. (1980) 'Soybean top and root response to temporary water tables imposed at three different stages of growth', *Agronomy Journal*, **72**, 341–6.

Stark, J. C. and Jarrell, W. M. (1980) 'Salinity-induced modifications in the response of maize to water deficits', *Agronomy Journal*, **72**, 745–8.

Stedman, O. J. (1980) 'Splash droplet and spore dispersal studies in field beans', *Agricultural Meteorology*, **21**, 111–27.

Stern, R. D. (1980a) 'The calculation of probability distributions for models of daily precipitation', *Archiv fur Meteorologie, Geophysik und Bioklimatologie*, Ser. B, **28**, 137–47.

Stern, R. D. (1980b) 'Analysis of daily rainfall at Samaru, Nigeria, using a simple two-part model', *Archiv fur Meteorologie, Geophysik und Bioklimatologie*, Ser. B, **28**, 123–35.

Stern, R. D. and Coe, R. (1982) 'The use of rainfall models in agricultural planning', *Agricultural Meteorology*, **26**, 35–50.

Stern, R. D., Dennett, M. D. and Dale, I. C. (1982a) 'Analysing daily rainfall measurements to give agronomically useful results. I. Direct Methods', *Experimental Agriculture*, **18**, 223–36.

Stern, R. D., Dennett, M. D. and Dale, I. C. (1982b) 'Analysing daily rainfall measurements to give agronomically useful results. II. A modelling approach', *Experimental Agriculture*, **18**, 237–53.

Stern, R. D., Dennett, M. D. and Garbutt, D. J. (1981) 'The start of the rains in West Africa', *Journal of Climatology*, **1**, 59–68.

Stern, W. R. (1967) 'Seasonal evapotranspiration of irrigated cotton in a low latitude environment', *Australian Journal of Agricultural Resources*, **18**(2), 259–69.

Stewart, J. B. (1983) 'A discussion of the relationships between the principal forms of the combination equation for estimating crop evaporation', *Agricultural Meteorology*, **30**, 111–27.

Stewart, J. I. and Hash, C. T. (1982) 'Impact of weather analysis on agricultural production and planning decisions for the semiarid areas of Kenya', *Journal of Applied Meteorology*, **21**, 477–94.

Stigter, C. J. (1980) 'Assessment of the quality of generalised wind functions in Penman's equations', *Journal of Hydrology*, **45**, 321–31.

Stocking, M. (1984) 'Rates of erosion and sediment yield in the African environment', in *Challenges in African Hydrology and Water Resources* ed. D. E. Walling, S. S. D. Foster and P. Wurzel, IAHS Publ. No. 144, 285–93

Stocking, M. (1985) 'Development projects for the small farmer: Lessons from eastern and central Africa in adapting conservation', in *Soil Erosion and Conservation*, eds S. A. El-Swaify, W. C. Moldenauer and A. Lo, Soil Conservation Society of America, pp. 747–58.

Stocking, M. A. and Elwell, H. A. (1973) 'Prediction of subtropical storm losses from field plot studies', *Agricultural Meteorology*, **12**(2), 193–201.

Stoeckenius, T. (1981) 'Interannual variations of tropical precipitation patterns', *Monthly Weather Review*, **109**, 1233–47.

Stol, Ph. Th. (1972) 'The relative efficiency of the density of raingauge networks', *Journal of Hydrology*, **15**(3), 193–208.

Suarez de Castro, F. (1951) 'Experimentos sobre la erosion de los suelos', *Fed. Nac. Carfeteros de Colombia*, Chinchina Bol. Tech. No. 6.

Subbaramayya, I., Babu, S. V. and Rao, S. S. (1984) 'Onset of the summer monsoon over India and its variability', *Meteorological Magazine*, **113**, 127–35.

Subbaramayya, I., Bhanu Kumar, O. S. R. U. and Babu, S. V. (1981) 'The summer monsoon rains over south and east Asia and west Pacific', *WMO Technical Conference on Climate – Asia and Western Pacific*, pp. 316–33.

Subrahmanyam, V. P. (1972) 'The concept and use of water balance in ecoclimate planning,' *Thornthwaite Memorial Volume I* in J. R. Mather (ed.), Publications in Climatology Vol. XXV, No. 2, 46–56. Laboratory of Climatology, Centerton, New Jersey.

Sumner, G. N. (1981) 'The nature and development of rainstorms in coastal East Africa', *Journal of Climatology*, 1, 131–52.

Sumner, G. N. (1984) 'The impact of wind circulation on the incidence and nature of rainstorms over Dar es Salaam, Tanzania,' *Journal of Climatology*, 4, 35–52.

Suppiah, R. and Yoshino, M. M. (1984a) 'Rainfall variations of Sri Lanka. Part 1: Spatial and temporal patterns', *Archiv für Meteorologie, Geophysik und Bioklimatologie*, Ser. B, 34, 329–40.

Suppiah, R. and Yoshino, M. M. (1984b) 'Rainfall variations of Sri Lanka. Part 2: Regional fluctuations', *Archiv für Meteorologie, Geophysik und Bioklimatologie*, Ser. B, 35, 81–92.

Sutcliffe, J. (1968) *Plants and Water*, Edward Arnold, London.

Sutcliffe, J. (1979) *Plants and Water*, Edward Arnold, 2nd edition, pp. 122.

Suwardjo, and Abujamin, S. (1985) 'Crop residue mulch for conserving soil in uplands of Indonesia', in *Soil Erosion and Conservation*, eds S. A. El-Swaify, W. C. Moldenhauer and A. Lo, Publication of the Soil Conservation Society of America, pp. 607–14.

Swift, J. (1973) 'Disaster and a Sahelian nomad economy', in Dalby and Harrison Church, *Report of the 1973 Symposium Drought in Africa*.

Tamiahe, R. V., Biswas, T. D., Das, B. and Naskar, G. C. (1959) 'Effect of intensity of rainfall on soil loss and runoff', *Indian Society of Soil Science*, 7(47), 231–38.

Tanwar, B. S. (1979) 'Effects of irrigation on the groundwater system in the semiarid zone of Haryana, India', in *The Hydrology of Areas of Low Precipitation*, IAHS Publ. No. 128, pp. 375–84.

Tavares, A.deS. and Ellis, J. (1980) *Chuva maxima em um dia no nordeste do Brasil*, B. tec Inst. Nac. Meteorol. No. 18, pp. 75.

Taylor, G. E. (1984) 'Hawaiian winter rainfall and its relation to the southern oscillation', *Monthly Weather Review*, 112, 1613–19.

Taylor, J. A. and Tulloch, D. (1985) 'Rainfall in the wet-dry tropics: extreme events at Darwin and similarities between years during the period 1870–1983', *Australian Journal of Ecology*, 10, 281–95.

Taylor, R. C. (1970) 'The distribution of rainfall over the tropical Pacific Ocean, deduced from island, atol and coastal stations', *Proceedings of the Symposium on Tropical Meteorology*, American Meteorological Society, World Meteorological Organisation, Hawaiian Institute of Geophysics, Honolulu.

Temple, P. H. (1972) 'Measurements of runoff and soil erosion at an erosion plot scale with particular reference to Tanzania', *Geografiska Annaler*, 54A, 3–4, 203–20.

Temple, P. H. and Rapp, A. (1972) 'Landslides in the Mgeta area, western Uluguru Mountains, Tanzania', *Geografiska Annaler*, 54A(3–4), 157–93.

Thomas, M. E. R. and de Bouvrie, C. (1973) 'Investigations into the magnitude of drought conditions in the Sahelian zone', *Food and Agricultural Organisation*, WS/D7284.

Thompson, B. W. (1957a) 'Some reflections on equatorial and tropical forecasting', East African Meteorological Department, Technical Memorandum 7.

Thompson, B. W. (1957b) 'The diurnal variation of precipitation in British East Africa', *East African Meteorological Department*, Technical Memorandum 8.

Thompson, B. W. (1965) 'Meteorological research in East Africa', *Proceedings of the Third Specialist Meeting of Applied Meteorology*, EAAFRO, Muguga, Kenya (discussion following the paper).

Thompson, G. D. and Boyce, J. P. (1967) 'Daily measurements of potential evapotranspiration from fully canopied sugar cane', *Agricultural Meteorology*, 4(4), 267–80.

Thompson, G. D. and De Robillard, P. J. M. (1968) 'Water duty experiments with sugar cane on 2 soils in Natal', *Experimental Agriculture*, 4, 295–310.

Thompson, G. D. and Wood, R. A. (1967) 'Wet and dry seasons and their effects on rain fed sugar cane in Natal', *Tropical Agriculture*, 44, 297–308.

Thompson, G. D., Gosnell, J. M. and De Robillard, P. J. M. (1967) 'Responses of sugar cane to supplementary irrigation on 2 soils in Natal', *Experimental Agriculture*, 3, 223–38.

Thompson, N. (1982) 'A comparison of formulae for the calculation of water loss from vegetated surfaces', *Agricultural Meteorology*, 26, 265–72.

Thornthwaite Associates, Laboratory of Climatology (1962, 1963, 1965) *Average Climatic Water Balance Data of the Continents*, Publications in Climatology, Vol. XV, No. 2 (1962), Vol XVI, No. 1 (1963), Vol. XVIII, No. 2 (1965). Laboratory of Climatology, Centerton, New Jersey.

Thornthwaite, C. W. (1948) 'An approach toward a rational classification of climate', *Geographical Review*, 38, 55–94.

Thornthwaite, C. W. and Holzman, B. (1939) 'The determination of evaporation from land and water surface', *Monthly Weather Review*, 67, 4–11.

Thornthwaite, C. W. and Mather, J. R. (1955) 'The water balance', *Drexel Institute of Technology*, Laboratory of Climatology Publications in Climatology, 8(1).

Tomar, V. S. and O'Toole, J. C. (1980a) 'Measurement of evapotranspiration in rice', in *WMO/IRRI Symposium on the Agrometeorology of the Rice Crop*, Los Banos, Philippines, pp. 87–95.

Tomar, V. S. and O'Toole, J. C. (1980b) 'Design and testing of a microlysimeter for wetland rice', *Agronomy Journal*, 72, 689–92.

Torrance, J. D. (1967) 'The nature of the rainy season in Central Africa', *First Rhodesian Scientific Congress*, pp. 13–43. Association of Scientific Societies in Rhodesia, Salisbury.

Trewartha, G. T. (1961) *The Earth's Problem Climates*, University of Wisconsin Press, Wisconsin.

Tumuhairwe, J. K. and Gumbs, F. A. (1983) 'Effects of mulches and irrigation on the production of cabbage (*Brassica oleracea*) in the dry season', *Tropical Agriculture*, 60, 122–7.

Turner, D. (1965) 'Investigations into the causes of low yield in late planted maize', *East African Agricultural and Forestry Journal*, **31**, 249–60.

Turner F. T., Cy-Chain Chen and McCauley, G. N. (1981) 'Morphological development of rice seedlings in water at controlled oxygen levels', *Agronomy Journal*, **73**, 566–70.

Tyson, P. D. (1981) 'Atmospheric circulation variations and the occurrence of extended wet and dry spells over southern Africa', *Journal of Climatology*, **1**, 115–30.

Tyson, P. D. (1984) 'The atmospheric modulation of extended wet and dry spells over South Africa, 1958–78', *Journal of Climatology*, **4**, 621–35.

Valentin, C. (1985) 'Effects of grazing and trampling on soil deterioration around recently drilled waterholes in the Sahelian zone', in *Soil Erosion and Conservation*, eds S. A El-Swaify, W. C. Moldenhauer and A. Lo, Soil Conservation Society of America, pp. 51–65.

van Alpen, J. G. (1984) 'Rice in the reclamation of salt affected soils', in *Ecology and Management of Problem Soils in Asia*, Food and Fertilizer Technology Center for the Asian and Pacific Regions, pp. 323–34.

Van Bavel, C. H. M. (1953) 'Chemical composition of tobacco leaves as affected by soil moisture conditions', *Agronomy Journal*, **45**, 611–14.

Van Bavel, C. H. M., Newman, J. E. and Hilgeman, R. H. (1967) 'Climate and estimated water use by an orange orchard', *Agricultural Meteorology*, **4**(1), 27–38.

Van Der Lingen, M. I. (1973) 'Lake Kariba: early history and south shore', in Ackerman, White and Worthington, *Man Made Lakes: Their Problems and Environmental Effects*, pp. 132–42.

Vasey, D. E., Hartis D. R., Olson, G. W., Spriggs, M. J. T. and Turner, B. L. (1984) 'The role of standing water and water-logged soils in raised-field, drained-field and island-bed agriculture', *Singapore Journal of Tropical Geography*, **5**(1), 63–72.

Vauclin, M., Haverkamp, R. and Vachaud, G. (1984) 'Error analysis in estimating soil water content from neutron probe measurements: 2 spatial standpoint', *Soil Science*, **137**(3), 141–8.

Vaughan, R. E. and Wiehe, P. O. (1947) 'Studies on the vegetation of Mauritius: (iv) Some notes on the internal climate of the upland climax forest', *Journal of Ecology*, **34**, 126–36.

Venkataramana, S. and Krishnamurthy, V. (1973) 'On the estimation of potential evaporation by the combination approach', *Archiv für Meteorologie, Geophysik und Bioklimatologie*, Ser. B, **21**(1), 1–10.

Venkataramana, S., Shunmugasundaram, S. and Naidu, K. M. (1984) 'Growth behaviour of field grown sugarcane varieties in relation to environmental parameters and soil moisture stress', *Agricultural and Forest Meteorology*, **31**, 251–60.

Veryard, R. G (1963) 'A review of studies on climatic fluctuations during the period of the meteorological record', *Arid Zone Research*, 20 Changes of Climate, Rome Symposium, pp. 3–15. UNESCO and World Meteorological Organisation, Paris.

Viesman, W., Knapp, J. W., Lewis, G. L. and Harbaugh, T. E. (1977) *Introduction to Hydrology*, Harper and Row, 2nd edition, pp. 704.

Virmani, S. M. (1975) *The Agricultural Climate of the Hyderabad Region in Relation to Crop Planning*, ICRISAT mimeo, pp. .61.

Virmani, S. M., Sivakumar, M. V. K. and Reddy, S. J. (1980) 'Climatic classification of the semi-arid tropics in relation to farming systems research' in *Climatic Classification in the Semi-arid Tropics*, ICRISAT, pp. 27–44.

von Leugerke, H. J. (1980) 'Heavy rainfall areas in peninsular India', *Archiv fur Meteorologie, Geophysik und Bioklimatologie*, Ser. B, **28**, 115–22.

Wada, H. (1971) 'Characteristic features of general circulation in the atmosphere and their relation to the anomalies of summer precipitation in Monsoon Asia', in M. M. Yoshino (ed.), *Water Balance of Monsoon Asia*, University of Hawaii Press, Hawaii.

Wade, M. K. and Sanchez, P. A. (1983) 'Mulching and green manure applications for continuous crop production in the Amazon Basin', *Agronomy Journal*, **75**, 39–45.

Wallace, J. S., Batchelor, C. H. and Hodnett, M. G. (1981) 'Crop evaporation and surface conductance calculated using soil moisture data from central India', *Agricultural Meteorology*, **25**, 83–96.

Wallin, J. R. (1967) 'Agrometeorological aspects of dew', *Agricultural Meteorology*, **4**(2), 85–102.

Walling, D. E., Foster, S. S. D. and Wurzel, P. (eds) (1984), *Challenges in African Hydrology and Water Resources*, IAHS Publ. No. 144, pp. 587.

Wallis, J. A. N. (1963) 'Water use by irrigated Arabica coffee in Kenya', *Journal of Agricultural Science*, **60**, 381–8.

Walton, K. (1969) *The Arid Zones*, Hutchinson & Co., London.

Wangati, F. J. (1984) 'Climate weather and plant production in sub-Saharan Africa: principles and contrasts with temperate regions', in *Advancing Agricultural Production in Africa*, Commonwealth Agricultural Bureau, pp. 293–8.

Ward, R. C. (1967) *Principles of Hydrology*. McGraw Hill, London.

Ward, R. C. (1971) 'Measuring evapotranspiration: a review', *Journal of Hydrology*, **13**(1), 1–21.

Ward, W. T. and Russell, J. S. (1980) 'Winds in southeast Queensland and rain in Australia and their possible long-term relationship with sunspot number', *Climatic Change*, **3**, 89–104.

Wartena, L. (1959) 'The climate and the evaporation from a lake in central Iraq', *Mededelingen van de Landbouwhogeschool*, Wageningen **59**(9), 1–59.

Watkins, L. H. and Fiddes, D. (1984) *Highway and Urban Hydrology in the Tropics*, Pentech Press, London, pp. 206.

Watts, I. E. M. (1955) *Equatorial Weather, With Particular Reference to South-East Asia*, University of London Press, London.

Watts, I. E. M. (1969) 'Climates of China and Korea', *Climates of Northern and Eastern Asia*, Arakawa, Amsterdam.

Weaver, D. C. (1968) 'The hurricane as an economic catalyst', *Journal of Tropical Geography*, **27**, 66–71.

Webb, E. K, (1975) 'Evaporation from catchments' in *Prediction in Catchment Hydrology*, eds T. G. Chapman and F. X. Dunin, Australian Academy of Science, pp. 203–36.

Webster, C. C. and Wilson, P. N. (1966) *Agriculture in the Tropics*.

Weisner, C. J. (1970) *Hydrometeorology*, Chapman & Hall, London.

Wendler, G. and **Eaton, F.** (1983) 'On the desertification of the Sahel zone', *Climatic Change*, **5**, 365–80.

Went, F. W. (1955) 'Fog, mist, dew and other sources of water', *The Yearbook of Agriculture*, US Department of Agriculture, Washington, pp. 103–8.

Wessling, W. H. (1966) 'Reaction of peanuts to dry and wet growing periods in Brazil', *Agronomy Journal*, **58**, 23–6.

Whipkey, R. Z. (1965) 'Subsurface stormflow from forested slopes', *Bulletin, International Association of Scientific Hydrology*, **10**, 75–85.

Whitehead, D., Okali, D. V. V. and **Fasehun, F. E.** (1981) 'Stomatal response to environmental variables in two tropical forest species during the dry season in Nigeria', *Journal of Applied Ecology*, **18**, 571–87.

Wiersum, K. F. (1985) 'Effects of various vegetation layers in an Acacia auriculiformis forest plantation on surface erosion in Java, Indonesia', in *Soil Erosion and Conservation*, eds S. A. El-Swaify, W. C. Moldenhauer and A. Lo, Soil Conservation Society of America, pp. 79–89.

Willheim, E. (1980) 'A legal regime for artificial cyclone modification', *Australian Meteorological Magazine*, **28**, 1–6.

Williams, C. N. and **Chew, W. Y.** (1980) *Tree and Field Crops of the Wetter Regions of the Tropics*, Longman, pp. 262.

Williams, C. N. and **Joseph, K. T.** (1970) *Climate, Soil and Crop Production in the Humid Tropics*, Kuala Lumpur. Oxford U.P., London.

Williams, K. (1970) 'Characteristics of the wind, thermal and moisture fields surrounding the satellite observed mesoscale trade wind cloud clusters of the western North Pacific', *Proceedings of the Symposium on Tropical Meteorology*, American Meteorological Society, World Meteorological Organisation, Hawaiian Institute of Geophysics, Honolulu.

Winkworth, R. E. (1970) 'The soil water regime of an arid grassland community in central Australia', *Agricultural Meteorology*, **7**, 387.

Wilson, J. M. and **Witcombe, J. R.** (1985) 'Crops for arid lands', in *Plants for Arid Lands*, eds G. E. Wickens, J. R. Goodin and D. V. Field, Allen and Unwin, London, pp. 35–52.

Winstanley, D. (1970) 'The North African flood disaster, September 1969', *Weather*, **25**(9), 390–403.

Wischmeier, W. H., Smith, D. D. and **Uhland, R. E.** (1958) 'Evaluation of factors in the soil-loss equation', *Agricultural Engineering*, **39**(8), 458.

WMO (1971) *Climatic Change*, Technical Note No. 79.

WMO (1979) *Operational Techniques for Forecasting Tropical Cyclone Intensity and Movement*, World Weather Watch, WMO Cyclone Project, Subproject No. 6, WMO Publication No. 528.

WMO (1981) *Guide to Hydrological Practices*, 4th edition, WMO Publication No. 168.

WMO (1982) *The Effect of Meteorological Factors on Crop Yields and Methods of Forecasting the Yield*, WMO Technical Note No. 174, WMO Publication No. 566.

WMO and **IRRI** (1980) *Proceedings of the Symposium on the Agrometeorology of the Rice Crop*, IRRI, Los Banos, Philippines, pp. 254.

Wollny, E. (1890) 'Untersuchungen uber die Beeinflussung der Fruchtbarkeit der Ackerkrume durch die Thatigkeit der Regenwurmer', *Forsch, Geb. Agric, Phys.*, **13**, 381–95.

Woodcock, A. H. and Jones, R. H. (1970) 'Rainfall trends in Hawaii', *Journal of Applied Meteorology*, **9**(4), August, 690–9.

Woodhead, T. (1970) 'Confidence limits for seasonal rainfall: their value in Kenya agriculture', *Experimental Agriculture*, **6**, 81–6.

Woodhead, T. (1982) 'Variability of seasonal and annual rainfall totals in East Africa', *East African Agricultural and Forestry Journal*, **45**(1), 74–82.

Woodhead, T., Waweru, E. S. and Lawes, E. T. (1970) 'Expected rainfall and Kenya agriculture confidence limits for large areas at minimum cost', *Experimental Agriculture*, **6**, 87–97.

Woodley, W. L. *et al.* (1982) 'Rainfall results of the Florida Area Cumulus Experiment, 1970–76', *Journal of Applied Meteorology*, **21**, 139–64.

Worthington, E. B. (ed.) (1977) *Arid Land Irrigation in Developing Countries: Environmental Problems and Effects*, Pergamon Press.

Wright, P. B. (1984) 'Relationships between indices of the Southern Oscillation', *Monthly Weather Review*, **112**, 1913–19.

Wright, P. B. (1985) 'The Southern Oscillation: an ocean-atmospheric feedback system?', *Bulletin American Meteorological Society*, **66**, 398–412.

Wrigley, G. (1969) *Tropical Agriculture, The Development of Production*, Faber & Faber, London.

Wustamidin, Douglas, L. A., Cumming, D. J. and Leslie, T. I. (1983). 'Comparison of the water-drop energy required to break down aggregates and soil loss caused by simulated rainfall', *Soil Science*, **136**(6), 367–70.

Yanuka, M., Leshem, Y. and Dovrat, A. (1982) 'Forage corn response to several trickle irrigation and fertilization regimes', *Agronomy Journal*, **74**, 736–40.

Yao, A. Y. M. (1981) 'A system for crop/climate assessment using meteorological and satellite data', *WMO Tech. Conference on Climate – Asia and Western Pacific* Publication No. 578, pp. 105–26.

Yarnal, B. and Kiladis, G. (1985) 'Tropical teleconnections associated with El Nino/Southern Oscillation (ENSO) events', *Progress in Physical Geography*, **9**(4), 524–58.

Yaron, D. (ed.) (1981) *Salinity in Irrigation and Water Resources*, Marcel Dekker, Inc., New York, pp. 432.

Yoshimura, M. (1971) 'Regionality of secular variation in precipitation over monsoon Asia and its relation to general circulation', *Water Balance of Monsoon Asia*, Yoshina, Hawaii, p. 195.

Young, H. M. (1982) *No-Tillage Farming*, No Till Farmer, Inc., pp. 202.

Zandastra, H. G. (1977) 'Cropping systems research for the Asian rice farmer' in *Symposium on Cropping Systems Research and Development for the Asian Rice Farmer*, IRRI, Los Banos, Philippines, pp. 11–30.

AUTHOR INDEX

SUBJECT INDEX